Amphibian Species in Environmental Risk Assessment Strategies

Issues in Toxicology

Editor-in-chief:
Diana Anderson, *University of Bradford, UK*

Series editors:
Michael D. Waters, *Michael Waters Consulting, USA*
Timothy C. Marrs, *Edentox Associates, UK*
Alok Dhawan, *CSIR-Indian Institute of Toxicology Research, India*

Titles in the Series:

1: Hair in Toxicology: An Important Bio-Monitor
2: Male-mediated Developmental Toxicity
3: Cytochrome P450: Role in the Metabolism and Toxicity of Drugs and other Xenobiotics
4: Bile Acids: Toxicology and Bioactivity
5: The Comet Assay in Toxicology
6: Silver in Healthcare
7: In Silico Toxicology: Principles and Applications
8: Environmental Cardiology
9: Biomarkers and Human Biomonitoring, Volume 1: Ongoing Programs and Exposures
10: Biomarkers and Human Biomonitoring, Volume 2: Selected Biomarkers of Current Interest
11: Hormone-Disruptive Chemical Contaminants in Food
12: Mammalian Toxicology of Insecticides
13: The Cellular Response to the Genotoxic Insult: The Question of Threshold for Genotoxic Carcinogens
14: Toxicological Effects of Veterinary Medicinal Products in Humans: Volume 1
15: Toxicological Effects of Veterinary Medicinal Products in Humans: Volume 2
16: Aging and Vulnerability to Environmental Chemicals: Age-related Disorders and their Origins in Environmental Exposures
17: Chemical Toxicity Prediction: Category Formation and Read-Across
18: The Carcinogenicity of Metals: Human Risk Through Occupational and Environmental Exposure
19: Reducing, Refining and Replacing the Use of Animals in Toxicity Testing
20: Advances in Dermatological Sciences
21: Metabolic Profiling: Disease and Xenobiotics
22: Manganese in Health and Disease
23: Toxicology, Survival and Health Hazards of Combustion Products
24: Masked Mycotoxins in Food: Formation, Occurrence and Toxicological Relevance
25: Aerobiology: The Toxicology of Airborne Pathogens and Toxins

26: Chemical Warfare Toxicology, Volume 1: Fundamental Aspects
27: Chemical Warfare Toxicology, Volume 2: Management of Poisoning
28: Toxicogenomics in Predictive Carcinogenicity
29: Human Stem Cell Toxicology
30: The Comet Assay in Toxicology, 2nd edition
31: Computational Systems Pharmacology and Toxicology
32: Ecotoxicology and Genotoxicology: Non-traditional Terrestrial Models
33: Ecotoxicology and Genotoxicology: Non-traditional Aquatic Models
34: Oxidative Stress and Redox Signalling in Parkinson's Disease
35: Nanotoxicology: Experimental and Computational Perspectives
36: Regulatory Toxicology in the European Union
37: Formaldehyde: Exposure, Toxicity and Health Effects
38: Chemical Health Threats: Assessing and Alerting
39: The Micronucleus Assay in Toxicology
40: Nanoparticle–Protein Corona: Biophysics to Biology
41: Big Data in Predictive Toxicology
42: Challenges in Endocrine Disruptor Toxicology and Risk Assessment
43: Bisphenol A: A Multi-modal Endocrine Disruptor
44: Marsupial and Placental Mammal Species in Environmental Risk Assessment Strategies
45: Bird and Reptile Species in Environmental Risk Assessment Strategies
46: Amphibian Species in Environmental Risk Assessment Strategies

How to obtain future titles on publication:
A standing order plan is available for this series. A standing order will bring delivery of each new volume immediately on publication.

For further information please contact:
Book Sales Department, Royal Society of Chemistry, Thomas Graham House, Science Park, Milton Road, Cambridge, CB4 0WF, UK
Telephone: +44 (0)1223 420066, Fax: +44 (0)1223 420247
Email: booksales@rsc.org

Visit our website at books.rsc.org

Amphibian Species in Environmental Risk Assessment Strategies

Edited by

Marcelo L. Larramendy
National University of La Plata, Argentina
Email: marcelo.larramendy@gmail.com

and

Guillermo Eli Liwszyc
University of Helsinki, Finland
Email: guillermo.liwszyc@kolumbus.fi

ROYAL SOCIETY OF CHEMISTRY

Issues in Toxicology No. 46

Print ISBN: 978-1-83767-152-6
PDF ISBN: 978-1-83767-200-4
EPUB ISBN: 978-1-83767-201-1
Print ISSN: 1757-7179
Electronic ISSN: 1757-7187

A catalogue record for this book is available from the British Library

The Royal Society of Chemistry is a charity, registered in England and Wales, Number 207890, and a company incorporated in England by Royal Charter (Registered No. RC000524), registered office: Burlington House, Piccadilly, London W1J 0BA, UK, Telephone: +44 (0) 20 7437 8656.

For further information see our website at www.rsc.org

Printed in the United Kingdom by CPI Group (UK) Ltd, Croydon, CR0 4YY, UK

Preface

This volume is the third in a series of four books published by the Royal Society of Chemistry. First on the list is *Marsupial and Placental Mammal Species in Environmental Risk Assessment Strategies*,[1] followed by *Bird and Reptile Species in Environmental Risk Assessment Strategies*.[2]

The chapters included in this book are a mere enumeration of some practical examples. There are many more species that can be used as non-conventional experimental models, and the list can certainly be expanded further by different research groups and academics all over the world. While animal laboratory models do have their merits, we certainly hope that many more scientists realise that it is more important to tackle subjects related to the status of the environment using autochthonous, or non-target species, which are truly exposed to locally used xenobiotic agents. In many cases, some wild and domestic household species also share the same physical space with humans, enabling us to identify which contaminants we are directly exposed to. This research plan would also increase the chances of independent scientists and research institutions getting access to grants and attracting the attention of local authorities, who would be more interested in financing projects that really are within their sphere of political interest and pride, not to mention their country or province/district.

Without running the risk of being repetitive, we would like to recap on some important concepts as previously mentioned in the two volumes already published by the Royal Society of Chemistry, *i.e.*, *Ecotoxicology and Genotoxicology – Non-traditional Terrestrial Models* and *Ecotoxicology and Genotoxicology – Non-traditional Aquatic Models*, by the same editor group.[3,4] We strongly recommend the perusal of both volumes, which are not overlapping subjects, in order to gain the full benefit of this series and have a more holistic and stimulating approach to the matter.

Issues in Toxicology No. 46
Amphibian Species in Environmental Risk Assessment Strategies
Edited by Marcelo L. Larramendy and Guillermo Eli Liwszyc

Published by the Royal Society of Chemistry, www.rsc.org

We share our planet with thousands of living species. Understanding how environmental stressors affect them will also help us in our quest for solutions to many of the problems humanity is facing at present, which will be more relevant in the near future. Thinking out of silos and cross-linking medical with applied science will lead to a faster pace of understanding many of the factors relating to the emergence of a disease and thus help us in our search for a cure.

New emerging studies in several reputable publications (*e.g.*, Worldwide Trends 1975–2015)[5] of common worldwide human diseases such as arterial hypertension, among others, seem to suggest that not only classical and lifestyle factors should be addressed in the search for adequate treatment. From a medical perspective, it is not only a disease of affluence. One in eight deaths worldwide is due to this condition and its corresponding factors (heart disease, kidney disease and stroke). These studies seem to suggest that a closer look at other, up until now unrelated factors is required. Early-life nutrition and exposure to air pollution, heavy metals and even noise are factors implicated that may push blood pressure up later in life.

Many researchers have contributed to the publication of this book. We hope that it serves as a herald in order to bolster enthusiasm for the use of native or easily available local species and widen our knowledge on the subject. We would like to especially thank the authors of the chapters for their positive response, their time and their contribution to making possible the compilation of this book. The task had been further complicated by the global COVID-19 pandemic, which we seem to have left behind, but not its long-term unforeseen consequences, which have only recently started to emerge. Many, if not most, of our institutions had to close their doors or cease all academic activity for long periods of time during the pandemic. Access to data banks, field trips, *etc.* were curtailed; nevertheless, against all odds, we were able to compile this piece of work, which we hope will be of interest and use, but above all, for the continuous nurturing of the appetite for future lines of research. Last but not least, our most sincere gratitude for the trust deposited upon us by Editor-in-Chief Diana Anderson, Series Editors Michael D. Waters, Timothy C. Marrs, Alok Dhawan and the Royal Society of Chemistry in this project. We hope to have fulfilled the reader's expectations.

Marcelo L. Larramendy
Universidad Nacional de La Plata, Argentina
Consejo Nacional de Investigaciones Científicas y Técnicas (CONICET), Argentina

Guillermo Eli Liwszyc
University of Helsinki, Helsinki, Finland

References

1. M. L. Larramendy and G. E. Liwszyc, in *Marsupial and Placental Mammal Species in Environmental Risk Assessment Strategies*, Royal Society of Chemistry, Cambridge, 2022.
2. G. E. Liwszyc and M. L. Larramendy, in *Bird and Reptile Species in Environmental Risk Assessment Strategies*, Royal Society of Chemistry, Cambridge, 2023.
3. M. L. Larramendy, in *Ecotoxicology and Genotoxicology: Non-traditional Terrestrial Models*, Royal Society Chemistry, Cambridge, 2017.
4. M. L. Larramendy, in *Ecotoxicology and Genotoxicology: Non-traditional Aquatic Models*, Royal Society of Chemistry, Cambridge, 2017.
5. NCD Risk Factor Collaboration, *Lancet*, 2016, **384**, 37–55.

Contents

Issues in Toxicology No. 46
Amphibian Species in Environmental Risk Assessment Strategies
Edited by Marcelo L. Larramendy and Guillermo Eli Liwszyc

Published by the Royal Society of Chemistry, www.rsc.org

CHAPTER 1

General Aspects – Current and Further Perspectives

GUILLERMO ELI LIWSZYC*[a] AND
MARCELO L. LARRAMENDY[b,c]

[a]Former Guest Scientist at the Department of Pathology, University of Helsinki, Helsinki, Finland; [b]Cátedra de Citología, Facultad de Ciencias Naturales y Museo, Universidad Nacional de La Plata, Calle 64 N° 3, B1904AMA La Plata, Argentina; [c]Consejo Nacional de Investigaciones Científicas y Técnicas (CONICET), Buenos Aires, Argentina
*E-mail: guillermoliwszyc@yahoo.com

By definition, it is accepted worldwide that environmental risk assessment is a scientific practice that identifies and evaluates the likelihood of an activity or substance threatening the environment in general and/or living organisms, natural habitats and ecosystems in particular. In other words, it is a term used to describe the overall process or method allowing (1) the identification of hazards and risk factors that have the potential to cause harm (hazard identification) and (2) the analysis and evaluation of the risk associated with that particular hazard (risk analysis and risk evaluation). It also suggests all the appropriate ways to eliminate or minimize the hazard or control the risk when the hazard cannot be eliminated (risk control).

The Food and Agriculture Organization (FAO) of the United Nations predicted a few years ago that the global population would be over 9.1 billion by the middle of the 21st century. Accordingly, food production will have to increase by about 70% above current levels to maintain pace with demand. Disruptions in transport and production of basic commodities

Issues in Toxicology No. 46
Amphibian Species in Environmental Risk Assessment Strategies
Edited by Marcelo L. Larramendy and Guillermo Eli Liwszyc

Published by the Royal Society of Chemistry, www.rsc.org

due to unforeseen circumstances, such as the present war in Ukraine, have revealed once again the caveats of this model. One plausible method for achieving this target would be to increase yields and the amount of arable land. However, the conversion of natural forests and/or other wild habitats engenders a number of well-known negative impacts on climate change and global biodiversity. Furthermore, it is accepted worldwide that such an expansion of agriculture could be responsible for approximately 12% of global warming. Several anthropogenic activities introduce extensive amounts of chemicals and pollutants into the environment on a daily basis. In this sense, fossil fuels, mining, manufacturing and construction industries, among others, are the main contributors to water, air and soil pollution. An increase in improper disposal of waste and chemicals, heavy usage of air-polluting equipment and machinery, asbestos handling, soil contamination and industrial noise also need to be factored in. Even our clothes, in many cases composed of artificial fibres that do not naturally degrade when thrown away or sent to some far-away country as a sort of "help" from more economically privileged nations, find their way to landfills. In some cases, these are located close to the seashores and eventually pollute the oceans. To this, we must add overfishing and overexploitation of marine resources, which complicate matters further.

Most xenobiotics in the different environmental compartments exert their effects through cytotoxicity and genotoxicity, interfering with metabolic pathways. Accordingly, current awareness of the real/potential hazards of pollutants in the environment has a high-ranking interest in using terrestrial invertebrate and vertebrate species as indicators for monitoring pollutant-induced environmental deleterious effects. Notwithstanding, it is known that agrochemicals, among others, not only affect target organisms but also concomitantly exert negative effects on non-target species as well.

In environmental risk assessment studies, there is an increasing interest in biomonitoring markers to provide practical indicators of biological exposure to pollutants. To achieve this goal, several end points for testing toxicity have been developed. Molecular techniques, plus cyto- and genotoxicity tests, have been employed on aquatic and terrestrial organisms to assess the impact of pollution on contaminated areas (*in situ* assays) and to screen xenobiotics after direct or indirect exposure (*in vivo* assays).

This book covers many non-traditional amphibian species models employed to analyse the jeopardizing effects of environmental pollutants, as well as examples of applications of models in integrated assessment solutions. We also hope it will provide answers, or at least some clues, to the inherent potential risk of the presence of these xenobiotics in our environment. To achieve this goal, consideration of exposure, hazards, risks, mitigation, and short- and long-term monitoring is required. This information will give all interested parties, *e.g.*, regulators and safety advisory bodies, the tools needed for a better understanding of the matter in a broader context.

Many important ecosystems around the world are constantly challenged due to growing human and industrial pressure exerted upon them. More than 1 million animals and plants are facing an unprecedented pattern of extinction. We are experiencing the largest biological disappearance wave since the extinction of the dinosaurs, and humans could be the main cause of this event.

The use of various biomarkers in local, easily available species can be useful to evaluate the response of the biota to such pollutants. Several biological parameters mirror the interactions between toxic agents and biotic matrices. These are powerful tools that can be applied to environmental monitoring tests and studies. Their responses may reveal general deleterious effects on the organism in general, pinpointing alterations at the cellular, biochemical and molecular levels, as well as at higher levels of organization.

The health of our planet and its ecosystems has severely deteriorated. There is no question about this statement. Humankind has reached an unprecedented level of food security and availability of consumer goods for the first time in its history. Unfortunately, this comes with a price tag. The different geographical areas, species and substances studied show that no region, even if far away from polluting centres, is spared from this unfortunate phenomenon. We hope that the contributions in this book will stimulate further work along these lines while at the same time raising several flags. As scientists, we have an obligation to present evidence of the damage that our planet, Earth, is being subjected to. A transversal approach is needed, even in matters that do not seem to be related but in reality, are. An example of this is light pollution in populated areas, which disrupts the normal behaviour of birds and insects, altering their feeding habits and the pollination role they play in the ecosystem. This information, in turn, will help launch the debate at a civil society level to change, in turn, many of its consumer patterns, press producers and manufacturers, while at the same time demanding more stringent action on behalf of politicians and lawmakers to protect this planet, our home, before it is too late.

Our global society needs to table down actions and set rules to evaluate and considerably reduce the real and potentially hazardous environmental factors that can, as previously stated, result in health risks for all forms of life (including *Homo sapiens sapiens*). Despite major positive contributions in the field of health, due to the immense progress achieved in science, technology and industrialization, the interaction between environmental risk and health is an often intricate, not self-evident equation that involves a variety of not only social, political and economic but also lifestyle factors. This cannot be emphasized enough. Health depends on the good quality of environmental "basic ingredients," such as air, water, soil and food. We believe that the ultimate challenge in this matter is to weigh in short-term positive gains while, at the same time, taking into account the long-term effects of the substances used. Available information about the toxicity of heterogeneous xenobiotics continuously released into human habitats, inadvertently, deliberately or by non-regulated industrial discharges on biological components of the environment is inconclusive.

There is no clear-cut definition of the concept of environmental health. Various openings help us understand this concept. According to the World Health Organization (WHO), it is defined as "... all the physical, chemical and biological factors external to a person and all the related factors impacting upon behaviours. It encompasses the assessment and control of those environmental factors that can potentially affect health. It is targeted towards preventing disease and creating health-supportive environments ...." For the National Environmental Health Association, this concept refers to "the protection against environmental factors that may adversely impact human health or the ecological balances essential to long-term human health and environmental quality, whether in the natural or man-made environment." A third definition by the National Institute of Environmental Health Science also involves the criterion that "the social environment encompasses lifestyle factors like diet and exercise, socioeconomic status, and other societal influences that may affect health."

In general terms, our health and the health of many other species are negatively affected by five broad categories of environmental hazards: electromagnetic fields (produced by high-power lines, electrical wiring, appliances, mobile phones, computers, TV sets, *etc.*), radiation (including nuclear fallout from weapons testing, fission materials from nuclear power plants and their respective accidents, leaking radioactive disposal sites, air travel and X-rays), toxic chemicals (some organochlorines, phthalates, polybrominated flame retardants, perfluorinated substances, bisphenol-A) and several toxic metals, among others, which have been shown to have endocrine-disrupting properties, and soil mineral depletion as a complex environmental hazard.

By definition, health risk assessment in its quantitative and/or qualitative determinations includes variants such as the type of risk involved and the severity of response, within or without a probabilistic context. In this regard, risk-based methods of analysis play a strategic role in identifying and ranking adverse responses or the structure of the effects of exposure vis-à-vis environmental factors.

Many compounds can be hazardous if not used appropriately and may present a real risk to the environment by contaminating soil, water and air. Most pollutants in the different environmental compartments exert their effects through cytotoxic, genotoxic and metabolically toxic mechanisms. In pollution studies, there is an increasing interest in biomonitoring markers of biological exposure to pollutants. To achieve this goal, several end points for the three above-mentioned factors have been applied to *in situ* and *in vivo* assays.

The use of species as indicators for monitoring pollutant-induced deleterious environmental effects will raise the current awareness of real and potential hazards. It is also known that most environmental pollutants not only affect target organisms but also concomitantly exert negative effects on other non-target species.

Invertebrate and vertebrate animal models have been used for decades in acute and chronic toxicity tests for hazard identification. They can be very efficient screening systems that have a major role to play in toxicity research because certain aspects of their biology, physiology and genetic characteristics make them suitable models in ecotoxicological and genotoxicological studies.

This volume aims to shed some light on the matter, offering relevant tools for evaluating risk and providing a framework for practical discussions. These will, in turn, foster the decisions and actions required to reduce the environmental health risk from environmental factors. This book presents some real-life examples, extending concepts (of hazardous factors) to living species that may stimulate new research lines and trends in the relevant fields.

Available information has been compiled from a diversity of sources in an attempt to achieve a representative global and geographical balance as far as possible while at the same time aiming for high-quality studies. We believe that this book is unique in this sense.

The scientific community is starting to envisage that the time has come for a shift in emphasis away from basic to applied research. While basic research is needed to shed more light on the fundamentals, applied research is required to find solutions to the many problems the world faces, *e.g.*, overpopulation, excessive use of the Earth's natural resources and pollution.

Many researchers from different parts of the world have contributed to the chapters of this book. We hope that this volume will meet the expectations and needs of all those interested in the field of environmental risk assessment by using widely available species worldwide. Finally, we also hope that the examples included in the different chapters of this book will awaken the ability to search for new organisms in local and regional ecosystems to pursue further studies in ecotoxicology and genotoxicology.

CHAPTER 2

Considerations of Amphibian Models and Testing Paradigms for Ecological Risk Assessment

MARK S. JOHNSON*[a] AND DAVID PILLARD[b]

[a]United States Defense Center for Public Health – Aberdeen, Toxicology Directorate, Aberdeen Proving Ground, MD, USA; [b]Ecotox Consulting, Ft. Collins, CO, USA
*E-mail: Mark.S.Johnson.civ@health.mil

2.1 Introduction

Amphibians are an important and diverse class of vertebrates, comprising over 7000 species distributed over numerous landscape features and performing multiple ecological functions.[1] In addition to being prey for aquatic, terrestrial and aerial predators, amphibians consume huge numbers of insects, helping to control mosquitoes, among other Arthropod species, and converting invertebrate biomass into nutrients utilized by primary producers.[2–4] The global decline in amphibian populations is well documented, representing the greatest loss in historical times of any single vertebrate group, with over 30% of amphibian species listed as vulnerable, endangered or even critically endangered.[5–10] Decline in amphibian diversity and density has been attributed to several mechanisms, including disease (such as chytrid fungus), UV radiation, habitat loss, temperature and moisture alterations and chemical toxicants.[8,11–13] Several of these factors (*e.g.*, habitat loss and disease range) are influenced by climate

Issues in Toxicology No. 46
Amphibian Species in Environmental Risk Assessment Strategies
Edited by Marcelo L. Larramendy and Guillermo Eli Liwszyc

Published by the Royal Society of Chemistry, www.rsc.org

change, which has been implicated as a major driver of global amphibian decline.[11,13–16] Given the vulnerability of amphibians to a variety of stressors, it is likely that much of the decline in the amphibian population can be attributed to multiple factors that vary regionally and locally, rather than to a single factor.[17]

Unlike other vertebrate tetrapods, amphibian life cycles remain dependent upon the presence of a minimal level of moisture for reproduction, as amphibian eggs do not possess the requisite membranes that form the key barriers against fatal water loss. As adults, thin, porous integument with little keratin allow amphibians to capitalize on dermal gas and water exchange, which can prove advantageous under adverse environmental conditions[18] but also makes amphibians more vulnerable to exposure to toxic chemicals, ultraviolet radiation and pathogens.[19] The sensitivity and relatively complex life histories of amphibians, bridging aquatic and terrestrial environments, suggest that they are important sentinels of environmental stress and ecosystem health.[20,21] An understanding of the risks posed by various stress agents can provide key information for assembling environmentally relevant risk analyses for application to risk-management decisions.

The potential effects of chemical contaminants on vulnerable amphibian populations are a major concern for ecotoxicologists. Aside from direct toxicological effects, sublethal exposure to pollutants can indirectly affect survival through reductions in predator surveillance and evasion.[22–29] Many other examples of sublethal effects from chemicals exist.[30,31] Clearly, test protocols using appropriate amphibian models are needed to assess the risk posed by chemical exposures, providing data that can be utilized to mitigate significant adverse impacts on vulnerable populations.

2.2 *In Situ* Monitoring of Impacts on Amphibians

In situ studies evaluate the direct effects of stress agents on organisms that are either naturally present in an ecosystem or are temporarily introduced in some manner to monitor a given set of metrics over time. The application of *in situ* techniques is not new for ecological risk assessments, and their use provides a way of establishing real-time causal relationships in ecotoxicological investigations by associating known stress agents with measurable biological impacts. This type of study may be completed alongside more traditional field studies and laboratory toxicity tests, providing a link to these different investigations.[32]

In situ methods tend to be more flexible than laboratory test protocols, allowing for more options in study design and execution.[32] Amphibian investigations, like those involving fish or invertebrates, may include comparison of native amphibian communities from contaminated sites to reference locations or using caged organisms where the test system is standardized by age and/or size of the surrogate species and the test duration is pre-determined for a given time or through a particular developmental stage.[33–37]

Caged amphibian studies have been conducted with several species, including tadpoles of the green frog (*Rana clamitans*), the Eastern American toad (*Bufo americanus*) and the Northern Pacific tree frog (*Pseudacris regilla*).[38,39] The general study design involves placing caged organisms at locations that represent a gradient of contamination either by a single chemical of concern or by a mixture of materials, any one of which may cause significant adverse effects. Data on survival and sublethal metrics (*e.g.*, growth, behavioral abnormalities and time to metamorphosis) are often collected; additional metrics may include DNA damage[38] and/or other biomarkers.[39]

The use of *in situ* monitoring techniques has been steadily evolving along with technological innovations and analytical resolution and precision. Non-invasive methods utilized for *in situ* assessments minimize or eliminate stress and/or destructive sampling while still obtaining quantifiable data. Narayan *et al.* monitored the presence of a glucocorticoid stress biomarker to assess the state of amphibian populations.[40] An increasing number of studies have capitalized on the use of environmental DNA (eDNA) to measure the presence and predominance of a variety of aquatic species, including amphibians.[41] As the accuracy of this method improves, eDNA methods may include both single-species markers and meta-genomic surveys of ecosystems that could potentially lend themselves to use in risk assessments. Saeed *et al.* developed an eDNA protocol to monitor the presence of Asian spiny frogs in the Himalayas.[42] Sakata *et al.* evaluated PCR primers for use in the eDNA metabarcoding of amphibian species.[43] While these techniques are currently used primarily to investigate species inventories, without taking into account specific events that may introduce risk from chemical stressors, eDNA methods coupled with traditional toxicity studies could prove valuable in risk assessment and ongoing monitoring. However, information is still needed on whether eDNA and similar non-invasive assessment methods can be correlated with significant community or population effects.

2.3 Laboratory Models and Testing Protocols

Laboratory animal models for amphibians are limited and, given the phyla's variety, may poorly reflect the responses of many taxa. Amphibians are represented by a wide array of species with differences in behavior (*e.g.*, seasonal niche partitioning), life history and environmental exposure. Dramatic physiological alterations during metamorphosis may affect response and recovery from exposure to environmental contaminants. These differences in environmental variables suggest a focused approach in the valuation of specific models when controlled testing regimes are needed. As biologists search for patterns in toxicity that may be associated with physiological, environmental, or behavioral attributes, refined reviews of the literature with regard to risk assessment are few, but essential. A robust ecotoxicological toolbox for amphibian assessment can reduce uncertainty and contribute to the applicability of risk assessment results.

2.4 Larval Models

2.4.1 Water Exposure Testing

The aquatic stage of larval amphibians provides the opportunity to follow the same general toxicity testing protocols used for invertebrates and fish, which have long been a cornerstone of the assessment of both single chemical compositions for product registration and whole effluent toxicity (WET) testing. Both embryos (developing eggs) and larvae are exposed to known chemical concentrations in test water at a defined temperature for a set duration. A standard guide for acute testing with amphibians, as well as macroinvertebrates and fish, is provided by the American Society for Testing and Materials International (ASTM, ASTM E729-96).[44] Amphibian acute studies, which are well-represented in the literature, are by definition short term (96 h is common) with a survival endpoint only, usually represented by a median lethal/effects concentration (LC50/EC50). A growing database indicates that chronic and subchronic assessments, rather than acute, are better suited for identifying toxic concentrations where significant ecological effects may occur. The frog embryo teratogenesis assay – *Xenopus* (FETAX) provides a standard, 96 h test protocol for detecting developmental toxicants (ASTM E1439-98).[45] The Organization for Economic Co-operation and Development (OECD) larval amphibian growth and development assay (LAGDA, OECD 241)[46] also utilizes the African clawed frog *Xenopus laevis* embryos at test initiation but extends test duration with several sublethal measurements, including thyroid histopathology. The OECD test 231, the amphibian metamorphosis assay (AMA),[47] uses the African clawed frog as the test model but begins with older larvae and includes sublethal endpoints similar to OECD 241. These test methods with *Xenopus* sp. have been widely employed, particularly when evaluating suspect teratogenic and endocrine-disrupting chemicals.[48,49]

Xenopus laevis is often the amphibian test organism of choice for toxicity testing as it is readily available commercially and easily cultured under laboratory conditions. *Xenopus* completes metamorphosis entirely in an aqueous medium, allowing for a simpler test system without the need for a froglet emergence strategy. While the *Xenopus* model has been extensively used and FETAX has been suggested as an applicable procedure for inclusion in ecological risk assessments,[50] there remain gaps in the toxicological database as to how *Xenopus* toxicity indicators compare to other aquatic organisms.[48] Some data suggest FETAX may be less sensitive to some chemicals than other aquatic species, including other amphibians.[48,51] Testing native or closely-related anurans is likely to generate more relevant toxicological data and reduce uncertainty in risk assessments.

Frequently utilized larval amphibian taxa include different anuran genera such as *Rana* (= *Lithobates*), *Bufo* and *Ambystoma*. An increasing volume of life history and testing data have shown that healthy individuals of these genera can be held for extended periods in the laboratory and will consistently yield acceptable control results. Organisms can be held and tested in standardized laboratory water with dry, commercial food (*e.g.*, ground fish flakes) as a food source.[52] Several toxicity studies suggest varying sensitivity at different life stages. Berrill *et al.* examined the toxicity of endosulfan on embryos and tadpoles of the wood frog *Rana sylvatica*, *B. americanus,* and the green frog *Rana clamitans*.[53] While two-week-old tadpoles of *B. americanus* were found to be more tolerant to the pesticide than either of the *Rana* spp., the larvae of any of the three species were more sensitive than embryos (eggs), possibly due to a protective factor provided by the gelatinous coating. In addition, newly-hatched tadpoles showed statistically significant growth effects that were not apparent in older tadpoles. Similar larval age-related sensitivity has been reported for the Northern leopard frog *Lithobates* (= *Rana*) *pipiens* tadpoles exposed to various non-lethal concentrations of 3-nitro-124-triazol-5-one (NTO),[52] Pb,[27] $CaCl_2$, NaCl and Cu.[54] The results of these studies indicate that at some sublethal exposure concentrations, young larvae are more likely to show significant effects (relative to older juveniles) but can recover from early growth inhibition prior to metamorphosis. The lower sensitivity of anuran embryos relative to larvae and the greater sensitivity of younger larvae were also reported by Biga and Blaustein.[55]

Sublethal metrics collected during amphibian tests include growth aspects [*e.g.*, body width, snout-to-vent length (SVL), weight], body uniformity and malformation (*e.g.*, spinal curvature, hind limb deformities, mouth deformities), development/metamorphic benchmarks (*e.g.*, time to metamorphosis) and behavior (*e.g.*, swimming speed). Clearly, many of these sublethal impacts are neither independent nor unidirectional as larval amphibians' progress steadily toward the completion of metamorphosis.

Metamorphic parameters appear to be the most sensitive in subchronic tests initiated with juvenile animals; they are certainly the most critical from an ecological perspective since emergence as an air-breathing, reproductively fit adult is essential to both survival and population continuity. Adverse impacts on a variety of physiological aspects (*e.g.*, spinal curvature, body mass, limb eruption and growth reduction) in several different anuran species, leading to significant metamorphic delays, have been reported following exposure to Pb,[27] triphenyltin,[24] fenpropimorph,[26] carbaryl[56] and nitrotriazolone (NTO).[52] In all these longer-term studies, accommodations are made in the test system to allow emerging adults to climb out of the water, often via a slanted ramp.[52] Duration may be a practical limiting factor in conducting one-generation metamorphic studies, as completion of emergence can occur 70+ days from the release of free-swimming larvae. Test termination usually involves anaesthetizing and euthanizing animals with buffered tricaine methanesulfonate (MS-222) and preserving them in dilute formalin or cold storage for external measurements (*e.g.*, limb length) and histopathology.[52]

2.4.2 Sediment Testing

Although sediments can act as sinks for inorganic and organic chemicals, sequestration is far from permanent. Sediments are a significant source of toxic materials, releasing parent and degradation products back into the water column as well as directly causing toxicity to obligate and casual benthic organisms.[57–61] For larval amphibians that may spend a considerable amount of their time loitering on the bottom and/or directly feeding on epipelic algae, sediment exposure can be a significant source of toxicity through both dermal and ingestion pathways. Sediment characteristics can also affect the bioavailability of contaminants (*e.g.*, by sorption onto various organic ligands) and, thus, significantly impact environmental risk.[62]

ASTM standard guide E2591-07 provides a protocol for testing the toxicity of freshwater sediments to larval anurans in a sediment–water study design.[63] This standard guide was based on research on spiked sediments using primarily *L. pipiens*, although other amphibian larvae were tested, including *B. americanus* and *Ambystoma tigrinum*.[54] Some of the aspects of the design of the E2591-07 test are similar to those used to assess sediment toxicity to freshwater benthic macroinvertebrates (the midge *Chironomus dilutus* and the amphipod *Hyalella azteca*),[64] with the intention of providing a relatively easy transition to amphibian sediment testing. The duration of this ASTM test is 10 days with survival and growth as endpoints, but it can be extended as additional endpoints can be added and other larval amphibians tested using the same basic design but adjusting where needed for temperature, food and loading density.

2.5 Adult Models

2.5.1 Salamanders

Several taxa have been utilized in laboratory toxicity testing, including species of the genus *Ambystoma* and *Plethodon*. The tiger salamander *A. tigrinum* and the spotted salamander *A. maculatum* have been used in soil exposures (predominantly dermal) where additional oral exposures were added by maintaining earthworm prey in soil with similar contaminant concentrations.[65–67] Individuals can be hand-fed and respond readily to food. Excess skin mucus production is considered a general indication of stress. Animals can be anesthetized in a buffered solution of tricaine (MS-222) and euthanized by decapitation/exsanguination. Test materials mixed into the soil included exposures for 14 days, but longer exposures are possible. As many species of Ambystomidae have relatively long life-spans in captivity (>30 years), these exposures were considered from subacute to subchronic.

The red-backed salamanders *P. cinereus* have been used in several studies where adults were exposed to contaminated soils.[68–70] Adults were captured and exposed to homogeneous spiked soils for up to 28 days (see Figure 2.1). The soils were characterized according to organic carbon content, proportions of silt, clay, sand, redox potential and pH. Animals were exposed in Petri dishes (60 × 15 mm) with sphagnum moss for environmental enrichment. Salamanders were fed with a mutant of the fruit fly *Drosophila melanogaster* (flightless) daily and accepted food readily. Limited blood work can be conducted (*e.g.*, smears for red and white blood cell counts, five-part differentials). Histopathological assessment is valuable and the entire sagittal portion can be sectioned, stained and contained on a single slide; however, these sections can miss vital organs. Therefore, various cross-sections are recommended.

Members of the woodland salamanders from the genus *Plethodon* are lungless, and exposures occur *via* oral and dermal routes. As water absorption occurs through ventral dermal surfaces, exposure to water-soluble compounds can be considerable. Gas exchange occurs through the skin, and as with *Ambystoma*, the excess dermal mucus production is an indication of stress (see Figure 2.2). Also like *Ambystoma*, members of the genus *Plethodon* are thought to be long-lived as reproductive maturity is attained after 4 years. Therefore, these exposures are considered subchronic in duration. No published standardized methods exist using adults of these species of salamanders.

2.5.2 Frogs and Toads

The African clawed frog *Xenopus laevis* and its close relative, the Western clawed *X. tropicalis,* live their entire lives in water and, therefore, can be tested in water at any life stage. As previously discussed, however, *Xenopus* spp. may be neither the most sensitive amphibian nor the most representative. Because measuring quantitative exposure to adult anurans is more complex, husbandry can be difficult, and because data strongly suggest young larvae are more sensitive, there are far fewer published toxicity studies with adult frogs and toads. Many published studies conducted with adult anurans measure mortality, some in conjunction with larval (tadpole) forms.[71–74] These are typically water exposures where larvae were found to be more sensitive than adults.[75]

As mentioned, the inclusion of an oral exposure pathway in adult anuran tests can be challenging. As adult forms are carnivorous, feeding cues often rely on the movements of live prey, making husbandry considerations difficult. Some studies have been conducted in mesocosm designs, where

Figure 2.1 Exposures of red-backed salamanders (*Plethodon cinereus*) to treated soils as part of a 28-day design.

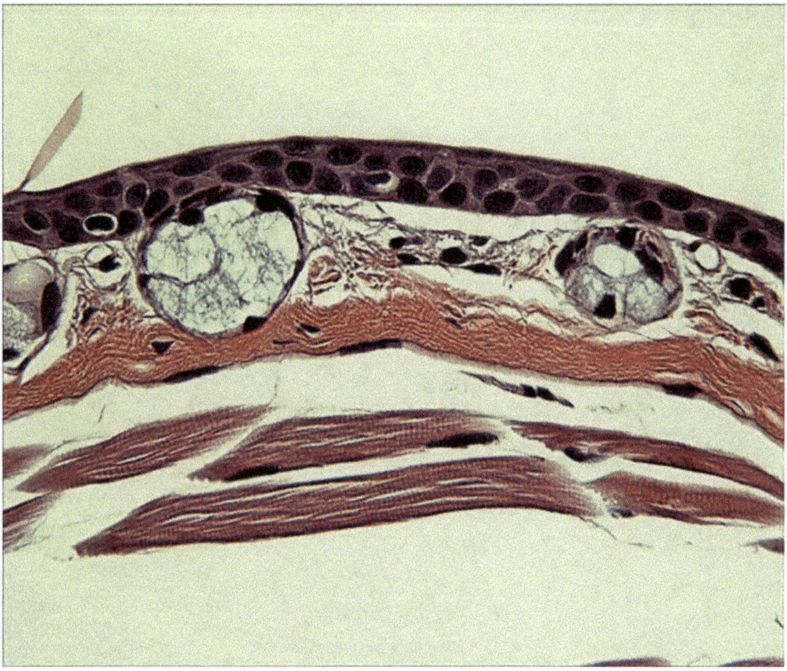

Figure 2.2 Photomicrohistograph of the skin of *P. cinereus* showing 2-4-cell thickness; predominant oval structures are mucus glands.

results are difficult to interpret due to interactions between abiotic and biotic factors.[76] Gavage, an oral exposure alternative to dosed food items for adult anurans, has been used successfully with the American bullfrog *Aquarama castebeiana*.[77] However, because gavage does not reflect natural eating habits and may require the use of light pre-dose sedation to prevent regurgitation, there remains a question as to its ecological relevance. As in salamanders, the dermal route of exposure in adult anurans may be toxicologically important as adults can be exposed both to water and to soil as they burrow or rest. Bidwell and Gorrie explored dermal exposure to glyphosate in three adult frogs using test chambers that forced post-metamorphic adults to remain in contact with water while still providing a small pocket of air.[78]

2.6 Considerations

The use of wildlife species, in general, can be problematic in a controlled laboratory environment, resulting mostly from complex husbandry considerations. Life history constraints make housing and feeding carnivorous adults challenging. Obtaining terrestrial adult amphibians is often relegated to wild-caught individuals, where sex and age can be confounding elements. A review of available toxicity data found that, for many substances, larval forms of amphibians appear to be more sensitive than adults, though the data are limited.[75] This seems not to be the case for compounds that act as endocrine disruptors, particularly those that can disrupt the thyroid-pituitary axis and affect metamorphosis. The reported sensitivity of larval forms may be due, in part, to an inappropriate comparison of exposure to substances in water relative to soil or sediments, where dermal and possible oral pathways occur. The type and magnitude of the observations for adverse effects must also be considered relative to life history challenges. For example, minor delays in metamorphosis can have significant survival implications for wild populations where larval water environments are transient.

2.7 Important Considerations for Amphibian Models

Physiology is an important determinant of exposure. Aquatic exposures for larval amphibians (*i.e.*, tadpoles) can be considered equivalent to those for fish, where water is the medium and exposure pathways can include dermal, oral and gill transfers. Adult terrestrial forms are varied, where gas exchange occurs predominantly through the skin, although some species do ventilate and have lungs (*e.g.*, terrestrial toads such as *Bufo* ssp.; red eft stage of the red-spotted newt *Notophthalmus viridescens*). Mucociliary clearance, however, is lacking. Many have a dorsal integument that is largely resistant to desiccation while also possessing ventral surfaces that function as water-uptake organs. Naturally, the dermal route of exposure could be important for compounds that are water soluble, though empirical data are lacking.

Many amphibians exist in habitats where bacterial and viral pathogens are prevalent. Declines in populations of some species have been attributed to infectious disease,[79] yet pathogen–animal–climate interactions can be complex.[80] An understanding of environmental stressors and how they affect amphibian immunity is an important consideration, particularly when stressors include exposure to chemicals that may suppress the immune system.

Environmental cues important in the initiation of metamorphosis have their basis in selective advantages in specific habitats. As previously mentioned, the timing of metamorphosis can be critical in survival and subsequent reproduction where aquatic breeding habitats are transitory. Ephemeral (and perennial) aquatic breeding habitats often offer larvae some refuge from predation events found in more stable water bodies but may not be present for extended periods of time. Typically, metamorphosis involves a niche shift in foraging preferences that are important in considering ecological consequences (*e.g.*, indirect effects if prey taxa are more sensitive to xenobiotic exposure). The relevance of ambient temperature on endotherms is also a consideration if toxicity is due to primary or secondary metabolites. However, we know of no published studies documenting such relationships.

2.8 Uncertainties

While ecological risk assessments can be a powerful tool, enabling risk managers to develop and/or modify activities designed to mitigate risk, there is always an inherent level of uncertainty associated with the application of data and the assumptions made to account for missing or unknown information. Recognition and acknowledgement of uncertainty are essential components of a risk assessment and must be considered when applying risk assessment results to risk management decisions. Uncertainty, though an inseparable part of a risk assessment, can be reduced by maximizing site-specific data and minimizing extrapolations. Sources of uncertainty in a risk assessment include the following:

- Incomplete characterization and formulation of the conceptual site model, including receptor species, exposure pathways and toxicity data.
- Poor understanding of chemical species, transformations and relative bioavailability.
- Cross-species extrapolations of toxicity and exposure data.
- Variability in environmental toxicity data and/or bias when collecting data.
- Incomplete life history of receptor species.

Typically, the risk to amphibian species is considered relative to media (soil, sediment and water) concentrations that align with the controlled toxicity study design. Although some are trophic-level predators following metamorphosis, insufficient life history information is often lacking to allow for any food web characterization in a quantitative sense. Food web transfer of compounds that biomagnify is particularly relevant for substances that are environmentally persistent and have intrinsically low acute toxicity, especially to invertebrate prey.

A key initial step in preparing for the execution of an ecological risk assessment is problem formulation, which provides a guide for the entire risk assessment process. A thorough problem formulation allows for a more accurate risk assessment and includes an *a priori* identification of uncertainties. Sample *et al.*[81] provide a review of problem formulation in wildlife risk assessment and identify four key problem formulation points to improve risk assessments, which are summarized as follows:

- Establishing concise, specific protection goals early in the risk assessment process.
- Assess how the collection of data and methods used will impact decisions.
- Ensure that relevant stakeholders are engaged when developing the conceptual site model that includes appropriate stress agents.
- Recognize that the problem formulation process is fluid and that multiple iterations may be needed to arrive at a feasible path forward.

Addressing uncertainty in an ecological risk assessment is a fundamental component of the process, regardless of whether the receptor species are amphibians, mammals, birds or other taxa. All fall under the same risk assessment framework; details will vary but will be described during problem formulation. The historical predominance of non-amphibian toxicological data can be a potential liability while contributing to uncertainty in amphibian-centric risk assessments. An appropriately designed risk assessment strategy will account for that uncertainty and establish the methods by which it can be minimized.

2.9 Research Needs and Recommendations

Recent research quantifying the toxicity of single contaminants, chemical combinations and complex environmental media has led to the development of a limited set of standardized toxicity models for amphibians. Both research and regulatory applications have proven the value of standardized testing with several invertebrate and vertebrate models and how they can be effectively used in risk assessment. The absence of amphibian toxicological data in aquatic, but especially terrestrial, environments will contribute to uncertainty in a risk assessment. While the extant data strongly support the use of chronic exposures and sublethal metrics, improvements to the resolution and applicability of ecological risk assessments for amphibian populations can be derived through additional studies on key species. Focused investigations on responses at different life stages are needed to better understand variations in sensitivity. Current data on long-term exposures indicate endpoints associated with metamorphosis, in particular time to metamorphosis, are often the most sensitive indicators of toxicity. Improved experimental designs that explicitly incorporate time to metamorphosis and adult transition can provide valuable information relevant to dispersal, reproduction and population success.[75]

Improving an understanding of exposure is important for adult carnivorous stages, particularly long-lived species (*e.g.*, Ambystomatid salamanders). Research is needed to help quantify daily food ingestion rates, prey types and metabolic considerations under various temperature conditions. As mentioned, food web modeling is most relevant for endocrine-disrupting and environmentally persistent substances likely to biomagnify given their chemical structures (*e.g.*, halogenated organics).

The paucity of toxicity data relative to adult amphibian forms can be partially attributed to the greater challenges in testing with adults but also in assumptions regarding the most sensitive stages. However, designing, implementing and confirming adult amphibian test models will generate critical ecotoxicological data that cannot be extrapolated from aquatic test data or avian or mammalian toxicity data on the same or similar compounds. A testing paradigm centered around consistent methods applied to key species and metamorphic stages is essential in identifying potential and actual risks to amphibian populations and implementing effective mitigation strategies.

Disclaimer

The views expressed in this document are those of the author(s) and do not necessarily reflect the official policy of the Department of Defense, Defense Health Agency or the U.S. The mention of any non-federal entity and/or its products is for informational purposes only, and is not to be construed or interpreted, in any manner, as federal endorsement of that non-federal entity or its products.

References

1. D. R. Frost, T. Grant, J. Faivovich, R. H. Bain, A. Haas, C. F. B. Haddad, R. De Sá, A. Channing, M. Wilkinson, S. C. Connellan, C. J. Raxworthy, J. A. Campbell, B. L. Blotto, P. Moler, R. C. Drewes, R. A. Nussbaum, J. D. Lynch, D. M. Green and W. C. Wheller, The amphibian tree of life, *Bull. Am. Mus. Nat. His.*, 2006, **297**, 1–291.
2. S. E. DuRant and W. A. Hopkins, Amphibian predation on larval mosquitoes, *Can. J. Zool.*, 2008, **86**, 1159–1164.
3. A. Valencia-Aguilar, A. M. Cortés-Gómez and C. A. Ruiz-Agudelo, Ecosystem services provided by amphibians and reptiles in Neotropical ecosystems, *Int. J. Biodiver. Sci., Ecosyst. Serv. Manage.*, 2013, **9**, 257–272.
4. D. J. Hocking and K. J. Babbitt, Amphibian contributions to ecosystem services, *Herpetol. Conserv. Biol.*, 2014, **9**, 1–17.
5. D. B. Wake, Declining amphibian populations, *Science*, 1991, **253**, 860.

6. R. A. Alford and S. I. Richards, Global amphibian declines: A problem in applied ecology, *Annu. Rec. Ecol. Syst.*, 1999, **30**, 133–165.
7. J. E. Houlahan, C. S. Findlay, B. R. Schmidt, A. H. Meyer and S. L. Kuzmin, Quantitative evidence for global amphibian population declines, *Nature*, 2000, **404**, 752–755.
8. S. N. Stuart, J. S. Chanson, N. A. Cox, B. E. Young, A. S. L. Rodrigues, D. L. Fischman and R. W. Waller, Status and trends of amphibian declines and extinctions worldwide, *Science*, 2004, **306**, 1783–1786.
9. A. Quaranta, V. Bellantuono, G. Cassano and C. Lippe, Why amphibians are more sensitive than mammals to xenobiotics, *PLoS One*, 2009, **4**, 1–4.
10. D. Rödder, J. Kielgast and S. Lötters, Future potential distribution of the emerging amphibian chytrid fungus under anthropogenic climate change, *Dis. Aquat. Org.*, 2010, **92**, 201–207.
11. J. Bosch, I. Martinez-Solano and M. Garcıia-Parıis, Evidence of a chytrid fungus infection involved in the decline of the common midwife toad (*Alytes obstetricans*) in protected areas of central Spain, *Biol. Conserv.*, 2001, **97**, 331–337.
12. J. M. Kiesecker, Global stressors and the global decline of amphibians: Tipping the stress immunocompetency axis, *Ecol. Res.*, 2011, **26**, 897–908.
13. Y. I. Yiming, J. M. Cohen and J. R. Rohr, Review and synthesis of the effects of climate change on amphibians, *Integrat. Zool.*, 2012, **8**, 145–161.
14. S. E. Williams, L. P. Shoo, J. L. Isaac, A. A. Hoffmann and G. Langham, Towards an integrated framework for assessing the vulnerability of species to climate change, *PLoS Biol.*, 2008, **6**, e325.
15. A. R. Blaustein, S. C. Walls, B. A. Bancroft, J. J. Lawler, C. L. Searle and S. S. Gervasi, Direct and indirect effects of climate change on amphibian populations, *Diversity*, 2010, **2**, 281–313.
16. W. B. Foden, S. H. M. Butchart, S. N. Stuart, J.-C. Vié, H. R. Akçakaya, A. Angulo, L. M. DeVantier, A. Gutsche, E. Turak, L. Cao, S. D. Donner, V. Katariya, R. Bernard, R. A. Holland, A. F. Hughes, S. E. O'Hanlon, S. T. Garnett, C. H. Sekercioğlu and G. M. Mace, Identifying the world's most climate change vulnerable species: a systematic trait-based assessment of all birds, amphibians and corals, *PLoS One*, 2013, **8**, e65427.
17. E. H. C. Grant, D. A. W. Miller, B. R. Schmidt, M. J. Adams, S. M. Amburgey, T. Chambert, S. S. Cruickshank, R. N. Fisher, D. M. Green, B. R. Hossack, P. T. J. Johnson, M. B. Joseph, T. A. G. Rittenhouse, M. E. Ryan, H. Waddle, S. C. Walls, L. L. Bailey, G. M. Fellers, T. A. Gorman, A. M. Ray, D. S. Pilliod, S. J. Price, W. Sadinski and E. Muths, Quantitative evidence for the effects of multiple drivers on continental-scale amphibian declines, *Sci. Rep.*, 2016, **6**, 25625.

18. H. B. Lillywhite, Water relations of tetrapod integument, *J. Exp. Biol.*, 2006, **209**, 202–226.
19. C. A. Brühl, S. Pieper and B. Weber, Amphibians at risk? Susceptibility of terrestrial amphibian life stages to pesticides, *Environ. Toxicol. Chem.*, 2011, **30**, 2465–2472.
20. H. H. Welsh Jr. and L. M. Ollivier, Stream amphibians as indicators of ecosystem stress: A case study from California's redwoods, *Ecol. Appl.*, 1998, **8**, 1118–1132.
21. D. W. Sparling, A review of the role of contaminants in amphibian declines, in *Handbook of Ecotoxicology*, ed. D. J. Hoffman, B. A. Rattner, A. G. Burton and J. Cairns, Lewis Publishers, Boca Raton, FL, 2003, pp. 1099–1128.
22. R. D. Semlitsch and J. W. Gibbons, Fish predation in size-structured populations of treefrog tadpoles, *Oecologia*, 1988, **75**, 321–326.
23. T. B. Watkins, Predator-mediated selection on burst swimming performance in tadpoles of the Pacific tree frog, (*Pseudacris regilla*), *Physiol. Zool.*, 1996, **69**, 154–167.
24. E. Fioramonti, R. D. Semlitsch, H.-U. Reyer and K. Fent, Effects of triphenyltin and pH on the growth and development of *Rana lessonae* and *Rana esculenta* tadpoles, *Environ. Toxicol. Chem.*, 1997, **16**, 1940–1947.
25. S. M. Raimondo, C. L. Rowe and J. D. Congdon, Exposure to coal ash impacts swimming performance and predator avoidance in larval bullfrogs, *J. Herpetol.*, 1998, **32**, 289–292.
26. C. Teplitsky, H. Piha, A. Laurila and J. Merilä, Common pesticide increases costs of antipredator defenses in *Rana temporaria* tadpoles, *Environ. Sci. Technol.*, 2005, **39**, 6079–6085.
27. T.-H. Chen, J. A. Gross and W. H. Karasoc, Sublethal effects of lead on northern leopard frog (*Rana pipiens*) tadpoles, *Environ. Toxicol. Chem.*, 2006, **25**, 1383–1389.
28. S. M. Hanlon and R. Relyea, Sublethal effects of pesticides on predator–prey interactions in amphibians, *Copeia*, 2013, **1**, 691–698.
29. N. Polo-Cavia, P. Burraco and I. Gomez-Mestre, Low levels of chemical anthropogenic pollution may threaten amphibians by impairing predator recognition, *Aquat. Toxicol.*, 2016, **172**, 30–35.
30. S. Yu, M. R. Wages, Q. Cai, J. D. Maul and G. P. Cobb, Lethal and sublethal effects of three insecticides on two developmental stages of *Xenopus laevis* and comparison with other amphibians, *Environ. Toxicol. Chem.*, 2013, **32**, 2056–2064.
31. S. A. Abercrombie, C. de Perre, M. Iacchetta, R. W. Flynn, M. S. Sepúlveda, L. S. Lee and J. T. Hoverman, Sublethal effects of dermal exposure to poly- and perfluoroalkyl substances on postmetamorphic amphibians, *Environ. Toxicol. Chem.*, 2020, **40**, 717–726.

32. K. Liber, W. Goodfellow, P. den Besten, W. Clements, T. Galloway, A. Gerhardt, A. Green and S. Simpson, *In situ*-based effects measures: considerations for improving methods and approaches, *Integr. Environ. Assess. Manag.*, 2007, **3**, 246–258.
33. D. J. Chappie and G. A. Burton Jr., Optimization if *in situ* bioassays with *Hyalella azteca* and *Chironomus tentans*, *Environ. Toxicol. Chem.*, 1997, **16**, 559–564.
34. B. J. Kater, J. F. Postma, M. Dubbeldam and T. H. J. Prins, Comparison of laboratory and *in situ* sediment bioassays using *Corophium volutator*, *Environ. Toxicol. Chem.*, 2001, **20**, 1291–1295.
35. E. R. Kline and M. S. Stekoll, Colonization of mine tailings by marine invertebrates, *Mar. Environ. Res.*, 2001, **51**, 301–325.
36. L. A. Courtney and W. H. Clements, Assessing the influence of water and substratum quality on benthic macroinvertebrate communities in a metal-polluted stream: An experimental approach, *Freshw. Biol.*, 2002, **47**, 1766–1778.
37. G. A. Burton Jr., M. S. Greenberg, C. D. Rowland, C. A. Irvine, D. R. Lavoie, J. A. Brooker, L. Moore, D. F. Raymer and R. A. McWilliam, *In situ* exposures using caged organisms: a multi-compartment approach to detect aquatic toxicity and bioaccumulation, *Environ. Pollut.*, 2005, **143**, 133–144.
38. S. Ralph and M. Petras, Caged amphibian tadpoles and *in situ* genotoxicity monitoring of aquatic environments with the alkaline single cell gel electrophoresis (comet) assay, *Mutat. Res.*, 1998, **413**, 235–250.
39. D. W. Sparling, J. Bickham, D. Cowman, G. M. Fellers, T. Lacher, C. W. Matson and L. McConnell, *In situ* effects of pesticides on amphibians in the Sierra Nevada, *Ecotoxicology*, 2015, **24**, 262–278.
40. E. J. Narayan, Z. R. Forsburg, D. R. W. Davis and C. R. Gabor, Non-invasive Methods for measuring and monitoring stress physiology in imperiled amphibians, *Front. Ecol. Evol.*, 2019, **7**, 431.
41. P. F. Thomsen and E. Willerslev, Environmental DNA – An emerging tool in conservation for monitoring past and present biodiversity, *Biol. Conserv.*, 2015, **183**, 4–18.
42. M. Saeed, M. Rais, A. Akram, M. R. Williams, K. F. Kellner, S. A. Hashsham and D. R. Davis, Development and validation of an eDNA protocol for monitoring endemic Asian spiny frogs in the Himalayan region of Pakistan, *Sci. Rep.*, 2022, **12**, 5624.
43. M. K. Sakata, M. U. Kawata, A. Kurabayashi, T. Kurita, M. Nakamura, T. Shirako, R. Kakehashi, K. Nishikawa, M. Y. Hossman, T. Nishijima, J. Kabamoto, M. Miya and T. Minamoto, Development and evaluation of PCR primers for environmental DNA (eDNA) metabarcoding of Amphibia, *Metabarcoding Metagenom.*, 2022, **6**, e76534.

44. American Society of Testing and Materials (ASTM), in Standard guide for conducting acute toxicity tests with fishes, macroinvertebrates, and amphibians, ASTM E729-96, American Society of Testing and Materials, West Conshohocken, PA, 2002.
45. American Society for Testing and Materials International (ASTM), in Standard guide for conducting the frog embryo teratogenesis assay-Xenopus (FETAX), ASTM E1439-98, American Society of Testing and Materials, West Conshohocken, PA, 2002.
46. Organization for Economic Co-operation and Development (OECD), Test no.241, The larval amphibian growth and development assay (LAGDA), Organization for Economic Cooperation and Development, Paris, France, 2015.
47. Organization for Economic Co-operation and Development (OECD), Test no.231, Amphibian metamorphosis assay, Organization for Economic Cooperation and Development, Paris, France, 2009.
48. R. A. Hoke and G. T. Ankley, Application of frog embryo teratogenesis assay - *Xenopus* to ecological risk assessment, *Environ. Toxicol. Chem.*, 2005, **28**, 2677–2690.
49. L. S. Ortego, A. W. Olmstead, L. Weltje, J. R. Wheeler, A. J. Bone, K. K. Coady, C. S. Banman, N. Burden and L. Lagadic, The extended amphibian metamorphosis assay: A thyroid-specific and less animal-intensive alternative to the larval amphibian growth and development assay, *Environ. Toxicol. Chem.*, 2021, **40**, 2135–2144.
50. J. A. Bantle, FETAX-s developmental toxicity assay using frog embryos, , in *Fundamentals of Aquatic Toxicology*, ed. G. M. Rand, Taylor and Francis, Washington DC, 1995, pp. 207–230.
51. A. N. Edginton, G. R. Stephenson, P. M. Sheridan, D. G. Thompson and H. J. Boermans, Effect of pH and Release® on two life stages of four anuran amphibians, *Environ. Toxicol. Chem.*, 2003, **22**, 2673–2678.
52. D. A. Pillard, W. S. Eck, M. S. Johnson and S. Packard, Effects of 3-nitro-1,2,4-triazol-5-one on survival, growth and metamorphosis in the Northern leopard frog, *Lithobates pipiens*, *Ecotoxicology*, 2017, **26**, 1170–1180.
53. M. Berrill, D. Coulson, L. McGillivray and R. Pauli, Toxicity of endosulfan to aquatic stages of anuran amphibians, *Environ. Toxicol. Chem.*, 1998, **17**, 1738–1744.
54. Naval Facilities Engineering Command (NAVFAC), in *Development of a standardized approach for assessing potential risks to amphibians exposed to sediments and hydric soils TR-2245-ENV*, ENSR International, Westford, MA, 2004, pp. 1–340.
55. L. M. Biga and A. R. Blaustein, Variations in lethal and sublethal effects of cypermethrin among aquatic stages and species of anuran amphibians, *Environ. Toxicol. Chem.*, 2013, **32**, 2855–2860.

56. K. M. Gaietto, S. L. Rumschlag and M. D. Boone, Effects of pesticide exposure and the amphibian chytrid fungus on gray treefrog (*Hyla chrysoscelis*) metamorphosis, *Environ. Toxicol. Chem.*, 2014, **33**, 2358–2362.
57. P. A. Lara-Martín, M. Petrovic, A. Gómez-Parra, D. Barceló and E. González-Mazo, Presence of surfactants and their degradation intermediates in sediment cores and grabs from the Cadiz Bay area, *Environ. Pollut.*, 2006, **144**, 483–491.
58. K. T. Ho and R. M. Burgess, What's causing toxicity in sediments? Results of 20 years of toxicity identification and evaluations, *Environ. Toxicol. Chem.*, 2013, **32**, 2424–2432.
59. D. J. Fairbairn, M. E. Karpuzcu, W. A. Arnold, B. L. Barber, E. F. Kaufenberg, W. C. Koskinen, P. J. Novak and D. L. Swackhamer, Sediment-water distribution of contaminants of emerging concern in a mixed use watershed, *Sci. Tot. Environ.*, 2015, **505**, 896–904.
60. L. Mustajärvi, E. Eek, G. Cornelissen, A.-K. Eriksson-Wiklund, E. Undeman and A. Sobek, *In situ* benthic flow-through chambers to determine sediment-to-water fluxes of legacy hydrophobic organic contaminants, *Environ. Pollut.*, 2017, **231**, 854–862.
61. A. C. Chiaia-Hernández, C. Casado-Martinez, P. Lara-Martin and T. D. Bucheli, Sediments: sink, archive, and source of contaminants, *Sci. Pollut. Res.*, 2022, **29**, 85761–85765.
62. L. Fuentes, L. J. Moore, J. H. Rodgers Jr., W. W. Bowerman, G. K. Yarrow and W. Y. Chao, Role of sediments in modifying the toxicity of two Roundup formulations to six species of larval anurans, *Environ. Toxicol. Chem.*, 2014, **33**, 2616–2620.
63. American Society for Testing and Materials International (ASTM), Standard guide for conducting whole sediment toxicity tests with amphibians, ASTM E2591-07, American Society of Testing and Materials, West Conshohocken, PA, 2013.
64. United States Environmental Protection Agency (U.S.EPA), *Methods for measuring the toxicity and bioaccumulation of sediment-associated contaminants with freshwater invertebrates*, Second Edition, EPA/600/R-99/064, Washington D.C., 2000, p. 213.
65. M. S. Johnson, L. S. Franke, R. B. Lee and S. D. Holladay, Bioaccumulation of 2,4,6-trinitrotoluene and polychlorinated biphenyls through two routes of exposure in a terrestrial amphibian: is the dermal route significant?, *Environ. Toxicol. Chem.*, 1999, **18**, 873–878.
66. M. S. Johnson, S. D. Holladay, K. S. Lippenholz, J. L. Jenkins and W. C. McCain, Effects of 2,4,6-trinitrotoluene in a holistic environmental exposure regime on a terrestrial salamander: *Ambystoma tigrinum*, *Toxicol. Pathol.*, 2000, **28**, 334–341.

67. M. S. Johnson, J. K. Vodela, G. Reddy and S. D. Holladay, Fate and the biochemical effects of 2,4,6-trinitrotoluene exposure to tiger salamanders (*Ambystoma tigrinum*), *Ecotoxicol. Environ. Saf.*, 2000, **46**, 186–191.

68. M. S. Johnson, H. I. Paulus, C. J. Salice, R. T. Checkai and M. Simini, Toxicologic and histopathologic response of the terrestrial salamander *Plethodon cinereus* to soil exposures of 1,3,5-trinitrohexahydro-1,3,5-triazine, *Arch. Environ. Contam. Toxicol.*, 2004, **47**, 496–501.

69. M. S. Johnson, J. S. Suski and M. A. Bazar, Toxicological responses of red-backed salamanders (*Plethodon cinereus*) to subchronic soil exposures of 2,4-dinitrotoluene, *Environ. Pollut.*, 2007, **147**, 604–608.

70. M. A. Bazar, M. J. Quinn Jr., K. Mozzachio and M. S. Johnson, Toxicological responses of red-backed salamanders (*Plethodon cinereus*) to subchronic soil exposures of 2,4,6-trinitrotoluene, *Environ. Toxicol. Chem.*, 2008, **27**, 1393–1398.

71. P. W. Beatty, M. A. Holscher and R. A. Neal, Toxicity of 2, 3, 7, 8-tetrachlorodibenzo-*p*-dioxin in larval and adult forms of *Rana catesbeiana*, *Bull. Environ. Contam. Toxicol.*, 1976, **16**, 578–581.

72. G. S. Schuytema, A. V. Nebeker, W. I. Griffis and K. N. Wilson, Teratogenesis, toxicity, and bioconcentration in frogs exposed to dieldrin, *Arch. Environ. Contam. Toxicol.*, 1991, **21**, 332–350.

73. R. Mann and J. Bidwell, The toxicity of glyphosate and several glyphosate formulations to four species of southwestern Australian frogs, *Arch. Environ. Contam. Toxicol.*, 1999, **36**, 193–199.

74. M. H. Bernal, K. R. Solomon and G. Carrasquilla, Toxicity of formulated glyphosate (glyphos) and cosmo-flux to larval and juvenile colombian frogs 2. Field and laboratory microcosm acute toxicity, *J. Toxicol. Environ. Health*, 2009, **72**, 966–973.

75. M. S. Johnson, C. Aubee, C. J. Salice, K. B. Leigh, E. Liu, U. Pott and D. Pillard, A review of ecological risk assessment methods for amphibians: comparative assessment of testing methodologies and available data, *Integr. Environ. Assess. Manag.*, 2017, **13**, 601–613.

76. C. L. Rowe and W. A. Dunson, The value of simulated pond communities in mesocosms for studies of amphibian ecology and ecotoxicology, *J. Herpetol.*, 1994, **28**, 346–356.

77. D. J. Fort, M. B. Mathis, F. Kee, P. Whatling, D. Clerkin, J. Staveley and C. Habig, Evaluation of an acute oral gavage method for assessment of pesticide toxicity in terrestrial amphibians, *Environ. Toxicol. Chem.*, 2018, **37**, 436–450.

78. J. R. Bidwell and J. R. Gorrie, Acute toxicity of a Herbicide to selected Frog species, Final Report, Western Australian Department of Environmental Protection, Technical Series 79, 1995.

79. P. Daszak, L. Berger, A. A. Cunningham, A. D. Hyatt, D. E. Green and R. Speare, Emerging infectious diseases and amphibian population declines, *Emerg. Infect. Dis.*, 1999, **5**, 735–748.
80. A. R. Blaustein, B. A. Han, R. A. Relyea, P. T. J. Johnson, S. S. Gervasi and L. B. Kats, The complexity of amphibian population declines: understanding the role of cofactors in driving amphibian losses, *Ann. N. Y. Acad. Sci.*, 2011, **1223**, 108–119.
81. B. E. Sample, M. S. Johnson, R. N. Hull, L. Kapustka, W. G. Landis, C. A. Murphy, M. Sorensen, G. Mann, K. A. Gust, D. B. Mayfield, J.-D. Ludwigs and W. R. Munns Jr., Key challenges and developments in wildlife ecological risk assessment: Problem formulation, *Integr. Environ. Assess. Manag.*, 2023, DOI: 10.1002/ieam.4710.

CHAPTER 3

Risk Assessment Studies in Amphibians: The Case of Nanoparticles in the Aquatic Medium

FLORENCE MOUCHET*[a] AND LAURY GAUTHIER*[a]

[a]Laboratoire Ecologie Fonctionnelle et Environnement, Université de Toulouse, CNRS, INPT, UPS, Toulouse, France
*E-mail: florence.mouchet@ensat.fr, laury.gauthier@univ-tlse3.fr

3.1 Introduction

Ecotoxicology is defined as the study of the effects of toxic chemicals on living organisms and their environment. This includes the interactions between different organisms and their ecosystems. This discipline results from pollution events that harmed human and environmental health by the mid-twentieth century.[1] It is considered a branch of environmental toxicology with more realism as it allows an overall framework for ecologic risk assessment for decision-making.[2]

In addition to a good knowledge of the exposome, the environmental risk assessment (ERA) process requires the use of biological tools to evaluate the hazards incurred by organisms subjected to chemical pollution or processes capable of disturbing the normal functioning of natural ecosystems. In aquatic environments, fish have historically been used as

Issues in Toxicology No. 46
Amphibian Species in Environmental Risk Assessment Strategies
Edited by Marcelo L. Larramendy and Guillermo Eli Liwszyc

Published by the Royal Society of Chemistry, www.rsc.org

relevant indicators to assess water quality,[3] even though other groups of organisms have subsequently proved to be just as relevant or even more so than fish (*e.g.*, various invertebrates, worms, molluscs and insect larvae), particularly because of their key role within ecosystems. Among other vertebrate groups, amphibians have received particular attention from scientists for both ecological and physiological reasons.[4,5] Indeed, from an ecological point of view, this diverse group, which is widespread throughout the world, remains strongly dependent on wetlands of all kinds, even at very high altitudes. Although, in most cases, the young stages remain restricted to aquatic environments, the adult stages often have a terrestrial life phase, which allows them to make the link between these two compartments of the biosphere, which gives them a more integrative character than fish. Their way of life, which is closely linked to their habitat, means that they have a very strong local indicator of the quality of the environment, which has led some authors to call them "plant-animals." Moreover, their particular physiology, linked in particular to the existence of a metamorphosis phase necessary for aquatic/terrestrial adaptation, gives them a natural receptivity to the integration of environmental changes in their general homeostasis. Thus, during their development, these bare-skinned organisms have different types of respiration (tegumentary, branchial and pulmonary), which, at the time of gas and metabolite exchanges, can make them particularly permeable to other constituents of their environment, including pollutants found therein. The consequences of this particular functioning have largely contributed to the general decline in amphibian populations, which has been well documented for several decades now.[6]

The particular sensitivity of amphibians to environmental pollution has led to their use as early indicators of the quality of aquatic environments in particular. Thus, bioassays have been developed in laboratories using amphibians in ecotoxicology to develop relevant tools for assessing the quality of these environments (for review, see A. Egea-Serrano *et al.*).[7] Most of them are addressed to the youngest developmental stages (eggs, embryos and larvae) directly related to water and recognized as the most sensitive. Several species of amphibians belonging to several groups have been used throughout the world to develop these bioassays (the American bullfrog *Aquarana catesbeiana,* the common frog or grass frog *Rana temporaria,* the African clawed frog *Xenopus laevis,* the Iberian ribbed newt *Pleurodeles waltl,* the tiger salamander *Ambystoma tigrinum,* the axolotl *Ambystoma mexicanum*, among others), but in the end, only a few of them have really been integrated into test batteries compatible with use in the context of a regulatory approach, such as the ERA approach.

The use of amphibians in the context of ERA requires the compilation of analytical results obtained during exposure to a small number of substances in the laboratory, most of the time from single-species tests. It is indeed difficult at this stage to use the numerous studies carried out in aquatic environments on natural populations in contamination situations due to the extreme complexity of possible interactions and the multiplicity of exposure products in these complex environments. The use of single-species laboratory tests makes it possible to isolate the problems to be investigated (a single substance, a single species) to have the most reproducible information possible, allowing a realistic inter-comparison of results. For this reason, most of the biological tests used are subject to standardization [*e.g.*, Association Française de Normalisation (AFNOR), Deutsches Institut für Normung (DIN), British Standards Institution (BSI), International Organization for Standardization (ISO) and Organisation for Economic Cooperation and Development (OECD)], which allows them to be integrated into test batteries as part of a regulatory ERA. The danger measured by these tools represents the potential toxicity of the substance under test, which is classically evaluated in terms of acute toxicity and chronic toxicity, depending on the type of parameter measured (mortality, development, reproduction, *etc.*) and the time of exposure (short or long) of the organism to the substance. Another class of toxicity, sometimes wrongly classified as chronic toxicity, concerns the genetic heritage of the organism. This is known as genetic toxicity or hidden toxicity because it is sometimes very difficult to establish a simple cause-and-effect relationship between the event causing the damage and its observation. This is the domain of mutagenesis, teratogenesis and carcinogenesis. Since the historic work of Rostand in 1958,[8] amphibians are well known to be particularly sensitive to this category of toxicity. But above all, much more than other toxicities, this one is of particular importance in terms of long-term prediction for populations, as the risk of increasing the levels of mutations within a population may prove to be much more deleterious in the long term in terms of fitness and adaptability than, for example, acute toxicity (mortality), which will certainly temporarily affect a significant proportion of the exposed organisms but which in the long term will have little or no impact on the population. This is why environmental regulations using biological tests do not attribute the same predictive value to the various toxicities measured by establishing uncertainty factor values specific to each of them to modulate the tolerance thresholds of substances in the environment.

In the laboratory, we have developed a bioassay using a variety of species of Urodeles and Anurans that can effectively detect genome damage in larvae exposed either to chemical substances added to the rearing water or to complex mixtures obtained from samples taken in the natural environment or from industrial processes after treatment. This test, known as the amphibian micronucleus (MN) assay, reveals genotoxic effects by assessing the levels of micronucleated red blood cells in the circulating blood of exposed larvae. The method has been standardized at the national level in France (AFNOR)[9,10] and then internationally (ISO and ISO AFNOR)[11] so that it can be used to assess the genotoxic nature of any type of water, included in the context of a specific ERA process. Furthermore, the methodology developed also allows the assessment of acute toxicity (mortality) and chronic toxicity (growth inhibition) resulting from the exposure of organisms. This tool, which is both sensitive and robust, has made it possible to reveal the potential genotoxic power of numerous categories of chemical substances used alone or in mixtures or to contribute to the knowledge of the impacts of numerous situations of contamination of natural and/or artificial environments, including solid fractions such as soils or sludge.[12–14]

Among the many substances studied, there is one category that has been the object of our attention in recent decades: Emerging Contaminants, for which the lack of knowledge leads to a lack of regulation allowing better control of their use and dissemination in the environment. Within this general framework of our recent work, we are particularly interested in the evaluation of the ecotoxic effects of nanoparticles.

This chapter aims to report on the knowledge acquired in this field over the last 15 years or so using amphibians as biological matrices, in particular through our contribution to the field of nanoecotoxicology.

3.2 Nanoecotoxicology, Nanoparticles and Environmental Context

Nanoecotoxicology is a relatively new field of study that emerged in the early 2000s. The first nanoecotoxicology papers came out in 2006, as reported by Kahru and Ivask in 2013.[15] As the potential risks and impacts of nanoparticlesnanoparticles and nanomaterials on the environment began to be recognized, researchers started investigating the effects of nanomaterials on various organisms and ecosystems.[16–18] Since then, nanoecotoxicology has rapidly developed into a multidisciplinary field that encompasses biology, chemistry, physics, environmental science and engineering. It focuses on understanding how nanoparticles interact with living organisms, such as plants and animals, and their impact on ecosystem health.[19] The ultimate goal of nanoecotoxicology is to develop strategies for minimizing the negative impact of nanoparticles on the environment while also utilizing their potential benefits with respect to safer by design objectives. In this field of research, it seems very relevant to study the potential genotoxic effects of nanoparticles, *i.e.*, how exposure to nanoparticles may cause damage to genetic material, such as DNA, leading to mutations, chromosomal abnormalities and other genetic changes in organisms. Therefore, "nano(geno)ecotoxicology" could be the new term to define, which has been absent from the scientific literature until now. This already exists in practise since it is a question of analysing the genotoxic potential of nanoparticles using comet assays, MN assays and DNA damage assays combined to investigate the physicochemical properties of nanoparticles, such as size, shape, surface charge and surface coating, that affect their genotoxicity.

Nanoparticles, are part of the so-called emerging contaminants, which are substances newly recognized as being harmful to the environment and human health but which are not yet regulated or well understood. Emerging contaminants include chemicals such as pharmaceuticals, personal care products and nanoparticles. Defining what nanoparticles are from a regulatory point of view is an obstacle course. There is sometimes confusion between nanoparticles and nanomaterials. Moreover, there is no single internationally accepted definition for nanomaterials and nanoparticles, which are commonly defined as having a diameter in the range of 1 to 100 nm. The recent paper by Jeevanandam and collaborators in 2018 summarizes the situation well.[20] The ISO has described nanomaterials as a "material with any external nanoscale dimension or having an internal structure or surface structure in the nanoscale"[21] with a size range between 1 and 100 nm. Nanoparticles can be naturally occurring or human-made, and can be found in various shapes, such as spheres, rods, cubes or tubes. Due to their small size and unique properties, nanoparticles have many potential applications in fields such as medicine, electronics, energy and environmental remediation. There are various types of nanoparticles, including mainly (1) inorganic metal nanoparticles such as silver (Ag), gold (Au), aluminium (Al), cadmium (Cd), copper (Cu), iron (Fe), zinc (Zn), lead (Pb) and organic metal oxide nanoparticles such as zinc oxide (ZnO), copper oxide (CuO), magnesium aluminium oxide ($MgAl_2O_4$), titanium dioxide (TiO_2), cerium oxide (CeO_2), iron oxide (Fe_2O_3), silica (SiO_2) and iron oxide (Fe_3O_4), among others; (2) quantum dots [semi-conductors with cadmium selenide (CdSe), cadmium sulphide (CdS), among others]; (3) carbon-based nanoparticles (fullerenes, carbon nanotubes, carbon fibres, graphenes, carbon black) and (4) some other organic nanoparticles formed from organic materials excluding carbon materials, such as dendrimers, lipid-based nanoparticles and polymeric nanoparticles.

Nanoparticles can be released into the environment in various ways at each step of their life cycle, including through manufacturing processes, product use and disposal. Once released, nanoparticles can interact with the environment depending on their physicochemical properties (size, shape, surface charge and surface coating), making them potentially hazardous to living organisms and the environment. Most of the nanoparticles released into the environment end up in the aquatic compartment, which is known to be a "concentrator" of all forms of pollution and, therefore, expected to be especially impacted. Among nanoparticles, engineered nanoparticles such as graphenes, nanodiamonds and carbon nanotubes corresponding to different allotropes of carbon are among the best candidates for applications in fast-growing nanotechnology.[22] This is why the scientific community has taken an interest in them since several years ago.

3.3 Amphibian Models Employed in Ecotoxicology

Amphibian models are very important in ecotoxicology for the study of emerging contaminants such as nanoparticles. Amphibians (*Amphibia*), formerly "Batrachians," are a class of tetrapod vertebrates. They are generally defined as a group that includes the "whole of the non-amniote tetrapods." The three main groups of amphibians currently living are *Anura* (toads and frogs), *Urodela* or *Caudata* (newts and salamanders) and *Gymnophiona* (*Caecilians*), which are quite different in both lifestyle and appearance. The first two are of interest in ecotoxicology because of their potential to serve as models for investigating the mechanisms of toxicity of pollutants on global health, including humans. Many of the physiological and developmental processes that occur in amphibians may occur in humans. Among amphibian species, the anuran *Xenopus laevis* (see Figure 3.1), called the African clawed frog, belonging to the family *Pipidae,* has been used as a model in laboratory studies due to its ability to easily breed and develop in captivity, its sensitivity and its capacity for reproducibility and facility to genomic analysis. *Pleurodeles waltl* (see Figure 3.2), also known as the Iberian ribbed newt and *Ambystoma mexicanum* (see Figure 3.3) are urodel amphibian species commonly used in ecotoxicology research because of their height sensitivity, just as for the *Xenopus. Pleurodeles* and *Ambystoma* are relatively large amphibian species, which makes it easier to work with than smaller species such as *X. laevis*. Their larger size also makes them easier to obtain tissue samples for analysis, including the size of the blood cells and the quantity of DNA. All three have a complex life cycle that involves metamorphosis from a larval stage to an adult stage, but they have some differences in terms of the timing and extent of their metamorphosis. Metamorphosis is determined by environmental

Figure 3.1 Photography of *Xenopus laevis* adult.

Figure 3.2 Photography of *Pleurodeles waltl* adult.

factors such as temperature and food availability for the three, but the metamorphosis of *Xenopus* is more rapid and completed within a few weeks, whereas it needs several months (or even years) to complete in both urodeles. Gradually, the tail of *Xenopus* retracts by apoptosis processes,

Figure 3.3 Photography of *Ambystoma mexicanum* adult.

the limbs and lungs develop, and the gills become baskets. The mouth changes shape to accommodate a carnivorous diet. The larvae of both urodeles develop limbs, gills and lungs, but external gills persist only in *Ambystoma*, whereas they are absorbed in *Pleurodeles*. In all cases, the timing of metamorphosis is influenced by internal hormonal changes.

In amphibian ecotoxicology, various stages of their life cycle may be analysed, depending on the specific research questions and goals. Eggs are studied to analyse the effect of the contaminants on the development and hatching success and allow evaluating future probability of malformations. Larvae are often used because they are the most susceptible stage to environmental contaminants due to their permeable skin and gills. Metamorphs are used to analyse the effects of contaminants on the metamorphosis of amphibian larvae, which involves the dramatic physiological changes mentioned above. Adults are less commonly studied and serve to investigate reproductive success, immune function and overall health. By analysing these different life stages, toxicity mechanisms may be better understood, including how contaminants affect amphibians at different points in their life cycle and how these effects impact the long-term viability of populations in the environment.

Among the four stages mentioned, this chapter will focus on larvae that are mostly used in our work because they are more exposed to the environment and, therefore, more likely to be affected by environmental contaminants, whereas eggs are protected by a jelly-like coating that can reduce the uptake of contaminants. Larvae have permeable skin and gills, which allow for a greater uptake of contaminants from the surrounding water or sediment. Additionally, they are often mobile and actively foraging, which can increase their exposure to contaminants. In our laboratory test, larvae are also preferred because they can be reared under controlled conditions after hormonal induction and provide a longer exposure window to environmental contaminants compared to their embryonic stage. Larvae from a single mating can be pooled together to obtain a larger sample size and especially to reduce the genetic variability, which is extremely important when conducting, *e.g.*, a genetic analysis. Their physiological advantage includes the formation of organs, such as the gills and primary lungs and the development of digestive and excretory systems that can lead to absorption, metabolization and excretion by the organisms. In the case of physical particles in the exposure media, the gills of aquatic larvae are privileged sites of gas exchanges and constitute the entry points of contaminants. Larvae have a functional circulatory system with a developed heart that transports contaminants through the organs. This heart is easily accessible, especially in *Xenopus*, *Pleurodeles* and *Ambystoma*, and allows obtaining a small volume of blood after an appropriate anaesthesia to ensure that the larvae do not experience pain during the procedure. Several biomarker tests are possible to be carried out on the blood, such as biochemical enzymes, including oxidative stress markers, enzymatic activities and biomolecules, such as proteins and lipids. From a genetic point of view, the analysis of the induction of micronuclei in the blood of the larvae proved to be relevant.

To obtain laboratory egg-laying in amphibians, sexually mature males and females were placed together in filtered tap water at 18 ± 2 °C for urodeles and 22 ± 2 °C for *Xenopus*. The rearing and breeding conditions of *X. laevis*, *P. waltl* and *A. mexicanum* were described by Mouchet and collaborators.[23,24] In the case of urodeles, fertilization occurs naturally and does not require hormonal injections as in *Xenopus*. *Xenopus* males are injected with 50 IU of PMSG 500 (Pregnant Male Serum Gonadotrophin) and females with 750 IU of HCG (Human Chorionic Gonadotropin) to induce spawning. Amplexus is the mating behaviour of amphibians and involves the male grasping the female tightly around the body (see Figures 3.4 and 3.5). During this period, the male releases sperm onto the eggs as they are laid by the female (external fecundation). A few hours later (24–48 h), the pair is separated and the eggs are collected and placed in tanks containing 5–6 L of filtered tap water (100 eggs per tank). Viable eggs were maintained at 18 °C for *Ambystoma* and 22 ± 2 °C for *Xenopus* until they reached the development stage appropriate for experimentation: the onset of formation of the two first digits for *Ambystoma* and *Pleurodeles*, and stage 50 of the *Xenopus* development table of Nieuwkoop and Faber of 1956

Figure 3.4 Photography of an amplexus between male and female *Pleurodeles waltl*.

Figure 3.5 Photography of an amplexus between male and female *Xenopus laevis*.

characterized by the hind limb bud being longer than broad, constricted at the base.[25] After hatching, the urodeles were fed on freshly hatched brine shrimp aquatic crustacean artemia (*Artemia salina*) and the *Xenopus* on dehydrated aquarium fish food as soon as they reached the feeding stage, *i.e.*, a few days after laying. One-third of the water is renewed daily before feeding the three. When the young axolotl and *Pleurodeles* reach around 30 mm, the food is switched to frozen chironomid larvae. *Xenopus* larvae are still herbivorous until they metamorphose, and urodeles are carnivorous. All the amphibians described in this piece of work originated from our certified breeding facilities (under the European approval number A31113002) and the experimental procedure was approved by an ethics committee (CEEA-073). All larvae were anaesthetized by bathing in a buffered solution of tricaine methanesulfonate (MS222) to perform further analysis, and APAFIS ("Autorisation de Projet utilisant des Animaux à des Fins Scientifiques" which can be translated as "Authorization for Projects using Animals for Scientific Purposes") was authorized. Managed by the French Ministry of Higher Education, Research and Innovation, APAFIS is the French national online platform for authorization before any animal experimentation can be conducted in research facilities.

3.4 Amphibian Studies on Carbon Nanoparticles

3.4.1 Carbon Nanotubes

Carbon nanotubes (CNTs), one allotrope of carbon, are one-dimension nanoscale objects (called 1D objects) with an extremely high aspect ratio (the ratio length/diameter ranges between 5000 and 50 000) and a very high specific surface area (800–1000 m^2 g^{-1}). CNTs have a diameter between 1 and 100 nm and a length of less than 1 µm up to tens of µm or more.[26] Their structure can be described as a graphene sheet rolled up to form a cylinder. Among them, two main types are well described. Double-walled carbon nanotubes (DWNTs) consist of two concentric cylinders of carbon atoms with a hollow space in between. They are similar in structure to single-walled carbon nanotubes (SWNTs), but with an extra layer of carbon atoms. DWNTs can have different diameters and lengths and also be functionalized with various molecules to make them suitable for specific applications in fields such as electronics, energy and medicine. Multi-walled carbon nanotubes (MWCNTs) are another type of CNT consisting of several layers of graphite cylinders nested within one another and forming a tubular structure.

The first of the studies dealing with the ecotoxicity of CNTS in amphibians concerns the larvae of the axolotl *A. mexicanum* and DWCNTs, published in 2007 in the *Nanotoxicology Journal*.[23] In this study, we employed a classical risk assessment approach to assess the potential environmental risk posed by a new contaminant. Our first step involved investigating the potential acute toxic effects of this new material on organisms, specifically by examining the biological effects using amphibian larvae. Due to non-existent data on the potential biological effects of CNTs on amphibians, we exposed the larvae to a broad range of DWNT concentrations, including very high concentrations (ranging from 1 up to 1000 mg L^{-1}) to observe any potential toxic effects on our test organism. Controlled laboratory conditions were used to expose axolotl larvae to CNTs for 12 days, with the aim of evaluating two different endpoints according to the French standard MNassay:[10] (1) acute toxicity on larvae and (2) genotoxicity as indicated by the appearance of micronucleated cells in erythrocytes of the running blood, which serves as an expression of clastogenic and/or aneugenic effects of CNTs. Additionally, the study aimed to characterize the DWNTs that were exposed to amphibian larvae to highlight any potential toxic effects of this material. The experimental exposure procedure was carried out following the AFNOR recommendations[9] and adhered to the French Standard AFNOR NF T90-325 (AFNOR 2000) and the international standard ISO 21427–1,[11] but with adaptations to ensure more realistic environmental exposure conditions that involved no daily water renewal. Reconstituted water (RW), composed of distilled tap water with added nutritive salts (294 mg L^{-1} $CaCl_2.2H_2O$, 123.25 mg L^{-1} $MgSO_4.7H_2O$, 64.75 mg L^{-1} $NaHCO_3$ and 5.75 mg L^{-1} KCl), was used to expose amphibian larvae. Exposure began when the hind limb buds of the larvae exhibited slight indentations, indicating the formation of the first two digits. The larvae were exposed in groups of 5 or 13 animals (100 mL $larva^{-1}$) in glass-crystallizing dishes containing either the control medium (negative control) or the test medium (CNTs at concentrations of 0.5, 1, 10, 100, 125, 250, 500 and 1000 mg L^{-1} in water). The larvae were kept at a temperature of 20 ± 0.5 °C with a 12-h light : 12-h dark cycle and fed with frozen chironomid larvae every day. This first present work revealed that DWCNTs deposited in water did not result in any acute toxicity or cause any induction of MNs in the erythrocytes of axolotl larvae, regardless of their concentration in water. However, upon visual inspection, it was observed that large black masses of ingested CNTs were present inside the gut even at the lowest concentration. In *Ambystoma* larvae, xenobiotics can enter the body through two main routes, the dermal and digestive routes, which may lead to high ingestion rates of particles, particularly those present in food or vehicle-agglutinated CNTs and water-containing suspended particles.

The second study aimed to compare the response obtained in larvae of *Xenopus* in the presence of the same DWCNTs according to the same strategy based on a classical risk assessment method.[24] Because of the absence of biological effects of DWNTs on *Ambystoma* larvae exposed to a wide range of DWNT concentrations in the previous study,[23] *Xenopus* larvae were also exposed to a wide range of concentrations (from 10 to 500 mg L^{-1}) under controlled laboratory conditions. The concentrations studied in the early days of nanoecotoxicology were very high and considered to mimic potential accidents, releasing the nanoparticles during their production or transport. The exposure conditions were later adapted to more realistic environmental conditions. According to the French and the International standards,[10,11] exposure was conducted without any renewal of the medium during test exposure instead of the semi-static exposure condition (daily renewal of the medium) of the standardized procedure and an air diffuser was placed in the water of each larvae group to promote DWNT dispersion in the media and compensate for the potential oxygen deficit in the water due to static exposure conditions. The reconstituted water of the exposure medium was the same as for the previous work in *Ambystoma*. The larvae were exposed in groups of 20 animals (100 mL larva^{-1}) in 2 L Pyrex™ crystallizing dishes containing either the control medium (RW) or the test containing the concentrations of DWNTs. The larvae were exposed to a natural light-dark cycle at 22.0 ± 0.5 °C during the 12 days of exposure. They were fed every day on dehydrated aquarium fish food. After 12 days of exposure, (1) acute toxicity on larvae (mortality and growth) and (2) genotoxicity as the expression of the clastogenic and/or aneugenic effects observed in the erythrocytes in the running blood were then analysed. Finally, the presence of DWNTs was investigated in the larvae using light and electron microscopy methods. The Raman spectrometry analysis was used to confirm the presence of DWNTs. Consistent with our previous findings in axolotl exposed to the same DWNTs and concentration range, our study found no evidence of MN induction in the erythrocytes of *Xenopus* larvae when exposed to rough DWNTs deposited in water, regardless of the concentration or exposure conditions. On the contrary, acute toxicity was observed in *Xenopus* larvae exposed to DWNTs, even at the lowest concentration (10 mg L^{-1}), with a high mortality rate of 85% at the highest concentration of 500 mg L^{-1}, suggesting that differences in sensitivity between these two species might be related to their biology, larval growth and behaviour, with *Xenopus* larvae often being more sensitive than urodel species to environmental substances.[27]

The third study contributed to the ecotoxicological assessment of the potential impact of DWNTs using the standardized method ISO 21427-1[11] but in the presence of a dispersant agent called Gum Arabic (GA), a natural polysaccharide.[28] *Xenopus* larvae were exposed in a group of 20 to concentrations ranging from 0.1 mg L^{-1} (to mimic potential environmental doses) to 50 mg L^{-1} (which may represent an accidental release and optimize the observation of the potential toxic effects), with and without GA in RW, under the same controlled laboratory conditions to evaluate the same two endpoints after 12 days of exposure: toxicity and genotoxicity. Then, the presence of DWNTs was also investigated in the larvae using traditional microscopy methods (photonic and electronic) and Raman spectroscopy to confirm the presence of DWNTs. The results showed no mortality of the larvae irrespective of the experimental condition (with or without GA), except for a small percentage at high concentrations of DWNTs without GA, where 5% and 15% of mortality were observed at 10 and 50 mg L^{-1} without GA, respectively. The larvae exposed to high concentrations of DWNTs (10 and 50 mg L^{-1}) were smaller, while those exposed to lower concentrations (0.1 and 1 mg L^{-1}) did not show any signs of toxicity. The growth rate decreased in a dose-dependent manner. Genotoxicity was not observed except for exposure to 1 mg L^{-1} of DWNTs with 3% GA. Raman spectroscopy showed the presence of CNTs in the intestine, but not all the black material seen in the microscope was necessarily DWNTs, as already suggested in the previous work.[24] The Raman signal decreased from the intestine lumen to the wall while the signal of the *Xenopus* biological matrix increased.

In both studies, toxicity could be influenced by various biological mechanisms. The first mechanism involved branchial obstruction and/or disturbance of gaseous exchanges in the gills. If the gills become obstructed or their normal function is disturbed, this could result in anoxia and contribute to acute toxicity. The second mechanism involves intestinal obstruction. If (nano)particles are ingested by an organism, they could potentially cause blockages in intestines, preventing food and other essential nutrients from being properly digested and absorbed. The third mechanism involves competition between CNTs and nutritive elements. If the organism is unable to obtain enough nutrients from its diet due to competition with the CNTs, this could also contribute to acute toxicity. Demonstrating the presence of CNTs in *Xenopus* larvae after exposure was necessary and initially accomplished through visual inspection of the larvae under a binocular, which revealed black-coloured and potentially blocked internal gills, indicating that anoxia may have contributed to the observed toxicity. To explore the second hypothesis, the larvae exposed to DWNTs were visually inspected under a binocular, revealing distended aspects in the intestine walls and black masses in the intestine lumen after dissection, which indicated potential absorption of the CNTs present in the exposure medium; in amphibian larvae, besides the gills, xenobiotics can enter the body through the dermal or digestive routes, with the latter potentially leading to high ingestion rates of particles, particularly those of food susceptible to vehicle-agglutinated CNTs and water with suspended particles. To determine if ingestion of CNTs caused toxic effects in *Xenopus* larvae, the presence of CNTs in their intestines was examined under microscopic observation and black masses were seen in the gut lumen of exposed larvae. However, it was difficult to identify CNTs in these masses. Transmission Electronic Microscopy (TEM) also did not confirm the presence of CNTs in the digestive tissue or lumen, making it impossible to distinguish between exposed and non-exposed animals. Additional investigations were conducted to confirm the presence of DWNTs in the black masses found in the *Xenopus* larvae. High Resolution of TEM (HRTEM) observations and Raman analysis were performed to verify the presence of CNTs. HRTEM observations revealed the presence of CNTs encapsulated in graphite shells, where circular features that could correspond to cross-sections of CNTs were also visible. Raman images confirmed the presence of CNTs in the intestines of the exposed larvae.[23] However, the image superposition analysis showed that not all black masses contained CNTs and some CNTs were not visible under light microscopy. More detailed analysis in Raman imaging allowed us to conclude their absence in the intestinal wall (villi and cells), suggesting that DWNTs do not cross the intestinal barrier.[24] The addition of GA helped to keep DWNTs well dispersed in the water when there were no larvae present.[24] However, when amphibian larvae were introduced to medium exposure, the DWNTs became unstable and clumped together. This was likely due to the larvae ingesting the GA, which reduced its concentration in the water and altered the pH of the solution.

The following study aimed to compare the effects of double and multi-walled CNTs in the same amphibian larvae of *Xenopus* to contribute to the ecotoxicological assessment of the potential impact of CNTs on the environment.[29] Due to the important differences in nature existing among both these CNTs, their potential toxic effects would be expected to be different and, therefore, must be investigated. *Xenopus* larvae were exposed in groups of 20 animals for 12 days, with a daily renewal of medium, according to the standardized method ISO 21427-1 (ISO, 2006) to 0.1, 1, 10 and 50 mg L^{-1} of MWNTs in RW. Once again, acute and genetic toxicities were observed, with a focus on the potential detection of these particles using observations of histological organs and TEM preparations. Carbon nanotubes were finally tracked using their Raman signal. As for DWCNTs in *Xenopus* and *axolotl*, no mortality at any concentration of DWCNTs was observed and the measurements of the larval size showed that the larvae exposed in the presence of 50 mg L^{-1} of MWNTs had also significantly reduced size compared to NC in a dose-dependent manner. In the same way, no genotoxicity via MN induction in the erythrocytes of *Xenopus* larvae was observed regardless of CNT concentration. Raw CNTs probably cannot enter the cells owing to their size and their strong aggregation in water. Macro observation of the gills showed no difference whatever the MWNT concentration, whereas the intestinal tract from exposed larvae presented black masses that were particularly visible through the thin peritoneal membrane of the larvae owing to their deep black colour. The proportion of black masses seemed to increase with the MWNT concentration. MWNTs were observed by TEM both in the lumen and in the villi. TEM observations were easier than in the DWNTs owing to their higher contrast and diameter. The Raman analysis confirmed that MWCTs do not cross the intestinal wall. MWNTs showed less acute and chronic toxicity compared to DWNTs in terms of concentration, with growth inhibition only observed in larvae at 50 mg L^{-1} of MWNTs compared to 10 mg L^{-1} of DWNTs in a previous study. This lower toxicity in the larvae may be due to the larger size or aggregate size of MWNTs, making it more difficult for them to enter organisms. Concentration units could be used only in the case of comparing the same kind of CNTs. The size of aggregates of CNTs in RW especially is thought to be a primary concern for toxicity. A few years later,[30] we demonstrated that a unit of comparison is essential. Mass concentration as a quantitative measure (metric) of exposure is not the most relevant descriptor at the nanoscale level. Surface area is defined as the most relevant descriptor of toxicity for different types of carbon allotropes.

In Bourdiol and collaborators in 2013,[31] a focus has been realized on the effects of the dispersion of the CNTs on the amphibian larvae after the publication of Hyung and Kim in 2008.[32] Their study demonstrated that raw CNTs have a hydrophobic surface that can be covered by natural organic matter or surfactants present in natural systems. The negative surface charge and large molecular size of CNTs can then cause charge repulsion and steric hindrance, respectively. This can enhance the dispersion and stabilization of CNTs at concentrations relevant to the environment, depending on water properties such as pH, ionic strength, total organic carbon content and salt concentration. For this reason, we designed a new exposure protocol to compare the biological responses of *X. laevis* larvae exposed in semi-static conditions to size-reduced agglomerates of MWCNTs in suspension in the water column and/or to larger agglomerates in order to maximize the exposure to individual CNTs/bundles and to limit agglomeration. Suspensions of CNTs were prepared using a combination of non-covalent functionalization with a non-toxic polymer [either carboxymethylcellulose (CMC) or GA] and mechanical dispersion methods (mainly ultrasonication). The same endpoint of growth inhibition and genotoxicity and the tracking of CNTS into organisms have been followed every time in the same exposure conditions according to the standardized method ISO 21427-1 to 0.1, 1, 10 and 50 mg L^{-1} of MWNTs in RW.[11] Agglomerates that settled down were found to cause disruption of intestinal transit and nutrient assimilation, leading to acute and chronic toxicities at higher concentrations. Suspended MWCNTs in the water column resulted in increased mortality and decreased growth rate and genotoxicity at low concentrations (1 mg L^{-1} in the presence of CMC). These effects were mitigated in the presence of GA, a potential antioxidant. It was suggested that MWCNTs may primarily affect the surface of the gills, acting as a respiratory toxicant and inducing indirect effects. Macro and TEM observations showed that MWCNTs were absorbed and moved within the digestive tract (*i.e.*, in the lumen) before being at least partially (if not totally) excreted in the exposure medium. For the first time, genotoxicity was revealed at 1 and 10 mg L^{-1} of MWCNTs in the presence of CMC. Larvae exposed to 50 mg L^{-1} of CMC-stabilized MWCNTs had a significantly lower mitotic index compared to the negative control. Although there was no significant induction of micronuclei erythrocytes (MNE) at this concentration, it is possible that the decrease in mitotic index could disrupt erythrocyte mitosis. Finally, results suggested that 1 mg L^{-1} of CMC-stabilized MWCNTs may be the threshold level at which genotoxic response is induced after 12 days of exposure. When MWCNTs were stabilized with GA, there was no induction of acute and chronic toxicity or genotoxic response at low concentrations. The concentration of MWCNTs was lower in the water column when stabilized with GA compared to CMC, which may have affected the biological responses. That is why it is possible that MWCNTs induce reactive oxygen species in the gills of larvae, leading to DNA damage and MN induction. However, the use of GA as a dispersing agent may have mitigated their genotoxicity, resulting in a non-significant induction of MN at the same concentration. The partitioning of CNTs is influenced not only by the presence of dispersants/stabilizers in natural water but also by the nature of the latter (*i.e.*, mixtures of amino acids, fulvic and humic substances, proteins, lipids, among others), make the accurate prediction of the ecotoxicological effects of CNTs on the aquatic organisms extremely difficult. The use of TEM to observe the internalization of MWCNTs in the intestines of larvae has been discussed. We used a contrast treatment to make the MWCNTs visible in the ultra-thin sections of the intestines and found that the MWCNTs were difficult to distinguish from the biological matrix in deeper layers. However, isolated MWCNTs in the apical region (microvilli) and the basal membrane region of enterocytes were observable using a different staining method. They looked randomly distributed, parallel to each other and oriented along the direction of the cutting. On the other hand, accumulations of MWCNTs were found on opposite sides of the preparation. We concluded that MWCNTs may have been artefacts caused by the cutting process used to prepare the samples. Finally, the internalization of MWCNTs in the intestines of larvae may be difficult to observe and requires careful sample preparation and analysis.

As this genotoxicity intrigued us, we focused our attention on the impact of MWCNTs using the amphibian model *X. leavis* at low concentrations and under certain experimental conditions to understand the mechanism of toxicity of these nanoparticles and whether they are able to induce an early response.[33] Amphibian larvae were then exposed for short times (2, 4, 8, 12 and 24 h) to three different concentrations of MWCNTs (0.1, 1 and 10 mg L^{-1}) to evaluate their potential toxicity in terms of oxidative stress with enzyme activities [hydrogen peroxide (H_2O_2) evaluation, glutathione reductase (GR), catalase (CAT) and super oxide dismutase (SOD) analyses] and DNA damage induction using the Comet Assay and MN assay. Results showed significant H_2O_2 production when larvae were exposed to 1 mg L^{-1} and 10 mg L^{-1} of MWCNTs after 2 and 4 h of exposure, respectively. The activity of certain antioxidant enzymes (CAT, GR and SOD) was also increased in response to exposure to MWCNTs, from only 2 h of exposure to 10 mg L^{-1} of MWCNTs. In the presence of 1 mg L^{-1} of MWCNTs, only GR and CAT activities were significantly induced at 4 h. Enzyme activities do not follow a simple dose-effect relationship, but the time of induction is shortened in relation to the tested concentration. Results of the Comet assay suggested that the larvae were experiencing oxidative stress. The profiles of DNA damages showed fluctuations in the course of time, which are characteristics of an oxidative stress response in relation to the continuous balance between damage and compensation processes. We observed the presence of agglomerated MWCNTs in the gills and in the intestine of tadpoles at various concentrations and time points, although agglomerated MWCNTs were rapidly excreted after 24 h. The entry pathway of MWCNTs in *X. laevis* larvae is through dermal exposure (integument and gills) and breeding exposure. The high ingestion rates of suspended particles by *Xenopus* larvae, especially of food susceptible to binding CNTs, could result in the adsorption of CNTs on the larval surface and gill surfaces, inducing gill clogging and gas exchange disturbance, as already suggested in previous works. This disturbance could promote larval hypoxia and oxidative stress, leading to primary DNA damages and induction of antioxidant enzymes. The presence of MWCNT in enterocytes, hepatocytes, branchial and blood cells using specific microscopy tools such as TEM and Raman spectroscopy has still not been demonstrated.

3.4.2 Graphenes

For a few years, researchers shifted their focus from studying CNTs to graphene in ecotoxicity studies. First, graphene is easier and less expensive to be produced in large quantities.[34,35] Its similar properties also make it an interesting material to study in terms of its environmental impacts. Graphene is more stable and less prone to agglomeration, making it easier to study. It is a two-dimensional allotrope of carbon that consists of a single layer of carbon atoms arranged in a hexagonal lattice.[36] It is the basic building block of other carbon-based materials such as graphite, carbon nanotubes and fullerenes. Graphene has unique properties such as high electrical conductivity, mechanical strength and thermal conductivity, making it useful for a variety of applications in areas such as electronics, energy storage and water filtration.[37] Graphene was first isolated in 2004 by researchers at the University of Manchester, who used sticky tape to peel off thin layers of graphite until they were left with a single layer of carbon atoms. The discovery of graphene earned Andre Geim and Konstantin Novoselov the Nobel Prize in Physics in 2010. Considering its popularity, a rapid increase in production and applications of graphene implied its potential release into the environment. The issue quickly became apparent to the entire environmental community of scientists. What is the environmental impact that can be expected after graphene's introduction into the aquatic environment? The study of Muzi *et al.* contributed to its toxicity assessment with the examination of the impact of Multi-Layer Graphene (MLG) belonging to the Graphene-based Nanomaterials (GBMs).[38] MLG is constituted by two to five graphene layers, while graphene oxide (GO) and reduced graphene oxide (rGO) are normally composed of a single layer. Exposure of *X. laevis* larvae was performed according to the International Standard 21427–1[11] in semi-static exposure conditions with the larvae at developmental stage 50.[25] They were exposed for 12 days in groups of 20 animals in crystallizing dishes containing 0, 0.1, 1, 10 and 50 mg L^{-1} of MLG in RW. No mortality in larvae exposed to MLG was observed, regardless of the concentration. However, larvae exposed to 10 and 50 mg L^{-1} of MLG had a significantly reduced size compared to the negative control. There was no significant reduction in larval size at 0.1 and 1 mg L^{-1}. The study also assessed whether MLG could induce genetic damage. However, no genotoxicity *via* MNs induction was found. Larvae exposed to MLG in RW showed black agglomerates in the gills and intestines as the concentration of MLGs in the media exposure increased, and larval excretion of MLGs agglomerated with excrements was easily observed. The study concluded that MLG is not toxic to *Xenopus* larvae. Nevertheless, only the growth rate was affected to some extent, with the larval size decreasing by about 40% upon exposure to the highest concentration of MLG.

Considering the *Xenopus* larvae as amphibian models, the following works investigated how different routes of dispersion of C-based nanoparticles (CbNPs) in water (covalent by grafting of oxygen-containing functions) or non-covalent by adsorption of a chemical dispersant may impact the potential chronic toxicity in amphibians.[30,39] Depending on the state of dispersion of each CbNPs, their bioavailability for *X. laevis* was modified. By using a dose-response model, this work compared the chronic toxicity observed on *Xenopus* larvae after a 12 day *in vivo* exposure to raw, oxidized Callotropes or in the presence of chemical dispersants. We showed that chemical dispersion does not influence the observed chronic toxicity, whether it is through surface chemistry (oxidation state) or the addition of a dispersant. As they are pelagic, larvae were more exposed to dispersed CbNPs in the water column than raw CbNPs. *Xenopus laevis* larvae are able to feed through both filtration and grazing. This particularity makes it indifferent to the dispersion state of CbNPs since it will be exposed both ways. So, whatever the kind of CbNPs and their state of dispersion, they are transferred via the general cavity to the basket gills and the digestive tract of larvae and finally excreted. As filter-feeder organisms, *Xenopus* larvae actively filter the water column containing particles in suspension and ions from RW and food, which will be found in basket gills, oral cavity, pharyngeal cavity and intestinal tract. Thus, it appears that regardless of their oxidation state or their load of organic matter, depending on their surface area, CbNPs could lead to larvae growth inhibition.

Numerous other works followed on the assessment of the risk of graphene in the *Xenopus*[40–44] and *X. laevis* is the biological model that has the privilege to study the effects of graphenes within the framework of the well-known European project FLAGSHIP (https://flagships.eu/). This project focuses on researching and developing applications for graphene and is funded by the European Union. The Graphene Flagship project aims to investigate and develop graphene and other 2D materials for a wide range of applications, including electronics, energy, biomedical devices and environmental sensors. The project is divided into several work packages (WP), each focusing on a specific area of research. WP4 of the Graphene Flagship project is focused on the environmental impact of graphene and 2D materials, and aims to develop safe and sustainable methods for the production, use and disposal of these materials. This includes studying the toxicity of graphene and its potential effects on human health and the environment and developing methods for monitoring and mitigating any potential risks. The ultimate goal of WP4 is to ensure that graphene and other 2D materials can be used safely and sustainably in a variety of applications.

The objective of these studies was to evaluate the genotoxicity of GO in *X. laevis* and gain insights into the toxicological mechanisms involved in the genotoxic response following exposure to a commercial form of the material. The study also investigated the role of oxidation degree and surface functions in the toxicological response by subjecting the material to thermal treatments in a hydrogen H_2 atmosphere at two different temperatures (200 °C and 1000 °C) to produce rGO with varying surface functions.[40] The materials were fully characterized to identify the contribution of functional groups to the genotoxic response. Groups of 20 larvae were exposed for 12 days under semi-static conditions with daily feeding and exposure media renewal (RW) following the international standard ISO 21427-1 procedure. GO and rGO tested concentrations ranged from 0.1 to 50 mg L^{-1}. The results showed that GO caused disruptions in the erythrocyte cell cycle, leading to cell accumulation in the G0/G1 phase and that low concentrations of GO (0.1 mg L^{-1}) induced genotoxicity through oxidative stress in exposed larvae. However, reduction of GO at 200 °C and 1000 °C eliminated genotoxicity at low concentrations. The analysis of the relative gene expression levels in the livers of larvae exposed to GO and rGO after 12 days of exposure revealed a significant induction of most of the studied genes to GO at 0.1 mgL^{-1}. Some genes involved in oxidative stress response and inflammation [gpx1, sod(Cu/Zn), sod(Mn), pparγ, 5-lox and cox1] were significantly induced from 2.6 to 5.84 times more than the negative control. Finally, some detoxification processes occurred, as shown by the induction of *cyp1a1* and *tap*. On the contrary, no significant modulation of gene expression was noticed in rGO conditions. X-ray photoelectron spectroscopy analysis suggested that epoxide groups may be responsible for the genotoxic effects of highly oxidized GO. The study based on *Xenopus* responses proposes that thermal reduction of GO could be a safer alternative for developing environmentally friendly materials.

Previoulsy,[43] we were interested in the endocrine disruption potential of the same GO and rGO using a triiodothyronine (T3)-induced *Xenopus* metamorphosis assay adapted from reported works.[45] This assay was used to detect chemical disruption of thyroid hormone (TH) signalling in *X. laevis* larvae. Unlike the spontaneous metamorphosis-based assay, this method involves early exposure to active T3 hormone to induce metamorphosis and assess quantitative morphological endpoints. Chemicals that inhibit or stimulate T3-induced changes are identified as TH signalling antagonists or agonists, respectively. The assay provides quick results with improved sensitivity compared to previous methods.[46,47] The objectives of this study were to go beyond the few existing data concerning the evaluation of GBM endocrine disruption potential in the African clawed frog. This work aimed to characterize possible endocrine disruption mechanisms associated with GBM exposure and to disentangle the toxicity-related effects of endocrine disruption. The main hypothesis was that the previously observed effects described in tadpoles of the species were associated with toxicity rather than with thyroid endocrine disruption. Eighteen *X. laevis* tadpoles at stage 52 were exposed to GBMs at concentrations of 0, 0.01, 0.1 and 1 mg L^{-1} in the presence or absence of T3.[25] The exposures were conducted for 96 h in RW. The exposure media, contamination and feeding were renewed daily. However, the exposure to the GBMs in the absence of the T3 hormone did not lead to the induction of T3-like effects. The results indicated that GBMs potentiated the effects of exogenous T3 with a more marked effect of GO compared to rGO. T3 quantifications in the exposure media indicated adsorption of the hormone on GBMs, increasing its bioavailability for organisms because it gets accumulated in the gut and gills of these amphibians. This study highlights that the tested GBMs do not disrupt the thyroid pathway in amphibians but indicates that the adsorption properties of these nanomaterials may increase the bioavailability and toxicity of other pollutants.

As GO is known to exert bactericidal effects, whereas the gut microbiota constitutes a compartment involved in host homeostasis regulation, it is important to determine if this microbial compartment constitutes a toxicological pathway involved in known GO-induced host physiological impairments. In this way, a recent study aimed to investigate the potential link between gut microbial communities and host physiological alterations induced by GO exposure.[44] For this purpose, *X. laevis* tadpoles were exposed during 12 days to GO. The growth rate was monitored every 2 days and the genotoxicity was assessed through the enumeration of MN erythrocytes. Genomic DNA was also extracted from the whole intestine to quantify gut bacteria and analyse the community composition. As previously observed in *Xenopus* larvae, a strong intestinal accumulation of GO was noticed in the gut of larvae exposed to any GO concentration after 2 days of exposure and up to the end of the experiment. At the end of the exposure, no developmental stage delay was observed. The metamorphosis process of tadpoles involves significant changes in their intestinal structure and complexity. In premetamorphic tadpoles, the intestine is a simple, thin tube that undergoes elongation, looping and rotation events. This results in increased gut complexity through crypt and villi structuration and shortening at climax, which occurs under the action of the T3 hormone. Larvae exposed to GO in this study did not exhibit significant differences in intestinal weight compared to unexposed larvae after 2 days of exposure. However, after 12 days of exposure to GO at 10 mg L^{-1}, a significant decrease in intestinal weight was observed, indicating impairment of intestinal development. No developmental stage delay was observed at the end of the experiment, but the shortening event that occurs during the climax stage was unlikely to occur due to GO exposure. Most of the tadpoles reached NF developmental stage 57 in all experimental conditions. The impairment of intestinal development in this study was less marked compared to the larval growth inhibitory effects determined based on larval length. GO exposure led to dose-dependent growth inhibition and genotoxic effects were detected following exposure to low doses. A transient decrease in total bacteria was noticed with a persistent shift in the gut microbiota structure in exposed animals. Genotoxic effects were associated with gut microbiota remodelling, characterized by an increase in the relative abundance of the anaerobic Gram-negative bacterium *Bacteroides fragilis*. The growth inhibitory effects would be associated with a shift in the Firmicutes/Bacteroidetes ratio, while metagenome inference suggested changes in metabolic pathways and upregulation of detoxification processes. These findings suggest that the gut microbiota is an important biological compartment that should be considered in ecotoxicological studies, as structural or functional impairments could lead to host fitness loss.

As standardized single-species-based assays fail to highlight toxicological pathways implying interactions between organisms, the use of micro/mesocosms appears to be a good solution to fill the gap of environmental realism inherent to these tests. In this work, microcosms were used to expose organisms interacting through a reconstituted trophic chain to GO at environmentally relevant concentrations (0.05 and 0.1 mg L^{-1}).[41] The trophic chain was composed of a consortium of algae and bacteria as primary producers, chironomids and *Xenopus* larvae as primary consumers and decomposers, while larvae of *P. waltl* constituted the secondary consumers. Both concentrations of GO were tested: 0.05 and 0.1 mg L^{-1}. Each condition was tested in triplicate with a random assignation of the microcosms. No endpoint was measured in *X. laevis* tadpoles that were all predated during the experiment by *Pleurodeles* larvae. For *Pleurodeles* larvae, mortality, growth and genotoxicity were studied after 10 days of exposure. No mortality in *Pleurodeles* larvae was observed during the whole experiment and larval growth was not affected by the GO exposure. In contrast, genotoxic effects were noticed in *Pleurodeles* larvae at the end of the exposure. This study highlighted the importance of using such experimental systems to better encompass the ecotoxic potential of GO through the determination of toxicological routes and consequences on the ecosystem's functioning. GO is able to induce genotoxic effects *in vivo* at low concentrations under realistic environmental conditions. This was also observed in *X. laevis*, which is consistent with the present results. The genotoxic response measured in *X. laevis* after exposure to 0.1 mg L^{-1} was associated with oxidative stress and a pro-inflammatory response, leading to an increase in circulating MN erythrocytes.[40] *Pleurodeles* larvae were exposed to GO simultaneously through direct exposure due to GO in the media and dietary exposure by feeding on contaminated chironomids. However, a Principal Component Analysis indicated that the marker of genotoxicity was independent of those associated with other trophic levels, which reinforces the hypothesis of effects associated with direct exposure over trophic effects. In the case of direct exposure, the contamination was performed sequentially and, according to previous studies indicating that at least 4 days of exposure to a genotoxic compound are necessary to induce the formation of MNs.[48] Accordingly, we can suggest that a concentration of 0.8 mg L^{-1} of GO is able to induce genotoxic effects (concentration reached 4 days before the sampling). It seems that the predation behaviour of the *Pleurodeles* was not influenced by GO exposure, as evidenced by the high decrease in chironomid larvae and the absence of differences in chironomid survival between the tested conditions at the end of the exposure.

3.5 Amphibian Studies on Metallic Nanoparticles

Nanoparticles of Ce dioxide (CeO_2 NPs) and Ti (TiO_2 NPs) have been investigated. In both cases, the aim was to provide wider information on CeO_2 and TiO_2 NP toxicity in aquatic environments at different trophic levels and environmentally relevant exposure conditions.

In the case of CeO_2 NP, two approaches were investigated. The first is the traditional method of studying ecotoxicity in monospecific assays[49] and the second is the use of mesocosms to simulate environmental conditions and study the impact of nanoparticles on ecological processes.[50,51] The first approach involves testing the effects of nanoparticles on a single species in a controlled environment, while the second approach allows for the examination of the broader impact of nanoparticles on ecosystems, including interactions between multiple species and the effects on ecosystem functions such as leaf litter decomposition. Two types of CeO_2 NPs were studied by Bour and collaborators.[49–52] Both CeO_2 NPs showed a size range of 2–5 nm and were coated with a tri-ammonium citrate layer. CeO_2 NP toxicity was studied on diatoms algae *Nitzschia palea*, on Diptera *Chironomus riparius* and amphibian larvae of *X. laevis* and *P. waltl* according to monospecific assays, *i.e.*, species by species and independently,[50] and mesocosms experimentation, *i.e.*, with all together.[51,52] Acute toxicity test was performed on *Xenopus* at larval stage 50 of the developmental table of Nieuwkoop and Faber after 96 h (adapted from the 96 h LC50 acute fish toxicity test) and after 12 days (ISO, 2006) of exposure to CeO_2 NPs from 0.1 up to 100 mg L^{-1} of RW.[25,53] Exposure of *Pleurodeles* larvae was also carried out according to the standardized conditions in RW at the same concentration of CeO_2.[11] Results showed that at the highest concentration of one type of NPs, 35% of *Xenopus* larvae died after 12 days of exposure but not after 96 h of static exposure, suggesting variation depending on the exposure mode. NPs do not act like traditional contaminants, such as metals or pesticides, and tend to agglomerate and settle, making exposure duration, medium renewal and other exposure factors crucial in terms of toxicity. Both types of NPs caused growth inhibition in *Xenopus* larvae at a concentration of 1 mg L^{-1}. On the other hand, only the highest concentration of one type of NPs caused growth inhibition in *Pleurodeles*. Additionally, while no genotoxicity was observed in *Xenopus*, *Pleurodeles* showed dose-dependent genotoxic effects when exposed to one type of NPs. It was suggested that the observed differences in toxicity between the two species may be due to the specific compartments, routes of exposure and differences in the NPs themselves. Finally, *Xenopus* larvae seemed to be more sensitive than *Pleurodeles* when assessing mortality or growth inhibition. The opposite response was observed for genotoxicity and may be explained by a difference in CeO_2 pharmacokinetics in the organism, associated with the physiology of these organisms (different feeding behaviour, external or internal gills) and by different haematopoiesis localization in *Xenopus* and *Pleurodeles*. MNs result from chromosome breaks (or mitotic anomalies) that require a passage through mitosis to be recognizable. So, the induction of MNs depends on the sensitivity of the organs where haematopoiesis takes place.

In the follow-up studies,[51,52] *Pleurodeles* larvae were used in mesocosm tests because they are particularly interesting for food chain studies thanks to their advantage of being carnivorous. The trophic chains contain a bacterial consortium and two diatom species as primary producers, *C. riparius* larvae as primary consumers and *Pleurodeles* larvae as predator species at the top of the trophic chain. In the first work,[51] significant mortality in *Pleurodeles* larvae was observed. The surviving larvae in the nanoparticle condition showed a disturbance in general health, with no hunting behaviour or response to mechanical stimulation. Due to the high level of intoxication in the larvae, genotoxicity was not assessed. TEM investigation did not reveal NP internalization in the gills or intestine, but Ce quantification indicated a significant accumulation of nanoparticles in the larvae, with a mean concentration of 13.5 ± 3.9 mg kg^{-1} of dry body weight. In mesocosms, *Pleurodeles* are likely to be exposed to NPs *via* direct exposure from the water column and dietary exposure via contaminated chironomids. Parallel experiments conducted on *Pleurodeles* according to a monospecific assay allowed a better understanding of the implications of these two routes of exposure for the toxicity observed in mesocosms.[11] By direct exposure, the absence of toxicity observed in this experiment indicates that direct exposure to CeO_2 NPs is not likely to be a major source of toxicity to *Pleurodeles*, at least in these exposure conditions. Indeed, as standardized and mesocosm exposure conditions are completely different, results cannot be directly extrapolated from one experiment to the other and are rather indicative of the general mechanism of action of CeO_2 NPs. By dietary exposure, the rapid ingestion of chironomid larvae suggests that they are not likely to depurate, the contact time with non-contaminated water being very short. This is confirmed by the Ce concentrations in the water below the detection limits, indicating that Ce is not released in the water by *Pleurodeles* too. These results confirm that *Pleurodeles* exposure occurred only via the trophic route during this experiment. Moreover, the important NP concentration measured in the larvae after depuration indicates that CeO_2 NPs are accumulated when ingested. The absence of toxicity in *Pleurodeles* larvae during this experiment indicates that the trophic route does not play a major role in the toxicity of CeO_2 NPs in this species, despite the observed accumulation. Complex phenomena occur in mesocosms and modulate NP toxicity in fine. The second work focused on the toxicity and genotoxicity of *Pleurodeles* larvae exposed to cerium in microcosms.[52] Neither mortality nor growth inhibition was recorded among amphibian larvae during this experiment. However, important genotoxicity was observed with both NP types, with a significantly higher MNE induction observed with bare NPs compared to coated citrated NPs. The higher genotoxicity observed with bare NPs would be due to the protective effect of citrate coating. Chironomid larvae, which feed on sediments, are concluded to be important vectors of CeO_2 NPs transfer due to their ability to ingest large amounts of contaminated sediment. Despite the absence of biomagnification, significant genotoxic effects were observed in the amphibian larvae, which are at the top of the trophic chain. In this study, *Pleurodeles* models were used as predators and their feeding behaviour played a crucial role in the transfer of CeO_2 NPs along the trophic chain. Their position as a carnivorous species in the food web made them an ideal model for studying the effects of NP accumulation and transfer through the food chain. In this case, the use of *Pleurodeles* as a model organism allowed for a better understanding of the fate and effects of CeO_2 NPs in environmental conditions.

In the case of TiO_2 NPs,[54] *X. laevis* larvae were used as a model to contribute valuable data in terms of the impact of TiO_2 NPs on environmentally relevant bioindicators of pollution and provide insight into implications for food safety and ecotoxicological risk assessment. TiO_2 NPs used consisted of 80% anatase and 20% rutile, with a nominal diameter of 25.0 ± 5.7 nm. Exposure media for amphibian larvae were prepared using spiked soil with the addition of 500 mg kg^{-1} of TiO_2 NPs to mimic several years of sewage sludge spreading on agricultural soils. According to the adapted static leaching procedure of the French norm XP 31-210138 without filtration,[55] the supernatant suspensions were collected by siphoning into glass flasks. Larvae in groups of 15 were exposed in semi-static conditions based on modified ISO 21427-1[11] for 12 days to these leachates and ICP equivalent concentrations of TiO_2 in control soil and spiked soil (3.6 and 12.3 mg L^{-1} in commercial mineral water, respectively). In the amphibian exposure media, Ti was finally dosed at 12.3 ± 0.15 mg L^{-1} in the higher concentrated leachate. There was no significant mortality observed in the larvae exposed to TiO_2 NPs. However, there was a significant difference in the length of the larvae exposed to the different concentrations of TiO_2 NPs and soil leachates, with larvae exposed to higher concentrations showing reduced length. In terms of genotoxicity, there was a significant genotoxic effect observed in larvae exposed to the soil leachates and the TiO_2 NPs, with the number of micronuclei increasing in these larvae compared to the control group. The study concludes that both soil leachates and TiO_2 NPs have the potential to inhibit larval growth and cause genotoxic effects in amphibian larvae. Thanks to the amphibian model, a downward transfer of Ti from soil to the amphibian exposure medium has been established.

Nanoparticles of B (BNNT) can be considered metallic nanoparticles because boron is classified as a metalloid, even if it is not a true metal. BNNT has unique properties that make it useful for a variety of applications, including nanoelectronics, energy storage and biomedical engineering.[56] BNNTs share a similar structure with CNTs but consist of boron and nitrogen instead of carbon atoms.[57] Due to their significant interest in various applications, large scale production in the near future is expected, which implies the evaluation of risk assessment of BNNT. This is the case with the *Xenopus* model in the work of Evariste *et al.*[42] It is an original approach to exploring the *in vivo* effects of BNNT exposure on *X. laevis* tadpoles by examining host physiology and gut bacterial community composition. Monitoring these endpoints would provide insights into potential metabolic changes induced by BNNT exposure. *Xenopus* larvae were exposed for 12 days under semi-static conditions with daily exposure media renewal according to the ISO procedure[11] in groups of 20 to 0, 0.1, 1 and 10 mg L^{-1} of BNNT in RW. Results suggested that BNNTs are generally biocompatible. Tadpole exposure to the nanoparticles did not lead to irreversible genotoxic effects expressed as MN induction regardless of the concentration and stimulated their growth since the significant increase of *Xenopus* tadpole growth in a dose-dependent manner. However, the mean weight/length ratio of larvae remained unaltered under the tested conditions. Unlike carbon-based nanomaterials (CBNs), the accumulation of BNNT in the tadpole gut was not as apparent, which is consistent with the absence of growth alterations observed. Although BNNT can absorb nutrients as per the previous study,[58] the hypothesis of nutrient intake limitation is unlikely to occur in the case of BNNT exposure. Previous studies have shown that a B deficiency (<0.003 mg L^{-1}), or excessive concentrations (>50 mg L^{-1}) can be harmful to the development of amphibian embryos and the reproduction of adults.[59] In line with these findings, the B concentrations measured in this study's exposure media

(0.32 mg L^{-1}) are not considered critical and do not appear to impair the physiology of the tadpoles. It has been suggested that B may play a role in the synthesis of T3, a hormone in the thyroid axis of *X. laevis*.[59] Therefore, the observed growth stimulation could be attributed, at least in part, to the stimulation of metamorphosis by this metalloid. This growth enhancement was associated with the remodelling of the host-associated gut microbiota, particularly benefiting the bacterial *Bacteroideta* phylum. The changes occurring in the gut microbial communities could be partly associated with the stimulation of the thyroid axis, leading to the acceleration of gut maturation. Measurement of B released in the exposure media and accumulated in the organisms suggest that the beneficial effects observed were due to the free metalloid associated with the nanoparticles. Finally, even if BNNT alone represents a minor threat for amphibians in aquatic environments, due to its sorption capacities, it would be necessary to evaluate the interactive effects of B with other common contaminants, such as PAHs, heavy metals or even pesticides, to complete the ecotoxic evaluation.

3.6 Final Remarks

Although it seems difficult to simply conclude on the different aspects mentioned in the preceding lines, we can attempt to draw some general conclusions in relation to the use of amphibians in the context of ERA related to exposure to nanoparticles.

Our initial work on CNTs showed the great difference in toxicity observed with the same particles using different amphibian species. Thus, the larvae of the urodele *A. mexicanum* seem particularly resistant to exposure to these nanoparticles, even at very high concentrations. This is not the case for the larvae of the anuran *X. laevis,* which showed signs of toxicity at much lower concentrations. The concentration of 1–10 mg L^{-1} seems to be a key concentration around which changes are commonly observed. In general, non-oxidized graphene carbon nanoparticles, such as FLGs, show relatively low biological activity compared to oxidized graphenes. In this sense, the work carried out on amphibians seems to confirm the effects already observed on other groups of organisms in relation to the level of oxidation of graphene compounds. In our work, genotoxic effects have been demonstrated during exposure of amphibian larvae to oxidized graphene, a reduction in these effects has been correlated with a partial reduction of the oxidized material. It is, therefore, understandable that amphibians should be used to assess the hazard, in this case genotoxic, of oxidized graphenes in the context of an ERA of these nanoparticles. Indeed, oxidized graphenes are much more interesting from an industrial point of view. They can, for example, be easily mixed with a dilution matrix, allowing them to be manipulated and integrated into a product whose characteristics are to be improved so that, in the end, it is mainly these oxidized materials that are produced at the industrial level. Even if today, the risk of these nanoparticles remains almost non-existent at the global level, insofar as the flows of these materials in the environment are still low, an *a priori* assessment of the danger they represent does not seem superfluous. The expected multiplication of future technologies integrating these new materials means that their significant dispersion in the environment may constitute an additional real risk in the near future. Whether CNTs or graphenes in general, the knowledge acquired on their potential toxicity towards a group of organisms as representative of aquatic environments as amphibians constitutes a fundamental step towards the notion of "Safer by Design." From now on, designing the products of the future without taking into account their fate during their entire life cycle no longer seems bearable for human societies that are increasingly concerned about their future in a world that is becoming more and more degraded every day. One way of contributing to this effort is to integrate, from the design process of new products, knowledge of the risks incurred and evaluate the means of reducing these risks, whether they relate to health or the quality of the environment.

While the use of amphibians has made it possible in some cases to highlight genotoxic effects, this is far from being the general case, particularly during our work on carbon nanoparticles. Thus, it is another toxicity that has proven to be much more effective in revealing the potential impacts of exposure to nanoparticles in a general way: the inhibition of larval growth. It seems that this parameter is often an early indicator of exposure to the nanoparticles that we have been able to study in our various approaches. It is often the only criterion that has made it possible to characterize chronic effects following exposure to nanoparticles. This integrating parameter seems well suited to revealing, in particular, the physical characteristic effects of exposure to nanoparticles related to their ingestion or their interaction with the respiratory functions of the amphibian. The systematic study of the evolution of larval growth during exposure allowed us to compare the potential effects of different carbonaceous nanoparticles and to propose new indices to better describe the notion of effect—concentration in the case of nanoparticles. Thus, it seems that the specific surface area is a much better descriptor of the biological interaction of the nanoparticle than the classic expression of the mass concentration used for chemical substances in particular.

ERA, in the case of nanoparticles, remains a real challenge. The physical and chemical methods for characterizing nanoparticles are still largely inadequate for many of them. The use of biological study models, such as amphibians, which constitute a key group from the point of view of ecosystems, allows for a global characterization of the interaction between particles and living organisms. The combination of ecotoxicological tools, such as monospecific tests, batteries of standardized tests and studies in microcosms or mesocosms could be used to contribute to the emergence of a specific regulatory approach to nanoparticles in the environment.

References

1. P. Vasseur, J.-F. Masfaraud and C. Blaise, Ecotoxicology, revisiting its pioneers, *Environ. Sci. Pollut. Res.*, 2021, **28**, 1–6.
2. P. M. Chapman, Integrating toxicology and ecology: putting the “eco” into ecotoxicology, *Mar. Pollut. Bull.*, 2002, **44**, 7–15.
3. A. Kumar and C. Jha, Fishes as environmental indicators of riverine ecosystem, *Life Sci. J.*, 2020, **11**, 62–67.
4. W. E. Duellman and L. Trueb, in *Biology of Amphibians*, Johns Hopkins University Press, McGraw-Hill Book Co, Baltimore, 1986, p. 670.
5. W. W. Burggren and S. Warburton, Amphibians as animal models for laboratory research in physiology, *ILAR J.*, 2007, **48**, 260–269.
6. W. A. Hopkins, Amphibians as models for studying environmental change, *ILAR J.*, 2007, **48**, 270–7.
7. A. Egea-Serrano, R. A. Relyea, M. Tejedo and M. Torralva, Understanding of the impact of chemicals on amphibians: a meta-analytic review, *Ecol. Evol.*, 2012, **2**, 1382–1397.
8. J. Rostand, in *Les anomalies des amphibiens anoures*, Société d'Édition d'Enseignement Supérieur, Paris, 1958, p. 100.
9. AFNOR, Association Française de Normalisation (the French National Organization for Quality Regulations), Norme NFT 90-325, Essais des eaux, Détection en milieu aquatique de la génotoxicité d'une substance vis-à-vis de larves de batraciens (*Pleurodeles waltl* et *Ambystoma mexicanum*), AFNOR Monograph, Paris, 1987, p.12.
10. AFNOR, Association française de normalisation (the French National Organization for Quality Regulations), Norme NFT 90-325, Qualité de l'Eau, Evaluation de la génotoxicité au moyen de larves d'amphibien (*Xenopus laevis, Pleurodeles waltl*), Paris, 2000, p. 17.
11. ISO, International Standard, Water quality – Evaluation of genotoxicity by measurement of the induction of micronuclei, Part 1: Evaluation of genotoxicity using amphibian larvae, ISO/DIS 21427-1, Draft International standard, ISO, Geneva CH, 2006, p. 15.
12. L. Gauthier, E. Tardy, F. Mouchet and J. Marty, Biomonitoring of the genotoxic potential (micronucleus assay) and detoxifying activity (EROD induction) in the river Dadou (France), using the amphibian *Xenopus laevis*, *Sci. Total Environ.*, 2004, **323**, 47–61.
13. F. Mouchet, L. Gauthier, C. Mailhes, M. J. Jourdain, V. Ferrier and A. Devaux, Biomonitoring of the genotoxic potential of draining water from dredged sediments, using the comet and micronucleus tests on

amphibian (Xenopus laevis) and bacterial assays (Mutatox® and Ames tests), *J. Toxicol. Environ. Health, A*, 2005, **68**, 811–832.

14. F. Mouchet, L. Gauthier, C. Mailhes, M. J. Jourdain, V. Ferrier, G. Triffault and A. Devaux, Biomonitoring of the genotoxic potential of aqueous extracts of soils and bottom ash resulting from municipal solid waste incineration, using the comet and micronucleus tests on amphibian (Xenopus laevis) and bacterial assays (Mutatox® and Ames tests), *Sci. Total Environ.*, 2006, **355**, 232–246.
15. A. Kahru and A. Ivask, Mapping the dawn of nanoecotoxicological research, *Acc. Chem. Res.*, 2013, **46**, 823–833.
16. J. J. Scott-Fordsmand, P. H. Krogh and J. R. Lead, Nanomaterials in ecotoxicology, *Integr. Environ. Assess. Manag.*, 2008, **4**, 126–128.
17. A. Kahruand and H. C. Dubourguier, From ecotoxicology to nanoecotoxicology, *Toxicology*, 2010, **269**, 105–119.
18. S. Patra, in *Nanotoxicology and Nanoecotoxicology, Vol. 1, Environmental impact and econanotoxicity of engineered nanomaterials*, ed. K. Vineet, G. Praveen, R. Shivendu, D. Nandita and E. Lichtfouse, Springer Nature, Switzerland AG, 2022.
19. H. C. Polonini and R. Brayner, in *Nanotechnologies in Food and Agriculture, Nanoecotoxicology: The State of the Art*, ed. M. Rai, C. Ribeiro, L. Mattoso and N. Duran, Springer, Cham, 2015.
20. J. Jeevanandam, A. Barhoum, Y. S. Chan, A. Dufresne and M. K. Danquah, Review on nanoparticles and nanostructured materials: history, sources, toxicity and regulations, *Beilstein J. Nanotechnol.*, 2018, **9**, 1050–1074.
21. ISO/TS 80004-1:2015 (en), in *Nanotechnology – Vocabulary – Part 1: Core Terms*, International Organization for Standardization, Geneva, Switzerland, 2015.
22. B.-V. Boros and V. Ostafe, Evaluation of ecotoxicology assessment methods of nanomaterials and their effects, *Nanomaterials*, 2020, **10**, 610.
23. F. Mouchet, P. Landois, E. Flahaut, E. Pinelli and L. Gauthier, Assessment of the potential *in vivo* ecotoxicity of Double-Walled Carbon Nanotubes (DWNTs) in water, using the amphibian *Ambystoma mexicanum*, *Nanotoxicology*, 2007, **1**, 149–156.
24. F. Mouchet, P. Landois, E. Sarremejean, G. Bernard, P. Puech, E. Pinelli, E. Flahaut and L. Gauthier, Characterisation and *in vivo* ecotoxicity evaluation of double-wall carbon nanotubes in larvae of the amphibian *Xenopus laevis*, *Aquat. Toxicol.*, 2008, **87**, 127–137.
25. P. D. Nieuwkoop and J. Faber, *Normal tables of Xenopus laevis (Daudin)*, North-Holland, Amsterdam, 1956.
26. M. Monthioux, P. Serp, E. Flahaut, M. Razafinimanana, C. Laurent, A. Peigney, W. Bacsa and J. M. Broto, Introduction to carbon nanotubes,

in *Handbook of Nanothechnology*, ed. B. Bushan, Springer, NY, 2nd edn, 2007, pp. 43–112.

27. L. Gauthier, The amphibian micronucleus test, a model for *in vivo* monotoring of genotoxic aquatic pollution, *Inter. J. Batrachol.*, 1996, **14**, 53–84.
28. F. Mouchet, P. Landois, V. Datsyuk, P. Puech, E. Pinelli, E. Flahaut and L. Gauthier, International amphibian micronucleus standardized procedure (ISO 21427-1) for *in vivo* evaluation of double-walled carbon nanotubes toxicity and genotoxicity in water, *Environ. Toxicol.*, 2011, **26**, 136–145.
29. F. Mouchet, P. Landois, P. Puech, E. Pinelli, E. Flahaut and L. Gauthier, Carbon nanotube ecotoxicity in amphibians: assessment of Multi Walled Carbon Nanotubes (MWNT) and comparison with Double Walled Carbon Nanotubes (DWNT), Special focus environmental toxicity of nanoparticles, *Nanomedicine*, 2010, 963–974.
30. A. Mottier, F. Mouchet, C. Laplanche, S. Cadarsi, L. Lagier, J.-C. Arnault, H. A. Girard, V. León, E. Vázquez, C. Sarrieu, É. Pinelli, L. Gauthier and E. Flahaut, Surface area of carbon nanoparticles: a dose metric for a more realistic ecotoxicological assessment, *Nano. Lett. Am. Chem. Soc.*, 2016, **16**, 3514–3518.
31. F. Bourdiol, F. Mouchet, A. Perrault, I. Fourquaux, L. Datas, C. Gancet, J. C. Boutonnet, E. Pinelli, L. Gauthier and E. Flahaut, Biocompatible polymer-assisted dispersion of multi walled carbon nanotubes in water, application to the investigation of their ecotoxicity using *Xenopus laevis* amphibian larvae, *Carbon*, 2012, **54**, 175–191.
32. H. Hyung and J. H. Kim, Natural organic matter (NOM) adsorption to multi-walled carbon nanotubes: effect of NOM characteristics and water quality parameters, *Environ. Sci. Technol.*, 2008, **42**, 4416–4421.
33. R. Saria, F. Mouchet, A. Perrault, E. Flahaut, C. Laplanche, J. C. Boutonnet, E. Pinelli and L. Gauthier, Short term exposure to multi-walled carbon nanotubes induce oxidative stress and DNA damage in *Xenopus laevis* tadpoles, *Ecotoxicol. Env. Saf.*, 2014, **107**, 22–29.
34. M. Segal, Selling graphene by the ton, *Nat. Nanotechnol.*, 2009, **4**, 612–614.
35. X. J. Lee, B. Y. Z. Hiew, K. C. Lai, L. Y. Lee, S. Gan, S. Thangalazhy-Gopakumar and S. Rigby, Review on graphene and its derivatives: Synthesis methods and potential industrial implementation, *J. Taiwan. Inst. Chem. Eng.*, 2019, **98**, 163–180.
36. C. Cao, X. Wu, X. Xi, T. Filleter and Y. Sun, Mechanical characterization of grapheme, in *Handbook of Nanomaterials Properties*, ed. B. Bhushan, D. Luo, S. R. Schricker, W. Sigmund and S. Zauscher, Springer, Berlin/Heidelberg, Germany, 2014, pp. 121–135.

37. Y. Zhu, S. Murali, W. Cai, X. Li, J. W. Suk, J. R. Potts and R. S. Ruoff, Graphene and graphene oxide: synthesis, properties, and applications, *Adv. Mater.*, 2010, **22**, 3906–3924.
38. L. Muzi, F. Mouchet, S. Cadarsi, I. Janowska, J. Russier, C. Ménard-Moyon, G. Risuleo, B. Soula, A. M. Galibert, E. Flahaut, E. Pinelli, L. Gauthier and A. Bianco, Examining the impact of multi-layer graphene using cellular and amphibian models, *2D-Nanomat.*, 2016, **3**, 025009.
39. L. Lagier, F. Mouchet, C. Laplanche, A. Mottier, S. Cadarsi, L. Evariste, C. Sarrieu, P. Lonchambon, E. Pinelli, E. Flahaut and L. Gauthier, Surface area of carbon-based nanoparticles prevails on dispersion for growth inhibition in amphibians, *Carbon*, 2017, **119**, 72–81.
40. L. Evariste, L. Lagier, P. Gonzalez, A. Mottier, F. Mouchet, S. Cadarsi, P. Lonchambon, G. Daffe, G. Chimowa, C. Sarrieu, E. Ompraret, A.-M. Galibert, C. M. Ghimbeu, E. Pinelli, E. Flahaut and L. Gauthier, Thermal reduction of graphene oxide mitigates its *in vivo* genotoxicity toward *Xenopus laevis* tadpoles, *Nanomaterials*, 2019, **9**, 584–600.
41. L. Evariste, A. Mottier, L. Lagier, S. Cadarsi, M. Barret, C. Sarrieu, B. Soula, F. Mouchet, E. Flahaut, E. Pinelli and L. Gauthier, Assessment of graphene oxide ecotoxicity at several trophic levels using aquatic microcosms, *Carbon*, 2020, **156**, 261–271.
42. L. Evariste, E. Flahaut, C. Baratange, M. Barret, F. Mouchet, E. Pinelli, A. M. Galibert, B. Soula and L. Gauthier, Ecotoxicological assessment of commercial boron nitride nanotubes toward *Xenopus laevis* tadpoles and host-associated gut microbiota, *Nanotoxicology*, 2020, **15**, 35–51.
43. L. Evariste, A. Mottier, E. Pinelli, E. Flahaut, L. Gauthier and F. Mouchet, Graphene oxide and reduced graphene oxide promote the effects of exogenous T3 thyroid hormone in the amphibian *Xenopus laevis*, *Chemosphere*, 2021, **281**, 130901.
44. L. Evariste, F. Mouchet, E. Pinelli, E. Flahaut, L. Gauthier and M. Barret, Gut microbiota impairment following graphene oxide exposure is associated to physiological alterations in *Xenopus laevis* tadpoles, *Sci. Total Environ.*, 2022, **857**, 159515.
45. M. Li, J. Zhu, M. Wang, H. Fang, G. Zhu and Q. Wang, Exposure to graphene oxide at environmental concentrations induces thyroid endocrine disruption and lipid metabolic disturbance in Xenopus laevis, *Chemosphere*, 2019, **236**, 124834.
46. X. Yao, X. Chen, Y. Zhang, Y. Li, Y. Wang, Z. Zheng, Z. Qin and Q. Zhang, Optimization of the T3-induced *Xenopus* metamorphosis assay for detecting thyroid hormone signaling disruption of chemicals, *J. Environ. Sci.*, 2017, **52**, 314–324.
47. Y. Wang, Y. Li, Z. Qin and W. Wei, Re-evaluation of thyroid hormone signaling antagonism of tetrabromobisphenol A for validating the T3-induced *Xenopus* metamorphosis assay, *J. Environ. Sci.*, 2017, **52**, 325–332.

48. A. Jaylet, P. Deparis, V. Ferrier, S. Grinfeld and R. Siboulet, A new micronucleus test using peripheral blood erythrocytes of the newt *Pleurodeles waltl* to detect mutagens in fresh-water pollution, *Mutat. Res.*, 1986, **164**, 245–257.

49. A. Bour, F. Mouchet, L. Verneuil, L. Evariste, J. Silvestre, E. Pinelli and L. Gauthier, Toxicity of CeO_2 nanoparticles at different trophic levels-Effects on diatoms, chironomids and amphibians, *Chemosphere*, 2015, **120**, 230–236.

50. A. Bour, F. Mouchet, S. Cadarsi, J. Silvestre, L. Verneuil, D. Baqué, E. Chauvet, J. M. Bonzom, C. Pagnout, H. Clivot, I. Fourquaux, M. Tella, M. Auffan, J. Silvestre, L. Gauthier and E. Pinelli, Toxicity of CeO_2 nanoparticles on a freshwater experimental trophic chain: a study in environmentally relevant conditions through the use of mesocosms, *Nanotoxicology*, 2016, **10**, 245–255.

51. A. Bour, F. Mouchet, S. Cadarsi, J. Silvestre, E. Chauvet, J. M. Bonzom, C. Pagnout, H. Clivot, L. Gauthier and E. Pinelli, Impact of CeO_2 nanoparticles on the functions of freshwater ecosystems: a microcosm study, *Environ. Sci. Nano.*, 2016, **3**, 830–838.

52. A. Bour, F. Mouchet, S. Cadarsi, J. Silvestre, D. Baqué, L. Gauthier and E. Pinelli, CeO_2 nanoparticle fate in environmental conditions and toxicity on a freshwater predator species: a microcosm study, *Environ. Sci. Pollut. Res. Int.*, 2017, **24**, 17081–17089.

53. OECD, in *Test no.203, Fish, Acute Toxicity Test, OECD Guidelines for the Testing of Chemicals, Section 2*, OECD Publishing, 1992.

54. V. Vijayaraj, C. Liné, S. Cadarsi, C. Salvagnac, D. Baqué, A. Elger, M. Barret, F. Mouchet and C. Larue, Transfer and ecotoxicity of titanium dioxide nanoparticles in terrestrial and aquatic ecosystems: A microcosm study, *Environ. Sci. Technol.*, 2018, **52**, 12757–12764.

55. AFNOR, Association française de normalisation, Norme NFT X31- 210, Déchets, Essai de lixiviation, 1992.

56. N. G. Chopra, R. J. Luyken, K. Cherrey, V. H. Crespi, M. L. Cohen, S. G. Louie and A. Zettl, Boron nitride nanotubes, *Science*, 1995, **269**, 966–967.

57. J. H. Kim, T. V. Pham, J. H. Hwang, C. S. Kim and M. J. Kim, Boron nitride nanotubes: Synthesis and applications, *Nano Converg.*, 2018, **5**, 17.

58. D. Farmanzadeh and S. Ghazanfary, Interaction of vitamins A, B1, C, B3 and D with zigzag and armchair boron nitride nanotubes: A DFT study, *C. R. Chim.*, 2014, **17**, 985–993.

59. D. J. Fort, R. L. Rogers, D. W. McLaughlin, C. M. Sellers and C. L. Schlekat, Impact of boron deficiency on *Xenopus laevis*, *Biol. Trace Elem. Res.*, 2002, **90**, 117–142.

CHAPTER 4

Evaluation of Genotoxicity and Mutagenicity in Amphibian Species that Inhabit Agroecosystem Environments

MARCELINO BENVINDO-SOUZA[a], RAQUEL F. SALLA[b] AND DANIELA MELO E SILVA*[c]

[a]Post-graduation Program in Natural Resources of Cerrado (RENAC), State University of Goiás, Anápolis, Goiás, Brazil; [b]Laboratory of Environmental Biotechnology and Ecotoxicology, Avenida Universitária, Setor Leste Universitário, Goiânia, Goiás, Brazil; [c]Mutagenesis Laboratory, Avenida Esperança, Campus Samambaia, Goiânia, Goiás, Brazil

*E-mail: danielamelosilva@ufg.br; silvadanielamelo@gmail.com

4.1 Introduction

Habitat modifications are causing a widespread decline in biodiversity and the homogenization of natural environments.[1] Agriculture is one of those activities that promotes these changes on a large scale. In addition to habitat fragmentation, some agricultural production systems use chemical inputs, such as synthetic fertilizers and pesticides. After their application in agricultural areas, pesticides are dispersed throughout the environment, causing contamination.[2] The fate and transport of pesticides in agroecosystems are determined by the chemical characteristics of the

Issues in Toxicology No. 46
Amphibian Species in Environmental Risk Assessment Strategies
Edited by Marcelo L. Larramendy and Guillermo Eli Liwszyc

Published by the Royal Society of Chemistry, www.rsc.org

formulations and soil attributes, in addition to the microbes, fauna and flora that inhabit the soil.[3] These contaminants may affect the agroecosystem's health, especially the fauna, such as anuran amphibians, that depend on water resources.[4]

Anurans are widely used as bioindicators of the effects of genotoxic and mutagenic substances, such as pesticide residues. These animals are closely associated with wetlands and have unique characteristics, such as unprotected eggs, permeable skin and low mobility, which make them highly susceptible to contamination and to the development of many biological damages, such as DNA damage.[5,6] Thus, the use of biomarkers has great relevance in ecotoxicological biomonitoring.[7] The comet or single gel electrophoresis assay (CA) and the micronucleus (MN) test are good examples of biomarkers for analyzing the effects of environmental chemical compounds. The CA assesses the integrity of the DNA structure and determines DNA strand breaks, among other DNA lesions, before repair mechanisms occur.[8] Many studies point to its efficiency in analyzing the effects of potentially toxic pesticides on anuran amphibians, such as imazethapyr,[9] Roundup,[10] glyphosate and POEA,[11] among many other chemicals from anthropogenic sources.

Since its first application in amphibians' research, an event that took place more than 30 years ago,[12] the application of MN has expanded among environmental stress studies with pesticides, heavy metals and radiation.[13] Mutagenic chemicals can affect DNA structure and cause strand DNA breaks, leading to the formation of MN.[14] The MN test also allows the detection of other nuclear lesions,[15] such as mitotic disorders and cytotoxicity. These biomarkers have been recognized as highly efficient in detecting the mutagenic effects of agrochemicals, as already reported in studies with the herbicide 2,4-D,[16] the insecticide chlorpyrifos[17] and the benzovindiflupyr/prothioconazole-based broad-spectrum fungicide Elatus®.[18]

Given the evidence that many agrochemicals can exert harmful ecological effects on organisms in agricultural landscapes,[19] this chapter compiled the findings concerning the main anuran species studied in agroecosystems in South America, with a special focus on genotoxicity and mutagenicity analyses. We chose South America, including countries like Brazil, Colombia, Ecuador and Mexico, as the main scenario due to their greater diversity of amphibian species in the world. Furthermore, South America has some of the main agricultural areas in the world, which reinforces the importance of conducting studies in these locations. In addition, this chapter also examined the applicability of the CA and MN tests to amphibians.

4.2 Methods

A systematic search was carried out using several databases, *i.e.*, Web of Science, Scopus and Google Scholar, concerning CA and MN tests, and anurans *in situ*/agroecosystems in South America. The investigation used a

Figure 4.1 Methodological steps in the selection of studies on frog species in agroecosystems in South America for the genotoxicity and mutagenicity analyses.

series of keyword combinations: TS = (anuran* OR frog* OR tadpole*) AND TS = (pesticide* OR agrochemical* OR agroecosystems) AND TS = (DNA damage* OR comet assay* OR single cell gel electrophoresis OR alkaline assay* OR genotoxicity* OR genotoxic* OR mutagenicity* OR micronuclei* OR micronucleus* OR erythrocytes*). As a criterion, only *in situ* papers that focused on agriculture (and pesticide use) were selected. Studies in which there were *in situ* experiments without the manipulation of chemical substances were also compiled. The research aimed to track species under the natural pressures of agroecosystems.

For each selected paper, a series of variables were compiled, such as the country where the research was carried out, the year of publication, the type of agricultural matrix, the type of agrochemical (analysis of water, sediments or tissues) and the main registered pesticides. We also considered the anuran species' developmental stage (whether tadpoles or adults) and the form of anesthesia/euthanasia to which the animals were submitted. We also compiled information on how the studies obtained the samples (with a cut at the base of the tail or blood puncture), the type of dye, and the parameters for detecting DNA damage (for the CA), nuclear abnormalities and the number of quantified cells (see Figure 4.1).

4.3 Results and Discussion

4.3.1 Overview of the *In Situ* Studies that Applied the CA and MN Tests in South America

We analyzed 256 research articles, but only 19 met the requirements for the agroecosystems/*in situ* studies. The articles were published between 2008 and 2022, although our search was carried out until February 18, 2023. The scarcity of studies was due to the filter of publications about agroecosystems and the effect of agrochemicals on fieldwork, which indicates the need for more research in natural conditions. Three countries have developed the main research: Argentina (n = 9), Brazil (n = 9) and Colombia (n = 1) (see Figure 4.2). Argentina and Brazil are the major agricultural producers in South America and are considered important producers and consumers of agrochemicals. Large areas of monocultures, such as soybeans, corn and sugarcane, routinely adopt

Figure 4.2 South American countries where research on frogs of agroecosystems was carried out for genotoxicity and mutagenicity between 2008 and 2022.

the intensive application of pesticides and chemical fertilizers,[20] which represent the major commodities in such countries. It is worth noting that 30% of pesticides applied in Brazil are banned in Europe,[21] which raises concerns about the ecosystem's health.

Although not all the studies quantified the presence of pesticides in agricultural fields, in the studies that included water quality analyses, there were different screening methodologies, such as water analysis from amphibian reproductive sites,[22–24] sediment collection,[4] chromatographic analysis of animal tissues[25] and verbal information reports with rural workers.[26] Alternatively, the studies were conducted before and after the application of a specific pesticide.[27] Indeed, occasionally, *in situ* studies do not necessarily aim to test the presence or quantify the concentrations of pesticides in the ecosystem, but they intend to verify the stress caused by agricultural activities due to the fact that many crops are associated with a complex mixture of pesticides applied.[28] Thus, more than 60 pesticides, including insecticides (endosulfan, carbaryl, dimethoate and trichlorfon), herbicides (atrazine, glyphosate and diuron) and fungicides (cyproconazole, pyraclostrobin, tebuconazole and trifloxystrobin), were the most registered products used in agricultural areas in the South American countries (see Figure 4.3).

Considering that pesticides and other aquatic contaminants, as well as various environmental disturbances (*e.g.*, habitat fragmentation, radiation

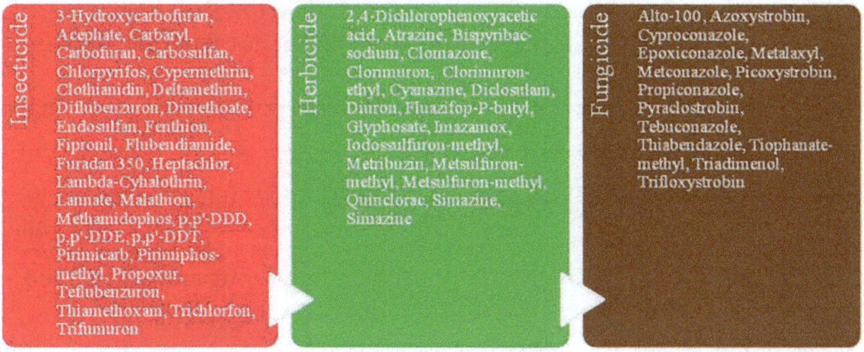

Figure 4.3 Pesticides registered in *in situ* studies on the genotoxic and mutagenic effects in frogs of South American countries.

and emerging diseases), are the conditioning factors for the amphibian population decline,[26,29] such dominance of environmental factors reinforces the need for more *in situ* studies. In addition to pesticides, fertilizers also represent a concern for aquatic fauna. In Argentina, scientists observed that tadpoles inhabiting a contaminated environment (with phosphorus and nitrate) had a higher frequency of MNs and other nuclear abnormalities.[30] The authors reported that the toxic action of nitrate occurs through the conversion of respiratory pigments into isoforms that are not capable of transporting and releasing oxygen. In addition, when phosphorus levels are higher (over 0.1 mg L^{-1}), they, in turn, generate eutrophication, producing a drastic decrease in dissolved oxygen and maximizing the toxic effects of the mixture.

4.3.2 Anuran Species in Agroecosystems

From the compilation of the South American studies, it was possible to identify the most common anuran species employed as biological matrices in studies employing the CA and MN tests as endpoints. We observed 26 frog species (see Table 4.1) investigated in the South American agroecosystems, representing 0.99% of the region's species richness. One species, the barker frog *Crossodactylus schmidti* (Hylodidae), was listed as "near threatened" on the International Union for Conservation of Nature (IUCN) and Natural Resources Red List, while others were considered as part of the "Least Concern" species. South America registered 2623 species of anurans until mid-2017,[31] which indicates a great diversity of animals with potential for further studies focused on agroecosystem, genotoxicity and mutagenicity, in addition to other relevant ecotoxicological biomarkers.

The three most studied species were the lesser tree frog *Dendropsophus minutus* (Hylidae), the barker frog *Physalaemus cuvieri* (Leptodactylidae)

Table 4.1 List of amphibian species recorded in agroecosystems studies on genotoxicity and mutagenicity in South America published between 2008 and 2022.

Families/species	Biomarker	Status	Distribution
Hylidae			
Boana albopunctata	CA, MN	LC	Argentina, Bolivia, Brazil, Uruguay and Paraguay
Boana curupi	MN	LC	Argentina, Brazil and Paraguay
			Brazil
Boana pulchella[a]	MN	LC	Argentina, Brazil, Paraguay and Uruguay
Bokermannohyla sp.	MN	LC	—
Pseudis minuta	MN	LC	Argentina, Brazil, Paraguay and Uruguay
Scinax montivagus	MN	DD	Brazil
Scinax nasicus	MN	LC	Argentina, Bolivia, Brazil, Paraguay and Uruguay
Scinax x-signatus	MN	LC	—
Scinax squalirostris	MN	LC	Argentina, Bolivia, Brazil, Paraguay and Uruguay
Scinax fuscovarius	MN	LC	Argentina, Bolivia, Brazil and Paraguay
Dendropsophus minutus	CA, MN	LC	Argentina, Bolivia, Brazil, Colombia, Ecuador, Paraguay, Peru, Trinidad, Uruguay and Venezuela
Leptodactylidae			
Leptodactylus chaquensis	MN	LC	Argentina, Bolivia, Brazil, Colombia, the Guianas, Paraguay and Uruguay
Leptodactylus fuscus	MN	LC	Argentina, Bolivia, Brazil, Panama and Paraguay
Leptodactylus latrans	MN	LC	Brazil
Leptodactylus luctator	MN	LC	Argentina, Bolivia, Brazil, Paraguay and Uruguay
Leptodactylus mystacinus	MN	LC	Argentina, Bolivia, Brazil, Paraguay and Uruguay
Leptodactylus vastus	MN	LC	Bolivia and Brazil

(*continued*)

Table 4.1 (*continued*)

Families/species	Biomarker	Status	Distribution
Rupirana cardosoi	MN	NT	Brazil
Physalaemus cuvieri	CA, MN	LC	Argentina, Bolivia, Brazil, Paraguay, Uruguay and Venezuela
Odontophrynidae			
Odontophrynus americanus	MN	LC	Argentina
Odontophrynus carvalhoi	MN	LC	Brazil
Odontophrynus cordobae	MN	LC	Argentina
Bufonidae			
Bufo arenarum[a]	MN	LC	Argentina, Bolivia, Brazil, East of the Andes, Paraguay and Uruguay.
Rhinella marina	MN	LC	Bolivia, Brazil, Colombia, Ecuador, the Guianan and Peru. Introduced widely worldwide (Antilles, Australia, Fiji, Guam, Hawaii, New Guinea, the Philippines, Ryukyu Islands (Japan), Solomon Islands, Taiwan, *etc.*
Hylodidae			
Crossodactylus schmidti	MN	NT	Argentina, Brazil, Paraguay and Paraná

[a]Species whose nomenclature was updated; Biomarkers: CA, comet assay; MN, micronucleus test; IUCN Red List Classification: DD, data deficient; LC, least concern; NT, near threatened.

and the South American common frog *Rhinella arenarum* (Bufonidae). *Dendropsophus minutus* is a tree frog of the Hylidae family considered widely distributed in South America. It occurs in Argentina, Bolivia, Brazil, Colombia, Ecuador, Paraguay, Peru and Uruguay.[32] This species is sensitive in agroecosystems with multiple pesticide uses, such as atrazine, glyphosate and malathion,[22,26,33] lives in arboreal habitats, and can be found in the leaves and branches of vegetation.[34] Its reproduction occurs in temporary water bodies, and the eggs are deposited in the water where the tadpoles develop.[34]

Physalaemus cuvieri is a frog that belongs to the Leptodactylidae family. It can be found in the Northeast, Central, Southeast and South regions of Brazil and possibly in Venezuela, Argentina and Paraguay.[32] This species is a generalist, reproducing in cattle ponds or temporary water bodies.[34]

The eggs are deposited in foam nests attached to grass stalks in the ponds.[34] Studies have reported that this species presented an increase in DNA damage (CA and/or MN) in animals that were sampled in agricultural areas in comparison to individuals sampled in the reference areas (free of agrochemicals).[22,26,28]

Finally, the third species, *R. arenarum*, from the Bufonidae family, is found in Southern Brazil, Argentina, Uruguay and Bolivia.[32] This species has adapted to environmental disturbance; its oviposition occurs in temporary water bodies, and the tadpoles feed and develop in the same place.[35] Unlike the two previously mentioned species, *R. arenarum* showed sensitivity to the MN occurrence and other nuclear abnormalities in agroecosystems,[30] as well as differences contrasting males and females.[7] Sex is a variable rarely considered in the evaluation of the genotoxicity and mutagenicity in amphibians.

4.3.3 Environmental Genotoxicity and Mutagenicity

Comet assay studies in anurans evaluating the effect of pesticides are more common in laboratory conditions, which explains the low number of *in situ* investigations. Among the countries in South America, Brazil was the only one that applied the test for animals on agricultural land (see Table 4.2). The authors described an increase in DNA damage in the three studies carried out in the region,[26,28,33] where crops were soybean and corn. The DNA damage was analyzed for three CA parameters (see Figure 4.4): tail length (TL), percentage of DNA in the tail (%DNA) and Olive tail moment (OTM). The tail length is directly related to the released DNA loops or the size of the DNA fragments. It is also proportional to the amount of DNA breaks and alkali-labile sites.[36] The percentage of DNA is the amount of DNA that has migrated out of the nucleus expressed as a percentage of the total cellular DNA content and is directly proportional to the amount of DNA damage.[36] In contrast, OTM is the product of the distance between head DNA, tail DNA and tail DNA distribution.[37]

MN and other nuclear abnormalities have been analyzed in frogs from agroecosystems (see Figure 4.5 and Table 4.3). In Central Cerrado, Brazil, the first *in situ* studies addressing the effect of pesticides have found an increase in the frequency of MNs in an agricultural area compared to a reference area.[26,33] Other scientists did not observe an increase in MNs but detected a significantly higher frequency of cells with a reniform nucleus and in the sum of nuclear abnormalities (nuclear buds, reniform nuclei, apoptotic cells, binucleated cells and anucleated cells) in animal samples from soy and corn cultures.[22] Other researchers also observed a higher frequency of MNs, lobed and reniform nuclei in the whistling frog *Leptodactylus fuscus* (Leptodactylidae) in the same region.[24] However, the water analysis did not detect pesticides in the breeding site.

In South Brazil, researchers reported that frogs coexist with mutagenic substances in a conservation unit due to indirect exposure to agricultural

Table 4.2 Summary of studies that used comet assay in studies of amphibian species in agricultural areas in South America.

Species	Fase	Culture	Pesticide	Parameters	Endpoint	Ref.
Dendropsophus minutus, Boana albopunctata and *Physalaemus cuvieri*	Tadpoles	Soybean and corn	Atrazine, Malathion, Furadan 350, Alto-100, Glyphosate, Lannate e Dimethoate	TL, %DNA, OTM	↑ %DNA for *D. minutus, B. albopunctata*	28
Physalaemus cuvieri	Tadpoles	Soybean and corn	—	TL, %DNA, OTM	↑ TL, %DNA, OTM	27
Dendropsophus minutus	Tadpoles	Soybean and corn	—	TL, %DNA, OTM	↑ TL, %DNA, OTM	26

↑, increase; %DNA, % DNA in tail; OTM, Olive tail moment; TL, tail length.

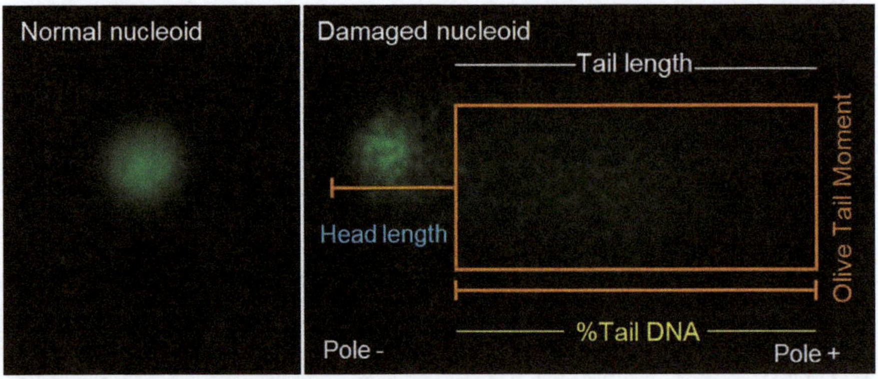

Figure 4.4 Images of anuran nucleoids observed in the comet assay. (A) Normal nucleoid. (B) Damaged nucleoid.

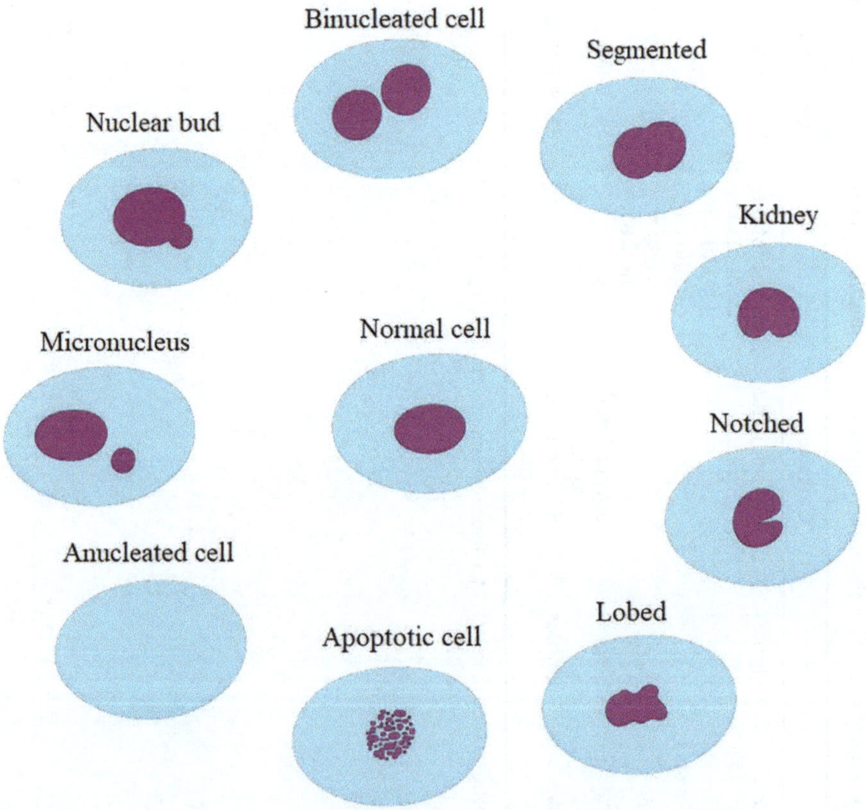

Figure 4.5 Micronucleus and other nuclear abnormalities analyzed in agricultural areas in South America.

Table 4.3 Summary of studies on the use of the MN test in frogs from agricultural areas in South America.

Species	Phase	Culture	Pesticides	MN and ENAs	Response	Ref.
Dendropsophus minutus, Boana albopunctata and *Physalaemus cuvieri*	Tadpoles	Soybean and corn	Atrazine, Malathion, Furadan 350, Alto-100, Glyphosate, Lannate and Dimethoate	MN	↑MN for *D. minutus, B. albopunctata*	28
Dendropsophus minutus	Tadpoles	Soybean and corn	—	MN	↑MN	26
Rhinella arenarum	Adults	Soybean	—	MN, notched, binucleated, bud, lobed nuclei	↑ MN in males relative to females	30
Pseudis minuta and *Leptodactylus gr latrans*	Adults	Multiple stressors, including rice planting	—	MN, kidney-shaped nuclei, lobed nuclei, notched nuclei and blebbed nuclei	↑MN and lobed nuclei in *P. minuta* than in *L. gr latrans* in UC	29
Leptodactylus luctator	Adults	Soybean	Multiple pesticides	MN, binucleated, notched, blebbed and lobed nuclei	Ø MN, ↑ of total ENAs	22
Leptodactylus fuscus	Adults	Soybean and corn	Not detected	MN, binucleated cells, nuclear buds, anucleated cells, lobed, notched, reniform and segmented nuclei	↑ MN and lobate and kidney-shaped nucleus	23
Boana curupi and *Crossodactylus schmidti*	Tadpoles	Soybean and corn around the UC	Multiple pesticides	MN	↑ MN in a less preserved area	24

(*continued*)

Table 4.3 (*continued*)

Species	Phase	Culture	Pesticides	MN and ENAs	Response	Ref.
Physalaemus cuvieri, Scinax fuscovarius and *Dendropsophus minutus*	Tadpoles	Soybean and corn	Atrazine	MN, nuclear bud, reniform, binucleated cell, apoptotic and anucleated	Ø MN, ↑ reniform nucleus; ↑ in total ENAs for *S. fuscovarius* from agricultural area	25
Rhinella marina	Adults	Soybean	*p,p'*-DDT, *p,p'*-DDE e *p,p'*-DDD	MN, nuclear eggs and pyknosis	↑ MN	4
Hypsiboas pulchellus	Adults	Agroecosystem	Glyphosate	MN	Ø MN 2 and 15 days after glyphosate application. But there was ↑ MN two months later	35
Rhinella arenarum[a]	Adults	Soybean	—	MN	↑ MN in crop area	34
Rhinella arenarum	Tadpoles	Agroecosystem	Nitrate and phosphorous	MN, lobed nuclei, binucleate or segmented nuclei, kidney-shaped, notched and pyknotic nuclei	↑ MN, ENAs in agroecosystem	31
Bokermannohyla sp., *Bokermannohyla oxente*, *Boana albopunctata*, *Scinax montivagus*, *Scinax x-signatus*, *Leptodactylus latrans*,	Tadpoles	Coffee, corn, vegetables and pumpkin	—	MN, lobed, binucleated	↑ MN	17

(*continued*)

Table 4.3 (*continued*)

Species	Phase	Culture	Pesticides	MN and ENAs	Response	Ref.
Leptodactylus mystacinus, *Leptodactylus vastus*, *Odontophrynus carvalhoi*, *Rupirana cardoso* and *Physalaemus cf. curvieri*						
Scinax nasicus	Tadpoles	Soybean	Heptachlor, Heptachlor epoxide, Endosulfan sulfate, Endosulfan II	MN	↑ Organochlorine residues; Ø MN, ↑Erythrocytes lysis	36
Leptodactylus chaquensis	Adults	Rice	Metamidofos, cipermetrina and Endosulfan	MN, lobed nuclei, binucleates, segmented nuclei, kidney-shaped and notched nuclei	Crops sprayed with endosulfan and cypermethrin potentiated the increase in immature erythrocytes and anisochromasia	32
Scinax squalirostris and *Leptodactylus mystacinus*	Tadpoles	Rice	Lambda-cialotrine	MN, lobed nuclei, binucleates, segmented nuclei, kidney-shaped and notched nuclei	Blood cell parameters of *S. squalirostris* and *L. mystacinus* from the agricultural site are similar	33

(*continued*)

Table 4.3 (*continued*)

Species	Phase	Culture	Pesticides	MN and ENAs	Response	Ref.
Odontophrynus cordobae and *O. americanus*	Adults	—	Nitrate and phosphate	MN, notched, binucleated, lobed and blebbed nuclei	↑MN and cytotoxicity in polyploid organisms and the highest frequency of ENAs in diploid organisms were recorded	30
Rhinella arenarum	Tadpoles	—	—	MN, lobed nuclei, multinucleate or segmented, kidney-shaped, notched and pyknotic nuclei	↑ MN, ENAs in agroecosystem	31

[a]Indicates species that have changed in the Genus.

Ø, there was no increase; ↑, increase; —, not investigated; ENAs, erythrocyte nuclear abnormalities; MN, micronucleus; UC, conservation unit.

activities.[24] Other researchers detected nine pesticides in frog liver samples in agricultural areas, of which deltamethrin was the most common and carbosulfan was recorded at the highest concentration.[25] Although they did not observe a significant increase in MNs, the total of other nuclear abnormalities (binucleated, notched, bubble and lobed nuclei) was significantly more frequent in the agricultural areas.

In Argentina,[38] investigations showed an increased frequency of MNs in populations in modified areas, such as agricultural environments, compared to control areas (reference areas). Indeed, the poor water quality of the agroecosystem ponds certainly has an impact on the health of the tadpoles, and this can, in turn, affect the persistence of the populations.[30] Water samples from ponds were analyzed for pesticide residues,[39] observing higher levels of organochlorine pesticide residues in agricultural ponds compared to the control area (a reference area). Several deleterious effects on survival, growth and development were found in the snouted tree frog, *Scinax nasicus* (Hylidae). In contrast, no increase in MN frequency was observed.[39]

In another study comparing 2 and 15 days after glyphosate application, there was no evidence of an increase in MNs in *Hypsiboas pulchellus*, but a difference was observed two months after these applications, possibly suggesting other multiple environmental stressors.[27] Researchers did not observe a mutagenic difference between the sex variable in *R. arenarum,* considering environments with different degrees of environmental alteration.[7] However, years later, they published a new study that evaluated the genotoxic and cytotoxic responses of diploid and polyploid organisms that inhabited agroecosystems.[40] As evidenced, polyploidy can provide greater genetic flexibility, increasing buffering against exogenous agents that damage the DNA of *Odontophrynus* sp. and thus conferring an advantage over diploids under certain environmental conditions. Finally, in Colombia, a study showed the presence of persistent organochlorine pesticides in sediment samples (*p,p'*-DDT, *p,p'*-DDE and *p,p'*-DDD) from agricultural soils and MN was statistically significant in individuals in these environments, indicating the toxic potential of persistent pesticides.[4]

Many physical and chemical factors can induce MNs, as well as those of an environmental nature.[14] MNs originate during anaphase from delayed chromosomes or acentric chromatid fragments caused by incorrect repair of DNA break or unrepaired DNA breaks. Erythrocyte nuclear abnormalities, such as buds, are considered as essential as MNs. They have the same morphology as an MN, except that they are connected to the nucleus by a narrow or wide rod of nucleoplasmic material, depending on the stage of the budding process.[41] Binucleated cells are due to cytokinesis failure on microfilament ring formation or cell cycle arrest due to incorrect chromosome segregation or telomere dysfunction.[42]

4.3.4 The Protocol: Obtaining the Samples and Slides Analyses

Samples for the CA and MN tests included tadpoles and adult specimens (see Figure 4.6). For the anesthesia/euthanasia of anurans, benzocaine and 5% xylocaine are usually applied,[24,26,28,33] as well as tricaine methanesulfonate[27] and chloroethylene solution.[30] However, the use of cold water has also been reported in the case of tadpoles.[22] The studies carried out with tadpoles indicate, in most cases, that to obtain blood samples, a cross-section at the base of the tail or the operculum is performed, and the smears for the MN test are immediately obtained by holding the animal with tweezers and gently rubbing the cut part on the clean slides. As for the CA, the tadpoles are placed after cutting in an Eppendorf with buffer solution.

In adult animals that are euthanized for tests other than the MN test, samples can be obtained through an abdominal incision or cardiac puncture. One of the studies cut the first 3 mm on the tip of finger three with a sterile scalpel.[4] Although animals were not euthanized for this study, less invasive procedures are strongly recommended whenever possible. The procedure is less invasive when the blood samples are obtained from the angularis vein, and then the animals are released.[38,40] Blood samples obtained from the angularis vein are also observed in German protocols.[4]

In the CA, slides were stained with 20 μL of ethidium bromide solution (0.02 mg mL^{-1}).[33,43,44] In the readings, the authors used a fluorescence microscope equipped with an appropriate filter combination and the Isis software (Metasystem, Heidelberg, Germany) with an excitation filter of 510–560 nm and a barrier filter of 590 nm with a magnification of 20×. DNA damage was observed using the TriTek Comet Score TM software, version 1.5. In this software analysis, pixel intensity is used to provide values corresponding to DNA damage, and the unit is an arbitrary unit for the parameters TL, %DNA and OTM (see Figure 4.4).[44] As for the MN test, Giemsa was the principal dye used to stain the anuran erythrocytes.

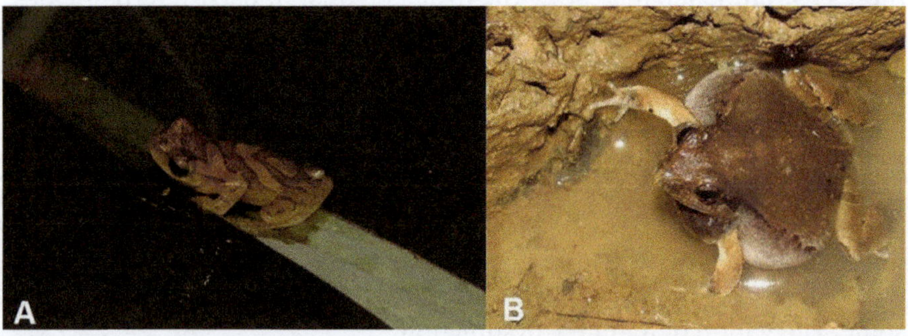

Figure 4.6 Images of (A) the lesser tree frog *Dendropsophus minutes* (Hylidae); (B) the barker frog *Physalaemus cuvieri* (Leptodactylidae). These images were gently given by Dr. R. P. Bastos (personal images).

The most common dye concentrations were 5% and 10%, with one study employing 15%. In addition to Giemsa, one study used Panoptic fast, a frequently used dye for erythrocytes.[6,10,11,13] The slide analysis was performed under an optical microscope with a 100× objective. Regarding the number of cells analyzed, out of the 19 articles selected, 14 evaluated 1000 cells per animal, although 4 quantified 2000 cells per specimen analyzed, and only 1 study considered 4000 cells for scoring.

4.3.5 Conclusions and Future Perspectives

A few studies have evaluated the genotoxicity and mutagenicity of frogs from agroecosystems in South America. Of the 26 species studied, the most frequent correspond to at least one representative of the Order Anura, such as the toad *R. arenarum* and the tree frogs *D. minutus* and *P. cuvieri*. Although most studies show genotoxicity and mutagenicity in agroecosystems, some studies did not observe an increase in genomic lesions. These responses demonstrate that many pesticides disperse in environments due to multiple environmental factors, such as wind, soil, water or the amount used in crops in the respective collection periods of animals, respectively.

Future studies may attempt to fill in the following gaps:

- Comparison before and after the application of pesticides in agroecosystems in different species and/or communities.
- Comparison of adult species of arboreal habit *vs* those of soil.
- Seek improved techniques for collecting blood samples without mutilating or sacrificing adult animals when applying only genotoxicity and mutagenicity tests.
- Seek tracking of pesticides applied in the study regions, either through analysis of water, sediments, chromatography of animal tissues, or through the collection of information from owners during the registration period.
- When possible, associate the genotoxic and mutagenic damage to "pesticides," and the landscape (land use and cover) where the animals were sampled, considering that many frog species have low mobility. Multivariate statistics can be used for this understanding. Thus, hypothetically it will be possible to generate associations between biological damage and environmental variables. The achievement of all these objectives will shed more light on the processes involved, help us on the road to a better understanding of the complexities involved, and help us in the search for a holistic risk assessment strategy.

Acknowledgements

The authors thank Dr R. P. Bastos for the original color images of *Dendropsophus minutes* and *Physalaemus cuvieri*. R. F. Salla and D. Melo e Silva thank the National Council for Scientific and Technological Development (CNPq) for their scholarships. The authors also thank FAPEG/CNPq (Fundação de Amparo à Pesquisa do Estado de Goiás (FAPEG)/CNPq: process 201710267000507-Pronex/Fapeg/CNPq). M. Benvindo-Souza thanks the Coordination for the Improvement of Higher Education Personnel Training (CAPES) and the Brazilian Fund for Biodiversity (FUNBIO) for their support.

References

1. A. J. Nowakowski, M. E. Thompson, M. A. Donnelly and B. D. Todd, Amphibian sensitivity to habitat modification is associated with population trends and species traits, *Glob. Ecol. Biogeogr.*, 2017, **26**, 700–712.
2. M. R. F. San Juan, S. M. L. Lavarías, V. Aparicio, K. E. Larsen, J. E. C. Lerner and A. Cortelezzi, Ecological risk assessment of pesticides in sediments of Pampean streams, Argentina, *Chemosphere*, 2023, **313**, 137598.
3. U. P. Dhuldhaj, R. Singh and V. K. Singh, Pesticide contamination in agro-ecosystems: toxicity, impacts, and bio-based management strategies, *Environ. Sci. Pollut. Res.*, 2022, **30**, 9243–9270.
4. Á. Cruz-Esquivel, J. Viloria-Rivas and J. Marrugo-Negrete, Genetic damage in *Rhinella marina* populations in habitats affected by agriculture in the middle region of the Sinú River, Colombia, *Environ. Sci. Pollut. Res.*, 2017, **24**, 27392–27401.
5. M. A. Albecker and M. W. McCoy, Adaptive responses to salinity stress across multiple life stages in anuran amphibians, *Front. Zool.*, 2017, **14**, 1–16.
6. R. E. Borges, L. R. de Souza Santos, R. A. Assis, M. Benvindo-Souza, L. Franco-Belussi and C. de Oliveira, Monitoring the morphological integrity of Neotropical anurans, *Environ. Sci. Pollut. Res.*, 2019, **26**, 2623–2634.
7. F. E. Pollo, C. L. Bionda, Z. A. Salinas, N. E. Salas and A. L. Martino, Common toad *Rhinella arenarum* (Hensel, 1867) and its importance in assessing environmental health: test of micronuclei and nuclear abnormalities in erythrocytes, *Environ. Monit. Assess.*, 2015, **187**, 1–9.
8. A. A. Abdel-Khalek, S. B. Dajem and K. Morsy, The potential use of rice husk for reducing the genotoxic effects of iron and aluminum oxides

nanoparticles in *Oreochromis niloticus*, *Water Air Soil Pollut.*, 2020, **231**, 1–11.

9. J. M. Pérez-Iglesias, G. S. Natale, S. Soloneski and M. L. Larramendy, Are the damaging effects induced by the imazethapyr formulation Pivot® H in *Boana pulchella* (Anura) reversible upon ceasing exposure?, *Ecotoxicol. Environ. Saf.*, 2018, **148**, 1–10.
10. W. F. Carvalho, C. de Arcaute Ruiz, J. M. Pérez-Iglesias, M. R. R. Laborde, S. Soloneski and M. L. Larramendy, DNA damage exerted by mixtures of commercial formulations of glyphosate and imazethapyr herbicides in *Rhinella arenarum* (Anura, Bufonidae) tadpoles, *Ecotoxicology*, 2019, **28**, 367–377.
11. A. T. C. Lopes, M. Benvindo-Souza, W. F. Carvalho, H. F. Nunes, P. N. de Lima, M. S. Costa, E. J. Benetti, V. Guerra, S. M. T. Saboia-Morais, C. E. Santos, K. Simões, R. P. Bastos and D. de Melo e Silva, Evaluation of the genotoxic, mutagenic, and histopathological hepatic effects of polyoxyethylene amine (POEA) and glyphosate on *Dendropsophus minutus* tadpoles, *Environ. Pollut.*, 2021, **289**, 117911.
12. A. Jaylet, P. Deparis, V. Ferrier, S. Grinfeld and R. Siboulet, A new micronucleus test using peripheral blood erythrocytes of the newt *Pleurodeles waltl* to detect mutagens in fresh-water pollution, *Mutat. Res. Genet. Toxicol. Environ. Mutagen.*, 1986, **164**, 245–257.
13. M. Benvindo-Souza, E. A. Santos Oliveira, R. A. Assis, C. G. Araújo-Santos, R. E. Borges, D. de Melo e Silva and L. R. de Souza Santos, Micronucleus test in tadpole erythrocytes: Trends in studies and new paths, *Chemosphere*, 2020, **240**, 124910.
14. J. Kwasniewska and A. W. Bara, Plant cytogenetics in the micronuclei investigation: The past, current status, and perspectives, *Int. J. Mol. Sci.*, 2022, **23**, 1306.
15. I. S. de Jesus, M. M. Cestari, M. D. A. Bezerra and P. R. A. D. de M Affonso, Genotoxicity effects in freshwater fish from a Brazilian impacted river, *Bull. Environ. Contam. Toxicol.*, 2016, **96**, 490–495.
16. C. Mesak, B. de Oliveira Mendes, R. de Oliveira Ferreira and G. Malafaia, Mutagenic assessment of *Lithobates catesbeianus* tadpoles exposed to the 2,4-D herbicide in a simulated realistic scenario, *Environ. Sci. Pollut. Res.*, 2018, **25**, 15235–15244.
17. M. B. Silva, R. E. Fraga, P. B. Nishiyama, N. L. B. Costa, I. S. S. Silva, T. S. Queiroz, M. A. Rocha and F. A. Juncá, Genotoxic effect of the insecticide Chlorpyrifos on the erythrocytes of *Odontophrynus carvalhoi* tadpoles (Amphibia: Odontophrynidae), *Ecotoxicol. Environ. Saf.*, 2020, **15**, 9–13.
18. R. A. Assis, M. Benvindo-Souza, C. G. Araújo-Santos, R. E. Borges, I. D. Santos-Filho, L. C. Oliveira, M. A. C. Mendonça and L. R. S. Santos, Mutagenic effect of a commercial fungicide on *Rana catesbeiana* and *Leptodactylus latrans* tadpoles, *An. Acad. Bras. Cienc.*, 2022, **94**, 20210161.

19. J. G. Zaller and C. A. Brühl, Editorial: Non-target effects of pesticides on organisms inhabiting agroecosystems, *Front. Environ. Sci.*, 2019, **7**, 75.
20. M. C. Triques, F. Ribeiro, D. de Oliveira, B. V. Goulart, C. C. Montagner, E. L. G. Espíndola and V. Menezes-Oliveira, The ecotoxicity of sugarcane pesticides to non-target soil organisms as a function of soil properties and moisture conditions, *Int. J. Environ. Res.*, 2022, **16**, 1–17.
21. E. C. Kuhn, M. T. Jacques, D. Teixeira, S. Meyer, T. Gralha, R. Roehrs, S. Camargo, T. Schwerdtle, J. Bornhorst and S. Ávila, Ecotoxicological assessment of Uruguay River and affluents pre- and post-pesticides' application using *Caenorhabditis elegans* for biomonitoring, *Environ. Sci. Pollut. Res.*, 2021, **28**, 21730–21741.
22. T. Ascoli-Morrete, N. M. G. Bandeira, E. Signor, H. A. Gazola, I. S. Homrich, R. Biondo, L. G. Rossato-Grando and N. Zanella, Bioaccumulation of pesticides and genotoxicity in anurans from southern Brazil, *Environ. Sci. Pollut. Res.*, 2022, **29**, 45549–45559.
23. R. A. Assis, W. R. Rezende, C. G. A. dos Santos, M. Benvindo-Souza, N. P. L. Amorim, R. E. Borges, L. Franco-Belussi, C. de Oliveira and L. R. de S. Santos, Habitat differences affect the nuclear morphology of the erythrocytes and the hepatic melanin in *Leptodactylus fuscus* (Anura) in the Brazilian Cerrado savanna, *Environ. Sci. Pollut. Res.*, 2021, **28**, 60741–60752.
24. M. C. da Rocha, M. B. dos Santos, R. Zanella, O. D. Prestes, A. S. Gonçalves and A. P. Schuch, Preserved riparian forest protects endangered forest-specialists amphibian species against the genotoxic impact of sunlight and agrochemicals, *Biol. Conserv.*, 2020, **249**, 108746.
25. R. E. Borges, L. R. de S. Santos, M. Benvindo-Souza, R. S. Modesto, R. A. Assis and C. de Oliveira, Genotoxic evaluation in tadpoles associated with agriculture in the Central Cerrado, Brazil, *Arch. Environ. Contam. Toxicol.*, 2019, **77**, 22–28.
26. M. W. Gonçalves, T. B. Vieira, N. M. Maciel, W. F. Carvalho, L. S. F. Lima, P. G. Gambale, A. D. da Cruz, F. Nomura, R. P. Bastos and D. de Melo e Silva, Detecting genomic damages in the frog *Dendropsophus minutus*: preserved *versus* perturbed areas, *Environ. Sci. Pollut. Res.*, 2015, **22**, 3947–3954.
27. M. W. Gonçalves, P. G. Gambale, F. R. Godoy, A. A. Alves, P. H.Almeida-Rezende, A. D.da Cruz, N. M. Maciel, F. Nomura, R. P. Bastos, P. de Marco-Jr. and D. de Melo e Silva, The agricultural impact of pesticides on *Physalaemus cuvieri* tadpoles (Amphibia: Anura) ascertained by comet assay, *Zool.*, 2017, **34**, 1–8.
28. M. W. Gonçalves, C. B. M. de Campos, F. R. Godoy, P. G. Gambale, H. F. Nunes, F. Nomura, R. P. Bastos, A. D. da Cruz and D. de Melo e Silva, Assessing genotoxicity and mutagenicity of three common amphibian species inhabiting agroecosystems, Environment, *Arch. Environ. Contam. Toxicol.*, 2019, **77**, 409–420.

29. M. E. Josende, A. M. Tozetti, M. T. Alalan, V. Mathies Filho, S. da Silva Ximenez, F. M. R. da Silva Júnior and S. E. Martins, Genotoxic evaluation in two amphibian species from Brazilian subtropical wetlands, *Ecol. Indic.*, 2015, **49**, 83–87.
30. F. E. Pollo, P. R. Grenat, M. A. Otero, S. Babini, N. E. Salas and A. L. Martino, Evaluation *in situ* of genotoxic and cytotoxic response in the diploid/polyploid complex *Odontophrynus* (Anura: Odontophrynidae) inhabiting agroecosystems, *Chemosphere*, 2019, **216**, 306–312.
31. M. S. Babini, C. de L. Bionda, N. E. Salas and A. L. Martino, Health status of tadpoles and metamorphs of *Rhinella arenarum* (Anura, Bufonidae) that inhabit agroecosystems and its implications for land use, *Ecotoxicol. Environ. Saf.*, 2015, **118**, 118–125.
32. A. M. Attademo, M. Cabagna-Zenklusen, R. C. Lajmanovich, P. M. Peltzer, C. Junges and A. Bassó, B-esterase activities and blood cell morphology in the frog Leptodactylus chaquensis (Amphibia: Leptodactylidae) on rice agroecosystems from Santa Fe Province (Argentina), *Ecotoxicology*, 2011, **20**, 274–282.
33. A. M. Attademo, P. M. Peltzer, R. C. Lajmanovich, M. C. Cabagna-Zenklusen, C. M. Junges and A. Basso, Biological endpoints, enzyme activities, and blood cell parameters in two anuran tadpole species in rice agroecosystems of mid-Eastern Argentina, *Environ. Monit. Assess.*, 2014, **186**, 635–649.
34. E. Caraffa, C. del L. Bionda, F. E. Pollo, N. E. Salas and A. L. Martino, Determinación de lafrecuencia de micronúcleos eneritrocitos de *Bufo arenarum* que habitan ambientes urbanizados, *Actatoxicológicaargentina*, 2013, **21**, 78–84.
35. J. C. Brodeur, J. Vera Candioti, S. Soloneski, M. L. Larramendy and A. E. Ronco, Evidence of Reduced Feeding and Oxidative Stress in Common Tree Frogs (*Hypsiboas pulchellus*) from an agroecosystem experiencing severe drought, *J. Herpetol.*, 2012, **46**, 72–78.
36. P. M. Peltzer, R. C. Lajmanovich, J. C. Sánchez-Hernandez, M. C. Cabagna, A. M. Attademo and A. Bassó, Effects of agricultural pond eutrophication on survival and health status of *Scinax nasicus* tadpoles, *Ecotoxicol. Environ. Saf.*, 2008, **70**, 185–197.
37. G. H. Carrasco, M. B. de Souza and L. R. de Souza Santos, Effect of multiple stressors and population decline of frogs, *Environ. Sci. Pollut. Res. Int.*, 2021, **42**, 59519–59527.
38. T. S. Vasconcelos, F. R. da Silva, T. G. dos Santos, V. H. M. Prado and D. B. Provete, in *Biogeographic Patterns of South American Anurans*, Springer Nature, Switzerland AG, 1st edn, 2019.
39. Amphibian Species of the World, Available from: https://amphibiansoftheworld.amnh.org. Accessed March 2023.

40. IUCN Red List of Threatened Species, Available from: https://www.iucnredlist.org/search?query=Dendropsophus%20minutus%20&searchType=species. Accessed March 2023.
41. *Rhinella arenarum*, Available from: https://www.iucnredlist.org/species/54576/11169255. Accessed March 2023.
42. T. S. Kumaravel and A. N. Jha, Reliable comet assay measurements for detecting DNA damage induced by ionising radiation and chemicals, *Mutat. Res. Genet. Toxicol.*, 2006, **605**, 7–16.
43. M. Fenech, M. Kirsch-Volders, A. T. Natarajan, J. Surralles, J. W. Crott and J. Parry, Molecular mechanisms of micronucleus, nucleoplasmic bridge and nuclear bud formation in mammalian and human cells, *Mutagenesis*, 2011, **26**, 125–132.
44. C. Bolognesi, S. Knasmueller, A. Nersesyan, P. Thomas and M. Fenech, The HUMNxl scoring criteria for different cell types and nuclear anomalies in the buccal micronucleus cytome assay - an update and expanded photogallery, *Mutat. Res.*, 2013, **753**, 100–113.

CHAPTER 5

Ecotoxicity and Risk Assessment Characterization of Veterinary Pharmaceuticals on Anuran Amphibian Larvae

PAOLA M. PELTZER*[a,b], ANA P. CUZZIOL BOCCIONI[a,b], CANDELA MARTINUZZI[a,b], AGUSTÍN BASSÓ[a,b], ANDRÉS M. ATTADEMO[a,b], MARÍA J. CULZONI[a,c], LESLY PARADINA-FERNANDEZ[a,c] AND RAFAEL C. LAJMANOVICH[a,b]

[a]Consejo Nacional de Investigaciones Científicas Técnicas (CONICET), Buenos Aires, Argentina; [b]Laboratorio de Ecotoxicología, Facultad de Bioquímica y Ciencias Biológicas, Universidad Nacional del Litoral (FBCB-UNL), Santa Fe, Argentina; [c]Laboratorio de Desarrollo Analítico y Quimiometría (LADAQ), Cátedra de Química Analítica I, Facultad de Bioquímica y Ciencias Biológicas, Universidad Nacional del Litoral (FBCB-UNL), Santa Fe, Argentina
*E-mail: paolapeltzer@hotmail.com

5.1 Introduction

Aquatic ecosystems are prone to contamination due to the synergies among high nutrients, pesticides, microplastics (MPs) and pharmaceutical compounds that, together or in isolated fashion, threaten their quality and biodiversity worldwide.

Issues in Toxicology No. 46
Amphibian Species in Environmental Risk Assessment Strategies
Edited by Marcelo L. Larramendy and Guillermo Eli Liwszyc

Published by the Royal Society of Chemistry, www.rsc.org

The widespread use of veterinary pharmaceuticals and their release into different environments, such as freshwater bodies and groundwater systems, has increased in the last two decades.[1,2] The occurrence of these compounds in aquatic systems has become a serious health and environmental concern and has been the subject of recent debate in scientific literature. Due to their massive environmental release, these contaminants of emerging concern (CECs) have been detected at ng L^{-1} to μg L^{-1} levels in several environmental sources, including surface water, groundwater, wastewater, and even drinking water.[3] Veterinary pharmaceuticals are generally used in livestock production for disease prevention and growth promotion. Significant quantities of antibiotics used as a feed supplement for growth enhancement of food animals have been found in animal waste manure (*e.g.*, poultry litter without any treatment) used as organic fertlizers.[4] The presence of these contaminants in aquatic environments poses an environmental risk for both human health and aquatic ecosystems.[5]

In most cases, the high risk of CECs is related to the potential bioaccumulation and ecotoxicological effects that could be triggered in aquatic organisms both after short- and long-term exposure. Likewise, since one of the major problems with the abovementioned contaminants is related to the lack of regulatory criteria and routine monitoring programs, data from screening in aquatic ecosystems and their effects at different hierarchical levels on wild fauna establish a powerful tool for future regulations and/or monitoring programs. Hence, controlling and/or monitoring these contaminants is essential for environmental sustainability, mainly if we consider that these contaminants may have synergistic, antagonistic, additive, or potentiation effects on aquatic ecosystems, wild fauna, and humans. In Argentina, the risk of the presence of CECs has increased in the last decades, mainly related to meat-producing agro-industries that have no treatment of their residues (*e.g.*, poultry litter and manure) and other anthropic activities, such as agrochemical application near aquatic systems, industrial effluent, wastewater, canalization or drought of rivers and/or tributaries, intentional fires, and urbanization in wetlands, which synergistically increase the extinction of amphibians and the well health of aquatic ecosystems.

5.2 Amphibians as Bioindicators

Amphibians are considered indicators of environmental quality worldwide, being model organisms in ecotoxicology. Given the similarity of the mechanisms of early development in amphibian embryos with that of human beings,[6] ecotoxicological studies have increased notably in recent years.[7] In these studies, the integration of different biological responses at different hierarchical levels and the search for new early warning and non-destructive biomarkers have become real challenges.[8,9]

This chapter aims to review different studies focused on the detection and quantification of such compounds in aquatic environments using different analytical techniques (ENVIRONMENTAL SCREENING) and how amphibian tadpoles respond (BIOLOGICAL BIOMARKERS) to different concentrations and compounds in mid-eastern Argentina. In this region, CECs are recorded in different aquatic systems and amphibian tissues, indicating their high risk and the potential collapse of aquatic systems if regulatory laws are not passed as soon as possible. A case in question is the handling of the so-called Wetlands Law by the Parliament of Argentina. These factors are threatening amphibian survival and are common in remnants of forests or altered wetlands, including drainage systems, fires, and drought, among others, producing an apocalyptic but unfortunately real scenario where not only amphibians are vulnerable but also all aquatic organisms.

5.3 Environmental Screening

The previous evidence endorses the relevance of developing sensitive, fast, and accurate analytical methodologies for the screening of CECs in biological and environmental matrices. For example, fluoroquinolones (FQs) are an important group of broad-spectrum antibiotics widely used to treat bacterial infections in humans, animals, poultry and fish.[10] Since these compounds are not only fully metabolized by the body but also not fully removed in wastewater treatment plants, they are discharged into surface water supplies as parent compounds or as metabolites. Besides considering that they are administered in large quantities, are highly resistant to biodegradation, and that the health exerted effects due to their persistence in the environment are not well known (even at very low levels), these compounds are of public and ecological health concern.[11] For all the abovementioned reasons, their evaluation and monitoring are imperative for human health protection and environmental control.

In this sense, studies performed in our group detected the occurrence of several FQs in rivers and wastewater in Argentina. A sensitive analytical method based on high-performance liquid chromatography coupled to fast scanning fluorescence detection was developed and used for detecting the presence of enrofloxacin, ciprofloxacin, and norfloxacin in rivers at 0.97–3.60, 0.40 and 1.00 $\mu g\ L^{-1}$ levels, respectively.[10,11] Besides, the occurrence of enoxacin, ofloxacin, ciprofloxacin, difloxacin, and enrofloxacin in wastewaters was observed at maximum concentrations of 22.1, 1.78, 1.14, 14.2 and 11.9 $\mu g\ L^{-1}$, respectively.[10]

In general, the FQ concentrations found in the present study are in accordance with the levels reported in the literature. Only for the case of enoxacin, the concentrations detected showed higher levels than those found by Adachi *et al.*[12] (*i.e.*, 0.02 $\mu g\ L^{-1}$) and Du *et al.*[13] (*i.e.*, 0.1 $\mu g\ L^{-1}$), which could constitute an important cause for concern. This persistence could probably be associated with the slow degradation process of enoxacin.[14] It is important to highlight that the presence of enrofloxacin was detected during different studies performed by Alcaraz *et al.*[11] and Teglia *et al.*[10] in the same river site location (Las Prusianas, Santa Fe, Argentina). This finding showed the prevalence of this emerging contaminant, which could suggest the continuous discharge of enrofloxacin to this river or its persistence in aquatic ecosystems. Many studies have noted the ubiquity of enrofloxacin worldwide. For example, enrofloxacin has been found in riparian water (4.42 $\mu g\ L^{-1}$) directly contaminated by large-scale animal farms,[15] in similar water bodies in China,[16] and in water courses at the end of aquaculture systems in Portugal,[17] at maximum concentrations of 0.979 $\mu g\ L^{-1}$ and 0.0232 $\mu g\ L^{-1}$, respectively. The occurrence of FQs in aquatic ecosystems could constitute a huge environmental issue related to the generation of harmful effects on native flora and fauna and the human population. In this regard, Peltzer *et al.*[18] have demonstrated the negative effects of enrofloxacin and ciprofloxacin on the development of the South American common toad *Rhinella arenarum*. In the *in vitro* study of biological endpoints (mainly development, growth and antioxidant enzyme activities), impairments were observed after exposure to enrofloxacin and ciprofloxacin, concluding that continuous exposure to these substances can lead to potential damage to the long-term maintenance of *R. arenarum* natural larvae populations (see Figure 5.1).[18]

Figure 5.1 The Cope's toad, also known as the "cururu" toad *Rhinella diptycha* male amplexus with a female of the South American common toad *Rhinella arenarum*.

5.4 Biological Biomarkers

5.4.1 Enzyme Biomarkers

The concept of a biomarker refers to a measurable change at the physiological, morphological, or histological level related to the exposure of an organism to one or more chemical or physical contaminants.[19] Variations in the metabolic responses of anuran amphibians exposed to emerging contaminants, such as human and/or veterinary pharmaceuticals, can be determined using different biomarkers, particularly enzymatic ones. In this context, research involving tadpoles of *R. arenarum* exposed to different agrochemicals, such as glyphosate and chlorpyrifos, and to antibiotics showed enzymatic changes in acetylcholinesterase (AChA) and glutathione *S*-transferase (GST).[20] Cuzziol Boccioni *et al.*[21] evaluated the GST activity to characterize the sublethal effects of chronic exposure to dexamethasone (1–1000 mg L^{-1}) at 96 h of exposure and found a significant increase in the GST activity level. GST also evidenced the effects of exposure to antiretrovirals in anuran tadpoles.[22]

Other contaminants of global emerging concern are MPs and chemical additives such as phthalates and bisphenols. MPs play a role as vectors for the transport of several toxic chemicals, enhancing their deleterious effects on environments and wild fauna.[23] In this sense, they might bind to human and/or veterinary pharmaceuticals.[24] Studies on the effects and toxicity of these MPs on aquatic organisms, especially amphibians, are scarce. Recently, Attademo *et al.*[25] found changes in carboxylesterases (CbEs) in tadpoles exposed to MPs. CbEs are a group of enzymes with an important role in the protection mechanisms against pesticides and other CECs, such as MPs.[26] Likewise, the enzyme alkaline phosphatase, which is involved in the removal of the phosphate group, was found to be affected by exposure to MPs.

5.4.2 Histological Biomarkers

Alterations in the number, structure, or function of cells are considered to be good indicators of pollutant effects.[27] In this sense, the configuration, composition, and appearance of some cells and tissues are frequently used as histological biomarkers to assess the effects of pollutants on amphibians. Histological changes in tissues can further reveal the effects of veterinary antibiotics on target organs, with histopathology reflecting the injuries according to the toxicity of these compounds, the duration and intensity of exposure of the organism, and the ability of tissues to respond and adapt.[28]

As the liver is involved in the biotransformation of toxic compounds, its histology gives relevant information about pollutants' effects.[29] A histopathological study can include the analysis of the general configuration of the organ or some specific cells. For example, a reduction of the whole organ size was reported on tadpoles of *Rana omeimontis* exposed to sulfamethoxazole and oxytetracycline.[30] Another study on tadpoles of *Aquarana catesbeiana* exposed to the same drugs described inflammatory infiltration as well as vascular congestion, blood cells, eosinophils and melanomacrophages.[31] Melanomacrophages are particularly interesting since their changes both in number and size and in the number of pigments phagocytized can indicate several physiological and immunological responses. For instance, a significant increase in number and size was reported for *R. arenarum* tadpoles exposed to dexamethasone.[21]

The appearance and configuration of the intestinal epithelium are also frequently evaluated as histological biomarkers of antibiotics exposure. Indeed, this tissue is sensitive to external agents and directly associated with physiological and metabolic processes involved in the absorption of nutrients. Therefore, changes in this tissue might affect those processes.[32] Tadpoles of *Pelophylax nigromaculatus* exposed to oxytetracycline showed severe histopathological gut injuries, including cytoplasmic damage, cell arrangement disorganization, and structure deformation.[33] Other effects of exposure to veterinary drugs such as dexamethasone on *R. arenarum* tadpoles include an increase in the gastrointestinal epithelium thickness and the presence of melanomacrophages in gut loops.[21]

Blood is also frequently used in the histological evaluation of veterinary drug effects. Since amphibians have nucleated erythrocytes, micronucleus (MNs) frequency is one of the most commonly used tests to detect the genotoxicity of chemical substances.[34] Alterations in the immune system cell count, due to exposure to contaminants, such as veterinary antibiotics are important in risk assessment studies because they are involved in the susceptibility of the organisms to pathogen infections, which is an important cause of amphibian population decline worldwide.[35] The antibiotics sulfamethoxazole and oxytetracyclyne in *A. catesbeiana* produced lymphocytes and neutropenia.[31]

Other less common organs used in the histological analysis of tadpoles for the evaluation of the effect of exposure to antibiotics include the gonads. These tissues are of particular interest for the study of drugs with the risk of endocrine disruption and their potential effects on sex differentiation. As suggested in a study on *Xenopus tropicalis* exposed to the fungicide clotrimazole (an aromatase inhibitor and consequently involved in sex differentiation), the gonadal study could include the number of germ cells as well as the cross-section area of ovaries, testes, and the ovarian cavity.[36]

Finally, whole-body sections of tadpoles are also useful for analysing not only the organs of interest but also their general and spatial configuration; thus, malformations observed at the macroscopic scale can be confirmed. The teratogenic effect of dexamethasone on *R. arenarum* development, dysplastic cells of digestive and central nervous systems, misplaced organs, abnormal gut coiling, heart function and shape, visceral congestion, and oedemas were also reported.[21] Similar findings were reported for diclofenac (DCF) on cross sections of *X. laevis* tadpoles, including developmental delay, lack of cell differentiation, abnormal organ shape and amorphous oedemas.[37]

5.4.3 Cellular and Nuclear Biomarkers

In amphibians, the leukocyte component is proposed as a cellular biomarker for the evaluation of individuals exposed to stressful situations, as a parameter indicative of populations coming from contaminated sites and as a measure of the immunomodulatory effect induced by a given pesticide.[38–41] On the other hand, various factors must be considered, such as the variability of the blood cells throughout the year and the age of the individuals. This is because normal cytological characteristics could be confused with pathological processes or responses to changes in the environment.[42,43] Besides, variations in the morphology of erythrocytes are associated with amoxicillin exposure, aromatic substances, metals and an altered agroecosystem.[40,44–46]

One of the most commonly used cellular biomarkers is the MN test, which reveals the early effect of genetic damage on tadpoles.[47] MNs are cytoplasmic bodies of a nuclear nature. They correspond to genetic material, from whole chromosomes to chromosomal fragments, not correctly incorporated into the daughter cells during cell division. In amphibians, as in other vertebrates, erythrocytes are the most abundant circulating cells. They are oval in shape, slender, with a central protrusion due to the nucleus. In most amphibian species, it is possible to observe anucleated erythrocytes (erythroplastids) of variable size and shape in low quantities (1–3% of the total red blood cells), with some species of salamanders in which the number of erythroplastids can account from 80% to 95% of the circulating cells.[48]

Genotoxicity of different pesticides has been determined by the MN test on anuran larvae exposed in the laboratory.[49–51] Similarly, the formation of nuclear aberrations in specimens in rice crops after field applications of pyrethroids (cypermethrin and lambda-cyhalothrin), organophosphates (methamidophos) and organochlorines (endosulfan),[52,53] and in restituted crop sediments contaminated by metals (As, Hg, and Cr) has been detected.[54] Nanomaterials have also been found to increase the MN frequencies in exposed larvae (*e.g.*, silicon dioxide, SiO_2 and Ni-Al nanoceramics).[55,56] Pharmaceutical drug mixtures in amphibian tadpoles (*i.e.*, carbamazepine, naproxen and sulfamethoxazole) produce increased toxicity compared to exposure to individual compounds.[57] Likewise, multiple nuclear abnormalities (*e.g.*, micronucleated erythrocytes, blebbed, kidney-shaped, binuclear, multilobulated and notched nuclei) have also been observed in the larvae exposed to a mixture of different pharmaceutical drugs.[58]

5.4.4 Microbial Biomarkers

The gut microbiota is known for its key roles in maintaining host health, including pathogen defense, food digestion, energy metabolism and storage, synthesis of vitamins and amino acid as well as ecological adaptation and behavioural regulation.[59,60] Pollutants can interact with the gut microbiota through different pathways and, therefore, alter its composition and homeostasis, leading to metabolic disorders, nutrient malabsorption and dysfunction of the immune system.[61] Although contaminants do not target the intestinal microbiota directly, some of them can disrupt the normal structure of intestinal microbiome communities (dysbiosis) and lead to an imbalance of bacterial species, which may, in turn, pose a risk to health.[62] The intestinal microbiota as a biomarker of the effects of veterinary pharmaceuticals has been poorly explored so far. Among the effects in the microbiotic community of the intestine due to exposure to pharmaceuticals, changes in relative abundance and dominance of genera and phyla are the most reported. For example, Ya *et al.*[33] described these effects on tadpoles of *P. nigromaculatus* exposed to oxytetracycline and Boccioni *et al.*[63] reported similar effects as well as diversity reduction on communities from *R. arenarum* tadpoles exposed to ciprofloxacin. Changes in the composition of the gut microbiota were also reported for *Rana dybowskii* exposed to gentamicin.[64] Moreover, decreased susceptibility and resistance acquisition to antibiotics are also being reported in the gut microbiota of amphibians exposed to pharmaceuticals.[33,65] In addition, amphibian tadpoles represent reservoirs of bacteria of sanitary interests, and their gut microbiota dysbiosis due to pollution exposure may lead to the emergence of resistant pathogens.[18,66]

5.5 Morphological Biomarkers

Numerous authors around the world have reported the finding of malformations in specimens of anuran amphibians, both in larvae and in adults. Johnson *et al.*[67] define morphological abnormality as any deviation from the normal range of anatomical variation, which can be caused by mutations, failures in embryonic development, or trauma. Although a small number of anomalies in natural amphibian populations are to be expected, this proportion is usually less than 5%. The causes of the appearance of these morphological anomalies in amphibians can be a consequence of both natural factors (*e.g.*, attacks by predators, parasitic infections, viruses and congenital malformations) and factors of anthropic origin (*e.g.*, pollutants of emerging concern, such as pesticides, hydrocarbons, metals, drugs and ultraviolet radiation).[68,69]

In this context, numerous morphological abnormalities were observed in *R. arenarum* larvae exposed under controlled laboratory conditions to different commonly used veterinary drugs (no published data). In the larvae exposed to different concentrations of dipyrone (0.1–1000 μg L^{-1}), we were able to observe a high prevalence of abnormal gut coiling (11%), gill chamber abnormalities (9%), abdominal oedema (9%) and abdominal organs atrophied (7%). Similarly, those larvae exposed to DCF (62.5–1000 μg L^{-1}) showed a high prevalence of abnormal gut coiling (18%), gill chamber abnormalities (10%), oral disc alteration (9%), abdominal organs atrophy (14%) and abdominal oedema (11%). Those larvae exposed to streptomycin (0.1–1000 μg L^{-1}) suffered a high prevalence of abnormal gut coiling (21%), abdominal organs atrophy (11%) and abdominal oedema (11%). Another antibiotic to which *R. arenarum* larvae were exposed was enrofloxacin (0.1–1000 μg L^{-1}) in which abdominal organs atrophy (19%), gill chamber abnormalities (15%), abnormal gut coiling (12%), abdominal oedema (12%) and oral disc alteration (11%) were observed. The glucocorticoid dexamethasone (2500–20 000 μg L^{-1}) caused abdominal oedema (33%), abdominal organs atrophy (14%) and abnormal gut coiling (8%) in larvae. Finally, *R. arenarum* larvae exposed to hyoscine butylbromide + dipyrone (0.1–1000 μg L^{-1}) suffered from abdominal organ atrophies (12%), abnormal gut coiling (11%) and gill chamber abnormalities (7%). Figure 5.2 shows images of the observed malformations.

Veterinary drugs tested have a high capacity to cause morphological alterations in the early stages of development in anuran amphibians. Peltzer *et al.*[8] described a high prevalence of morphological abnormalities in *Trachycephalus typhonius* and *Physalaemus albonotatus* larvae chronically exposed to DCF, finding abnormalities in the heart, viscera, oral disc organ, chondrocranium and hyobranchial skeleton. Similarly, Cardoso-Vera *et al.*[70] described malformations in *X. laevis* and *A. catesbeiana* larvae exposed to DCF, revealing the presence of abdominal, cephalocranial and cardiac oedema, hypopigmentation, microcephaly and atrophy of abdominal organs. An investigation carried out by Boccioni *et al.*[21] in *R. arenarum* larvae exposed to veterinary dexamethasone described the presence of abnormalities that included gills alterations, bilateral asymmetry, cranial and oral disc abnormalities, variation in the shape of the eyeball, abdominal oedema, abnormal gut coiling and visceral hypertrophy. Similarly, Lorenz*et al.*[71] observed that tadpoles of *X. laevis* exposed to dexamethasone had shorter body and limb lengths compared to unexposed larvae.

The abnormal tadpoles exposed to different CECs have high rates of early mortality or extensive development. In general, they do not reach metamorphosis. In addition, the larvae face an ecological death; that is, the ability of the organisms to carry out their basic ecological functions is reduced (impaired swimming ability, less ability to escape from prey, difficulty obtaining food and completing metamorphosis, among others). Thus, the increase in the frequency of morphological abnormalities in anuran amphibians could exacerbate the current and progressive decline of these populations and represent a new challenge for conservation biology.

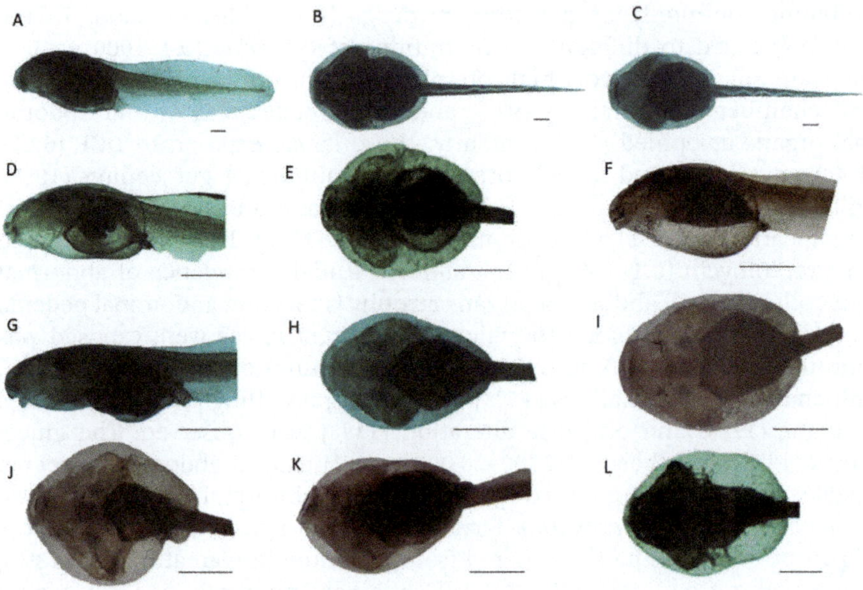

Figure 5.2 Morphological abnormalities in *R. arenarum* larvae exposed to different veterinary drugs. A: control larvae lateral view; B: control larvae dorsal view; C: control larvae ventral view; D: abnormal gut coiling, abdominal organs atrophy and abdominal oedema; E: abnormal gut coiling, atrophy of abdominal organs, gill chamber abnormalities and abdominal oedema; F: abdominal organs atrophy and abdominal oedema; G: abdominal organs atrophied; H: abdominal organs atrophy and oral disc alteration; I: oral disc alteration, gill chamber abnormalities and abdominal oedema; J: abnormal gut coiling, abdominal organs atrophy, abdominal oedema, oral disc alteration and gill chamber abnormalities; K: oral disc alteration; gill chamber abnormalities and abdominal organs atrophy; L: gill chamber abnormalities, abdominal oedema and abdominal organs atrophy. Scale bar: 1 mm.

5.6 Behavioural Biomarkers

Behavioural biomarkers constitute a group of noninvasive methods of assessing the general health of animals.[8] Since behaviour is the outcome of the integration of countless physiological processes, changes in physiology such as those produced by stress, pesticides, or medications have the potential to affect animal behaviour.[72]

Similar to other biomarkers, in behavioural biomarkers, it is essential to determine the profile of the activity of the selected parameters and to compare the changes that have occurred in them due to the challenging conditions in the studied medium. These basal levels are species-specific and may be affected by developmental stage, temperature, and other physical, chemical, or biological variables that must be considered.[8] Behavioural assessments can be done by direct observation, which can be time-consuming, subject to bias and/or suffer from a poorly defined link between behaviour and well-being, in general.[72] However, recent behavioural software analyses offer a solution to many of these issues and provide an objective and time-efficient method of quantifying behaviour.[73]

Although anurans played an essential role in the development of classical embryology, the behaviour of anuran larvae, in general, has been largely neglected,[74] particularly in ecotoxicology. Recently, some authors focused on the effect of widely used agrochemicals on the behaviour of the anuran larvae alone, in groups, or in a predator-prey scenario.[75,76] Recorded effects of pharmaceuticals on anuran larvae are even more scarce.

To increase the information about pharmaceuticals effects on anuran larvae, we tested streptomycin, enrofloxacin, dipyrone + hyoscine, dipyrone, DCF and dexamethasone.[77] For that purpose, ten larvae of *R. arenarum* per treatment were individually filmed to determine possible changes in behaviour. The larvae were individually placed in circular Petri dishes (Ø 10 cm, height 1 cm) containing the solutions to be tested, recorded for 3 min with a previous adaptation period of 1 min and analysed with Smart® motion analysis software. The ethological parameters determined were mean speed (MV, cm s^{-1}), total distance covered (DTR, cm), global activity (GA) and total immobility time (TI, s). Finally, each variable was evaluated by means of an ANOVA test and the corresponding Dunnett's post-test. In general, results showed that exposure to DCF, dipyrone + hyoscine, dexamethasone, streptomycin and enrofloxacin caused a linear dose-response relationship with DTR and GA decreasing with higher treatment concentrations, while dipyrone showed a hormetic relationship with an inhibitory response at low doses and a stimulatory response at high dipyrone concentrations.[77]

Moreover, Peltzer *et al.*[8] described significant effects of DCF on the swimming distance, speed and global activity of *T. typhonius* (see Figure 5.3) and *P. albonotatus* that had been exposed to low concentrations (125–250 μg L^{-1}). On the contrary, these parameters increased at higher concentrations (1000–2000 μg L^{-1}). These behavioural effects were linked to a neurotoxic effect due to acetylcholinesterase inhibition.[8] Other anti-inflammatory drugs, such as ibuprofen and anti-depressants like fluoxetine, were tested by Aliko *et al.*,[78] who described behavioural impairments and a significant increase of unresponsiveness to different stimuli in *Bufo bufo* tadpoles exposed. Nassef *et al.*[79] found that carbamazepine and DCF affected the feeding behaviour, while swimming speed was altered by exposure to carbamazepine and triclosan in the medaka fish *Oryzias latipes*. In addition, some authors have tested a mix of different types of medication (antibiotic, anti-inflammatory, antidepressant, anxiolytic, analgesic and antacid drugs) at environmentally relevant concentrations and found changes in

Figure 5.3 Adult specimen of the White-lipped tree frog *Trachycephalus typhonius*.

social aggregation and antipredatory defensive response in *A. catesbeiana*.[80] Melvin found no significant effect on swimming performance, but a trend of decreased swimming speed coupled with an increasing concentration of pharmaceutical drugs mix.[81] All these effects are directly related to an organism's fitness and survival. In summary, behavioural biomarkers are sensitive to a myriad of bioactive substances despite their mechanisms of action and are an accessible tool to complement ecotoxicological studies.

In conclusion, the abovementioned findings highlight the importance of pharmaceuticals as CECs for the aquatic environment and wildlife. Moreover, more attention needs to be given in relation to the toxic effects and potential risks these compounds pose to environments and all living beings, including humans. Paradigms in biological measurements that investigate in terms of biological relevance the toxicity of different CECs are extremely necessary in the face of the collapse of aquatic systems and they should be based on ecotoxicological task I-biological evidence or task II-bioaccumulation[82] using early warning organisms (bioindicator species), such as amphibians.[18] Environmental managers and decision-makers should be aware that human and veterinary pharmaceutical residues and their metabolites have been contributing to the current contamination of aquatic and terrestrial ecosystems all over the world. Treatment and regulation of these compounds are in urgent need. In addition, it is also important to note that these compounds occur with pesticides and nutrients that together may be lethal for amphibians and contribute to their decline.

Acknowledgements

We thank H. Beldoménico, H. Goicochea, C. Teglia, S. Seib, L. Curi, T. Sandoval, C. Colussi, E. León and R. Lajmanovich Jr. for their collaboration in different field and laboratory studies. We also thank CONICET, SECyT FONCyT and CAI + D UNL from Argentina for the provision of research grants.

References

1. V. Osorio, A. Larrañaga, J. Aceña, S. Pérez and D. Barceló, Concentration and risk of pharmaceuticals in freshwater systems are related to the population density and the livestock units in Iberian Rivers, *Sci.Total Environ.*, 2016, **540**, 267–277.
2. I. Tlili, G. Caria, R. B. Sghaier, S. Net, I. Ghorbe-Abid, R. Ternane, B. Ouddane and M. Trabelsi-Ayadi, Occurrence of 28 human and veterinary antibiotics residues in waters, North–Eastern Tunisia by liquid chromatography–tandem mass spectrometry, *Chem. Afr.*, 2022, **5**, 2163–2172.
3. L. Paradina Fernández, R. Brasca, M. R. Repetti, A. M. Attademo, P. M. Peltzer, R. C. Lajmanovich and J. Culzoni, Bioaccumulation of abacavir and efavirenz in *Rhinella arenarum* tadpoles after exposure to environmentally relevant concentrations, *Chemosphere*, 2022, **301**, 134631.
4. C. M. Teglia, P. M. Peltzer, S. N. Seib, R. C. Lajmanovich, M. J. Culzoni and H. C. Goicoechea, Simultaneous multi-residue determination of twenty–one veterinary drugs in poultry litter by modeling three-way liquid chromatography with fluorescence and absorption detection data, *Talanta*, 2017, **167**, 442–452.
5. M. L. Larramendy and S. Soloneski, (eds.), in *Emerging Pollutants in the Environment – Current and Further Implications*, InTech, Rijeka, Croatia, 2015.
6. D. R. Buchholz, More similar than you think: Frog metamorphosis as a model of human perinatal endocrinology, *Dev. Biol.*, 2015, **408**, 188–195.

7. L. S. Liu, L. Y. Zhao, S. H. Wang and J. P. Jiang, Research proceedings on amphibian model organisms, *Zool. Res.*, 2016, **37**, 237–245.
8. P. M. Peltzer, R. C. Lajmanovich, C. S. Martinuzzi, A. M. Attademo, L. M. Curi and M. T. Sandoval, Biotoxicity of diclofenac on two larval amphibians: Assessment of development, growth, cardiac function and rhythm, behavior and antioxidant system, *Sci. Total Environ.*, 2019, **683**, 624–637.
9. J. Peluso, C. M. Aronzon, M. Acquaroni and C. S. Pérez Coll, Biomarkers of genotoxicity and health status of *Rhinella fernandezae* populations from the lower Paraná River Basin, Argentina, *Ecol. Indic.*, 2020, **117**, 106588.
10. C. M. Teglia, F. A. Perez, N. Michlig, M. R. Repetti, H. C. Goicoechea and M. J. Culzoni, Occurrence, distribution, and ecological risk of fluoroquinolones in rivers and wastewaters, *Environ. Toxicol. Chem.*, 2019, **38**, 2305–2313.
11. M. R. Alcaraz, M. J. Culzoni and H. C. Goicoechea, Enhanced fluorescence sensitivity by coupling yttrium-analyte complexes and three-way fast high-performance liquid chromatography data modeling, *Anal. Chim. Acta.*, 2016, **902**, 50–58.
12. F. Adachi, A. Yamamoto, K. I. Takakura and R. Kawahara, Occurrence of fluoroquinolones and fluoroquinolone-resistance genes in the aquatic environment, *Sci. Total Environ.*, 2013, **444**, 508–514.
13. J. Du, H. Zhao, S. Liu, H. Xie, Y. Wang and J. Chen, Antibiotics in the coastal water of the South Yellow Sea in China: Occurrence, distribution and ecological risks, *Sci. Total Environ.*, 2017, **595**, 521–527.
14. M. Sturini, A. Speltini, F. Maraschi, L. Pretali, E. N. Ferri and A. Profumo, Sunlight-induced degradation of fluoroquinolones in wastewater effluent: Photoproducts identification and toxicity, *Chemosphere*, 2015, **134**, 313–318.
15. R. Wei, F. Ge, M. Chen and R. Wang, Occurrence of ciprofloxacin, enrofloxacin, and florfenicol in animal wastewater and water resources, *J. Environ. Qual.*, 2012, **41**, 1481–1486.
16. Y. Jiang, M. Li, C. Guo, D. An, J. Xu, Y. Zhang and X. Beidou, Distribution and ecological risk of antibiotics in a typical effluent–receiving river (Wangyang River) in North China, *Chemosphere*, 2014, **112**, 267–274.
17. A. M. Pereira, L. J. G. Silva, L. M. Meisel and A. Pena, Fluoroquinolones and tetracycline antibiotics in a Portuguese aquaculture system and aquatic surroundings: occurrence and environmental impact, *J. Toxicol. Environ. Health, A*, 2015, **78**, 959–975.
18. P. M. Peltzer, R. C. Lajmanovich, A. M. Attademo, C. M. Junges, C. M. Teglia, C. S. Martinuzzi, L. Curi, M. J. Culzoni and H. C. Goicoechea, Ecotoxicity of veterinary enrofloxacin and ciprofloxacin antibiotics on

anuran amphibian larvae, *Environ. Toxicol. Pharmacol.*, 2017, **51**, 114–123.

19. C. Cossu-Leguille and P. Vasseur, Aquatic Biomarkers, in *Encyclopedia of Aquatic Ecotoxicology*, ed. J. F. Férard and C. Blaise, Springer, Dordrecht, 2013, pp. 49–66.
20. A. P. Cuzziol Boccioni, R. C. Lajmanovich, P. M. Peltzer, A. M. Attademo and C. S. Martinuzzi, Toxicity assessment at different experimental scenarios with glyphosate, chlorpyrifos and antibiotics in *Rhinella arenarum* (Anura: Bufonidae) tadpoles, *Chemosphere*, 2021, **273**, 128475.
21. A. P. Cuzziol Boccioni, P. M. Peltzer, C. S. Martinuzzi, A. M. Attademo, E. J. León and R. C. Lajmanovich, Morphological and histological abnormalities of the neotropical toad, *Rhinella arenarum* (Anura: Bufonidae) larvae exposed to dexamethasone, *J. Toxicol. Environ. Health, B*, 2020, **56**, 41–53.
22. L. Paradina Fernández, R. Brasca, A. M. Attademo, P. M. Peltzer, R. C. Lajmanovich and M. J. Culzoni, Bioaccumulation and glutathione S-transferase activity on *Rhinella arenarum* tadpoles after short-term exposure to antiretrovirals, *Chemosphere*, 2020, **246**, 125830.
23. R. C. Lajmanovich, A. M. Attademo, G. Lener, A. P. Cuzziol Boccioni, P. M. Peltzer, C. S. Martinuzzi, L. Demonte and M. R. Repetti, Glyphosate and glufosinate ammonium, herbicides commonly used on genetically modified crops, and their interaction with microplastics: Ecotoxicity in anuran tadpoles, *Sci. Total Environ.*, 2022, **804**, 150177.
24. T. Atugoda, M. Vithanage, H. Wijesekara, N. Bolan, A. K. Sarmah, M. S. Bank, S. You and Y. S. Ok, Interactions between microplastics, pharmaceuticals and personal care products: Implications for vector transport, *Environ. Int.*, 2021, **149**, 106367.
25. A. M. Attademo, A. P. Cuzziol Boccioni, P. M. Peltzer, V. G. Franco, M. F. Simoniello and M. C. G. Passeggi, Effect of microplastics on the activity of carboxylesterase and phosphatase enzymes in *Scinax squalirostris* tadpoles, *Environ. Monit. Assess.*, 2022, **194**, 1–10.
26. M. Solé, A. Bassols and V. Labrada-Martagón, Plasmatic B-esterases as potential biomarkers of exposure to marine plastics in loggerhead turtles, *Environ. Res.*, 2022, **213**, 113639.
27. E. C. López González, A. Larriera, P. A. Siroski and G. L. Poletta, Micronuclei and other nuclear abnormalities on *Caiman latirostris* (Broad-snouted caiman) hatchlings after embryonic exposure to different pesticide formulations, *Ecotoxicol. Environ. Saf.*, 2017, **136**, 84–91.
28. A. C. L. Albinati, E. L. T. Moreira, R. C. B. Albinati, J. V. Carvalho, A. D. De Lira, G. B. Santos and L. V. O. Vidal, Biomarcadores histológicos: toxicidade crônica pelo Roundup em piauçu (*Leporinus macrocephalus*), *Arq. Bras. Med. Vet. Zootec.*, 2009, **61**, 621–627.

29. I. F. Fernandes, H. S. M. Utsunomiya, B. S. de Lacerda Valverde, J. V. C. Ferraz, G. H. Fujiwara, D. M. Gutierres, C. de Oliveira, L. F. Belussi, M. N. Fernandes and C. D. S. Carvalho, Ecotoxicological evaluation of water from the Sorocaba River using an integrated analysis of biochemical and morphological biomarkers in bullfrog tadpoles, *Lithobates catesbeianus*, *Chemosphere*, 2021, **275**, 130000.
30. W. Zhu, D. Yang, L. Chang, M. Zhang, L. Zhu and J. Jiang, Animal gut microbiome mediates the effects of antibiotic pollution on an artificial freshwater system, *J. Hazard. Mater.*, 2022, **425**, 127968.
31. C. F. Rutkoski, S. C. Grott, N. G. Israel, F. E. Carneiro, F. de Campos Guerreiro, S. Santos, P. A. Horn, A. A. Trentini, E. Barbosa da Silva, C. A. Coelho de Albuquerque, T. C. Alves and E. Alves de Almeida, Hepatic and blood alterations in *Lithobates catesbeianus* tadpoles exposed to sulfamethoxazole and oxytetracycline, *Chemosphere*, 2022, **307**, 136215.
32. R. C. Lajmanovich, C. S. Martinuzzi, C. L. Colussi, P. M. Peltzer, A. Bassó, A. M. Attademo and L. Curi, in *The ecological risks of BT crops in amphibians: possible effects of insect-resistant intacta RR2 PRO® soybean diets on Leptodactylus gracilis*, Nova Science Publishers, 2017.
33. J. Ya, W. Kong, W. Chen, G. Wang and H. Zhao, Exposure to environmentally relevant oxytetracycline induced gut histopathological damages and microbiota alterations of *Pelophylax nigromaculatus* larvae, *Aquacult. Res.*, 2021, **52**, 4526–4537.
34. S. Feng, Z. Kong, X. Wang, L. Zhao and P. Peng, Acute toxicity and genotoxicity of two novel pesticides on amphibian, *Rana* N. Hallowell, *Chemosphere*, 2004, **56**, 457–463.
35. T. B. Hayes, V. Khoury, A. Narayan, M. Nazir, A. Park, T. Brown, L. Adame, E. Chan, D. Buchholz, T. Stueve and S. Gallipeau, Atrazine induces complete feminization and chemical castration in male African clawed frogs (*Xenopus laevis*), *Proc. Natl. Acad. Sci. U.S.A.*, 2010, **107**, 4612–4617.
36. I. Gyllenhammar, H. Eriksson, A. Söderqvist, R. H. Lindberg, J. Fick and C. Berg, Clotrimazole exposure modulates aromatase activity in gonads and brain during gonadal differentiation in *Xenopus tropicalis* frogs, *Aquat. Toxicol.*, 2009, **91**, 102–109.
37. J. P. Chae, M. S. Park, Y. S. Hwang, B. H. Min, S. H. Kim, H. S. Lee and M. J. Park, Evaluation of developmental toxicity and teratogenicity of diclofenac using *Xenopus* embryos, *Chemosphere*, 2015, **120**, 52–58.
38. A. K. Davis, D. L. Maney and J. C. Maerz, The use of leukocyte profiles to measure stress in vertebrates: a review for ecologists, *Funct. Ecol.*, 2008, **22**, 760–772.
39. M. Cabagna, R. C. Lajmanovich, G. Stringhini and P. M. Peltzer, Hematological studies in the common toad (*Bufo arenarum*) in agrosystems of Argentina, *Appl. Herpetol.*, 2005, **2**, 373–380.

40. S. Barni, E. Boncompagni, A. Grosso, V. Bertone, I. Freitas, M. Fasola and C. Fenoglio, Evaluation of *Rana sculenta* blood cell response to chemical stressors in the environment during the larval and adult phases, *Aquat. Toxicol.*, 2007, **81**, 45–54.
41. D. Forson and A. Storfer, Effects of atrazine and iridovirus infection on survival and life-history traits of the long-toed salamander (*Ambystoma macrodactylum*), *Environ. Toxicol. Chem.*, 2006, **25**, 168–173.
42. S. K. Arserim and A. Mermer, Hematology of the Uludağ frog, *Rana macrocnemis* Boulenger, 1885 in Uludağ National Park (Bursa, Turkey), *E.Ü. Su ÜrünleriDergisi*, 2008, **25**, 39–46.
43. S. M. Marques, S. C. Antunes, H. Pissarra, M. L. Pereira, F. Gonçalves and R. Pereira, Histopathological changes and erythrocytic nuclear abnormalities in Iberian green frogs (*Rana perezi* Seoane) from a uranium mine pond, *Aquat. Toxicol.*, 2009, **91**, 187–195.
44. P. W. Krauter, Micronucleus incidence and hematological effects in bullfrog tadpoles (*Rana catesbeiana*) exposed to 2-acetylaminofluorene and 2-aminofluorene, *Arch. Environ. Contam. Toxicol.*, 1993, **24**, 487–493.
45. S. I. Perí, N. E. Fink and A. Salibian, Hematological parameters in *Bufo arenarum* injected with sublethal dose of lead acetate, *Biomed. Environ. Sci.*, 1998, **11**, 70–74.
46. Z. A. Salinas, N. E. Salas, M. Baraquet and A. L. Martino, Biomarcadores hematológicos del sapo común *Bufo (Rhinella) arenarum* en ecosistemas alterados de la provincia de Córdoba, *Acta Toxicol. Argent.*, 2015, **23**, 25–35.
47. P. W. Krauter, S. L. Anderson and F. L. Harrison, Radiation-induced micronuclei in peripheral erythrocytes of *Rana catesbeiana*: an aquatic animal model for in vivo genotoxicity studies, *Environ. Mol. Mutagen.*, 1987, **10**, 285–296.
48. C. A. Glomski, J. Tamburlin, R. Hard and M. Chainani, The phylogenetic odyssey of the erythrocyte IV. The amphibians, *Histol. Histopathol.*, 1997, **12**, 147–170.
49. M. Cabagna, R. C. Lajmanovich, P. M. Peltzer, M. Attademo and E. Ale, Induction of micronuclei in tadpoles of *Odontophrynus americanus* (Amphibia: Leptodactylidae) by the pyrethroid insecticide cypermethrin, *Environ. Toxicol. Chem.*, 2006, **88**, 729–737.
50. R. C. Lajmanovich, M. Cabagna, P. M. Peltzer, G. A. Stringhini and A. M. Attademo, Micronucleus induction in erythrocytes of the *Hyla pulchella* tadpoles (Amphibia: Hylidae) exposed to insecticide endosulfan, *Mutat. Res.*, 2005, **587**, 67–72.
51. R. C. Lajmanovich, M. Cabagna-Zenklusen, A. M. Attademo, C. M. Junges, P. M. Peltzer, A. Bassó and E. Lorenzatti, Induction of micronuclei and nuclear abnormalities in common toad tadpoles (*Rhinella

arenarum) treated with Liberty® and glufosinate-ammonium, *Mutat. Res.*, 2014, **769**, 7–12.

52. A. M. Attademo, P. M. Peltzer, R. C. Lajmanovich, M. Cabagna and G. Fiorenza, Plasma B-esterase and glutathione S-transferase activity in the toad *Chaunus schneideri* (Amphibia, Anura) inhabiting rice agroecosystems of Argentina, *Ecotoxicology*, 2007, **16**, 533–539.
53. A. M. Attademo, M. Cabagna-Zenklusen, R. C. Lajmanovich, P. M. Peltzer, C. Junges and A. Bassó, B-esterase activities and blood cell morphology in the frog *Leptodactylus chaquensis* (Amphibia: Leptodactylidae) on rice agroecosystems from Santa Fe Province (Argentina), *Ecotoxicology*, 2011, **20**, 274–282.
54. P. M. Peltzer, R.Lajmanovich, M. Attademo, C. Junges, M. Cabagna-Zenklusen, R. Repetti, and H. Beldoménico, Effect of exposure to contaminated pond sediments on survival, development, and enzyme and blood biomarkers in veined treefrog (*Trachycephalus typhonius*) tadpoles, *Ecotoxicol. Environ. Saf.*, 2013, **98**, 142–151.
55. R. C. Lajmanovich, P. M. Peltzer, C. S. Martinuzzi, A. M. Attademo, C. L. Colussi and A. Bassó, Acute toxicity of colloidal silicon dioxide nanoparticles on amphibian larvae: emerging environmental concern, *Int. J. Environ. Res.*, 2018, **12**, 269–278.
56. G. Svartz, C. Aronzon, S. Pérez Catán, S. Soloneski and C. Pérez Coll, Oxidative stress and genotoxicity in *Rhinella arenarum* (Anura: Bufonidae) tadpoles after acute exposure to Ni-Al nanoceramics, *Environ. Toxicol. Pharmacol.*, 2020, **80**, 103508.
57. S. D. Melvin, M. C. Cameron and C. M. Lanctôt, Individual and mixture toxicity of pharmaceuticals naproxen, carbamazepine, and sulfamethoxazole to Australian striped marsh frog tadpoles (*Limnodynastes peronii*), *J. Toxicol. Environ. Health, A*, 2014, **77**, 337–345.
58. D. F. do Amaral, M. F. Montalvão, B. de Oliveira Mendes, A. P. da Costa Araújo, A. S. de Lima Rodrigues and G. Malafaia, Sub-lethal effects induced by a mixture of different pharmaceutical drugs in predicted environmentally relevant concentrations on *Lithobates catesbeianus* (Shaw, 1802) (Anura, ranidae) tadpoles, *Environ. Sci. Pollut. Res.*, 2019, **26**, 600–616.
59. S. M. Jandhyala, R. Talukdar, C. Subramanyam, H. Vuyyuru, M. Sasikala and D. Nageshwar Reddy, Role of the normal gut microbiota, *World J. Gastroenterol.*, 2015, **21**, 8787–8803.
60. C. Milani, S. Duranti, F. Bottacini, E. Casey, F. Turroni, J. Mahony, C. Belzer, S. Delgado Palacio, S. Arboleya Montes, L. Mancabelli, G. A. Lugli, J. M. Rodriguez, L. Bode, W. de Vos, M. Gueimonde, A. Margolles, D. van Sinderen and M. Ventura, The first microbial colonizers of the human gut: Composition, activities, and health implications of the infant gut microbiota, *Microbiol. Mol. Biol. Rev.*, 2017, **81**, e00036-17.

61. L. Zhang, R. G. Nichols, J. Correll, I. A. Murray, N. Tanaka, P. B. Smith, T. D. Hubbard, A. Sebastian, I. Albert, E. Hatzakis, F. J. Gonzalez, G. H. Perdew and A. D. Patterson, Persistent organic pollutants modify gut microbiota-host metabolic homeostasis in mice through aryl hydrocarbon receptor activation, *Environ. Health Perspect.*, 2015, **123**, 679–688.
62. S. Pindling, D. Azulai, B. Zheng, D. Dahan and G. G. Perron, Dysbiosis and early mortality in zebrafish larvae exposed to subclinical concentrations of streptomycin, *FEMS Microbiol. Lett.*, 2018, **365**, fny188.
63. A. P. Cuzziol Boccioni, G. García-Effron, P. M. Peltzer and R. C. Lajmanovich, The effect of glyphosate and ciprofloxacin exposure on enteric bacteria of tadpoles, *Rev. Argent. Microbiol.*, 2023, In Press, DOI: 10.1016/j.ram.2022.08.004.
64. Q. Tong, L. Y. Cui, J. Bie, X. Y. Han, Z. F. Hu, H. B. Wang and J. T. Zhang, Changes in the gut microbiota diversity of brown frogs (*Rana dybowskii*) after an antibiotic bath, *BMC Vet. Res.*, 2021, **17**, 1–13.
65. A. P. Cuzziol Boccioni, G. García-Effron, P. M. Peltzer, C. S. Martinuzzi and R. C. Lajmanovich, Susceptibility of intestinal bacteria to emerging contaminants: assessment of chronic exposure of glyphosate and ciprofloxacin on tadpoles of *Rhinella arenarum*, *XXXIX Jornadas Interdisciplinarias de Toxicología, Acta Toxicol. Argent.*, 2022, **30**, 39.
66. R. C. Lajmanovich, F. Emiliani and P. M. Peltzer, Bacterias coliformes y otras bacterias de interés sanitario en larvas de *Bufo arenarum* Hensel, 1887 (Anura, Bufonidae) en Santa Fe (Argentina), *Alytes*, 2001, **18**, 197–208.
67. P. T. Johnson, K. B. Lunde, R. W. Haight, J. Bowerman and A. R. Blaustein, *Ribeiroia ondatrae* (Trematoda: Digenea) infection induces severe limb malformations in western toads (*Bufo boreas*), *Can. J. Zool.*, 2001, **79**, 370–379.
68. S. E. Haas, M. K. Reeves, A. E. Pinkney and P. T. J. Johnson, Continental-extent patterns in amphibian malformations linked to parasites, chemical contaminants, and their interactions, *Glob. Change Biol.*, 2018, **24**, e275–e288.
69. K. Henle, A. Dubois and V. Vershinin, A review of anomalies in natural populations of amphibians and their potential causes, *Mertensiella*, 2017, **25**, 57–164.
70. J. D. Cardoso-Vera, H. Islas-Flores, N. SanJuan-Reyes, E. I. Montero-Castro, M. Galar-Martínez, S. García-Medina, A. Elizalde-Velázquez, O. Dublán-García and L. M. Gómez-Oliván, Comparative study of diclofenac-induced embryotoxicity and teratogenesis in *Xenopuslaevis* and *Lithobatescatesbeianus*, using the frog embryo teratogenesis assay: *Xenopus* (FETAX), *Sci. Total. Environ.*, 2017, **574**, 467–475.
71. C. Lorenz, R. Opitz, I. Lutz and W. Kloas, Corticosteroids disrupt amphibian metamorphosis by complex modes of action including

increased prolactin expression, *Comp. Biochem. Physiol., C, Toxicol. Pharmacol*, 2009, **150**, 314–321.

72. A. B. Lawrence, Applied animal behaviour science: Past, present and future prospects, *Appl. Anim. Behav. Sci.*, 2008, **115**, 1–24.
73. A. C. Arkwright, E. Archibald, A. Fahlman, M. D. Holton, J. L. Crespo-Picazo, V. M. Cabedo, C. M. Duarte, R. Scott, S. Webb, R. M. Gunner and R. Wilson, Behavioral biomarkers for animal health: A case study using animal-attached technology on loggerhead turtles, *Front. Ecol. Evol.*, 2020, **7**, 1–19.
74. K. D. Wells, in *The ecology and behavior of amphibians*, University of Chicago Press, Chicago, 2010.
75. L. M. Curi, A. P. Cuzziol Boccioni, P. M. Peltzer, A. M. Attademo, A. Bassó, E. J. León and R. C. Lajmanovich, Signals from predators, injured conspecifics, and pesticide modify the swimming behavior of the gregarious tadpole of the Dorbigny's Toad, *Rhinella dorbignyi* (Anura: Bufonidae), *Can. J. Zool.*, 2022, **100**, 19–27.
76. A. Bassó, S. Devin, P. M. Peltzer, A. M. Attademo and R. C. Lajmanovich, The integrated biomarker response in three anuran species larvae at sublethal concentrations of cypermethrin, chlorpyrifos, glyphosate, and glufosinate-ammonium, *J. Environ. Sci. Health, B*, 2022, **57**, 687–696.
77. C. Martinuzzi, P. M. Peltzer and R. C. Lajmanovich, Ecotoxicidad de contaminantes emergentes farmacológicos: efectos en el desarrollo y comportamiento en anfibios anuros, IV Report of the Doctoral thesis, ESS-FBCB-UNL, 2022, p. 19.
78. V. Aliko, R. S. Korriku, M. Pagano and C. Faggio, Double-edged sword: Fluoxetine and ibuprofen as development jeopardizers and apoptosis' inducers in common toad, *Bufo bufo,* tadpoles, *Sci. Total Environ.*, 2021, **776**, 145945.
79. M. Nassef, S. Matsumoto, M. Seki, F. Khalil, I. J. Kang, Y. Shimasaki, Y. Oshima and T. Honjo, Acute effects of triclosan, diclofenac and carbamazepine on feeding performance of Japanese medaka fish (*Oryzias latipes*), *Chemosphere*, 2010, **80**, 1095–1100.
80. D. Ferreira do Amaral, M. F. Montalvão, B. de Oliveira Mendes, A. P. da Costa Araújo, A. S. de Lima Rodrigues and G. Malafaia, Sub-lethal effects induced by a mixture of different pharmaceutical drugs in predicted environmentally relevant concentrations on *Lithobates catesbeianus* (Shaw, 1802) (Anura, ranidae) tadpoles, *Environ. Sci. Pollut. Res.*, 2019, **26**, 600–616.
81. S. D. Melvin, Oxidative stress, energy storage, and swimming performance of *Limnodynastes peronii* tadpoles exposed to a sub-lethal pharmaceutical mixture throughout development, *Chemosphere*, 2016, **150**, 790–797.

82. European Medicines Agency, Veterinary Medicines and Inspections, Committee for Medicinal Products for Veterinary Use, Guideline on Environmental Impact Assessment for Veterinary Medicinal Products, Phase II, Scientific Guidance, VICH Topic GL, **38** (CVMP/VICH/790/03-Final), 2004.

CHAPTER 6

Adult African Toads as Bioindicators of Environmental Xenobiotic-induced DNA Damage and Systemic Toxicity Provide Insight into the Decline of Amphibian Populations

CHIBUSI GIDEON ALIMBA*[a]

[a]Cell Biology and Genetics Unit, Department of Zoology, University of Ibadan, Ibadan, Oyo State, Nigeria
*E-mail: chivoptera@yahoo.com

6.1 Introduction

In 2004, Stuart *et al.* published a novel article that described how amphibian populations are declining at a faster rate than any other group of vertebrates.[1] Alterations in the composition of the natural environment, including discharge of contaminants, radiations (UV), acid rain, and climate change, among others, were implicated as contributing factors to the rapid decline.[2] These events negatively affect reproduction and increase morbidity and mortality of amphibians. Also of interest is the occurrence

Issues in Toxicology No. 46
Amphibian Species in Environmental Risk Assessment Strategies
Edited by Marcelo L. Larramendy and Guillermo Eli Liwszyc

Published by the Royal Society of Chemistry, www.rsc.org

of infectious diseases. For instance, Chytridiomycosis was described as the most studied infectious disease of toads and probably contributes greatly to anuran decline around the world.[3] This chapter aims to discuss the use of African toads as a bioindicator for the assessment of geno- and systemic toxicity induced by environmental xenobiotics common in the African environment. This may increase the understanding of the possible mechanisms involved in the decline in toad populations on the continent.

6.2 Genotoxicity Biomarkers as Indicators of Genome Instability in Xenobiotic-exposed Toads

Adult African toads have been rarely used in genotoxicity studies despite their suitability as bioindicators for monitoring environmental contamination.[4,5] Adult toads are exposed to xenobiotics *via* absorption through permeable dermal contact with xenobiotics in water, sediments and during hibernation in soils.[6,7] They are also exposed to these pollutants *via* the ingestion of contaminated food.[8] Significant genotoxicity induction is associated with genome instability and genetic-related diseases, which are important factors in amphibian decline.[9,10]

Seid *et al.* exposed specimens of the African common or Egyptian toad *Amietophrynus regularis* (=*Sclerophrys regularis*) to varying lead (Pb) concentrations and examined the peripheral blood of the toad for micronuclei (MN) formation and the induction of nuclear morphological abnormalities (NAs).[11] The authors observed a significant increase in MN and NA formation, which is positively correlated with Pb accumulation in the liver and kidney of exposed toads. Khattab *et al.* captured free-ranging adult *A. regularis* that were potentially exposed *in situ* to a mixture of metals such as Pb, Cd, Cr and Mn and analysed peripheral blood for MN and NA frequency induction.[12] The findings revealed that toads collected from a high metal-contaminated site (El Giza) had significantly higher frequencies of MNs and NAs compared to those collected from an uncontaminated site (El Wahat). Besides, metal analysis of soils collected from the sampled sites showed a significant correlation with MN and NA formation in the circulating erythrocytes of the toads. The authors justified the use of the MN test as a suitable biomarker for *in situ* genotoxic assessment of metal-polluted sites that may suggest one of the mechanisms involved in the amphibian decline and extinction. Said *et al.* exposed premetamorphic tadpoles at developmental Gosner stages 24–27 of *A. regularis* to sub-lethal concentrations of the herbicide atrazine and the fertilizer sodium nitrate ($NaNO_3$) in an individual and binary fashion in order to examine peripheral blood for MN and NA formation.[5] The individual and binary mixture of atrazine and $NaNO_3$ solutions induced a significant increase in MNs and NAs in the peripheral erythrocytes of exposed toads. Furthermore, the interactive effects of the binary mixture induced larger disruptive effects on the genetic material of the exposed toads. This finding suggests that mixtures of these previously mentioned xenobiotics in the environment possibly interact additively and/or synergistically, but not antagonistically, to induce a greater deleterious health impact on toads. This effect may increase mortality and, hence, lead to amphibian decline. Osman *et al.* exposed adult *A. regularis* to 1/10 of the application rate of copper sulphate ($CuSO_4$), organophosphate larvicide temephos, herbicide glyphosate and pyrethroid insecticide bifenthrin for 46 h and examined the peripheral erythrocytes of toads for MNs and NAs formation.[13] At tested concentrations of the aforementioned pesticides, there was a significant increase in the induction of MNs and NAs in peripheral erythrocytes in the exposed Egyptian toads. These findings further showed the sensitivity of the MN test conducted in adult African toads for the genotoxic assessment of pesticides.

Bone marrow and spleen are the predominant erythropoietic organs for most African toads, including the European toad *Bufo bufo*, the Egyptian green toad *B. viridis* and the toad *B. regularis*.[14] Therefore, it is important to employ a genotoxic biomarker (MN assay) in the erythropoietic system (bone marrow and spleen) to understand the mechanisms of MNs and NAs formation during exposure to xenobiotics. In view of this, Alimba *et al.* exposed *A. regularis* to various concentrations of Cd and Pb compounds in an aqueous solution for 96 h to determine the LC_{50} values.[15] Toads were subsequently exposed to sub-lethal concentrations equivalent to the 96 h LC_{50} values for Cd and Pb for 14 consecutive days to assess the genotoxicity potentials of the metal solutions in the bone marrow cells and peripheral erythrocytes of the exposed toads. Similar exposure duration and conditions were given to toads exposed to 0.01 mL benzene L^{-1}, a haematopoietic genotoxic inducer and dechlorinated tap water as positive and negative controls, respectively.[16] Using the micronucleus cytome assay (MN-cyt) for the first time in toads, the cytotoxicity and genotoxicity of Cd and Pb in the differentiating bone marrow cells and peripheral erythrocytes of adult toads were evaluated. Figure 6.1 shows a concentration-dependent increase in the frequency of MNs induced in bone marrow cells and peripheral erythrocytes of toads exposed to 5–75% equivalent to the respective 96 h LC_{50} values of Pb and Cd for 14 days. An obvious finding is that peripheral erythrocytes contain higher micronucleated cells and NAs compared to the bone marrow cells in both Pb- and Cd-exposed toads (see Figure 6.1). These observations suggest that these chemicals induce haematopoietic dysfunction and DNA damage in the toad's bone marrow cells. Physiologically, when blood cells are formed, they are released into the body circulation. This may account for the higher frequency of micronucleated erythrocytes in circulating peripheral erythrocytes.

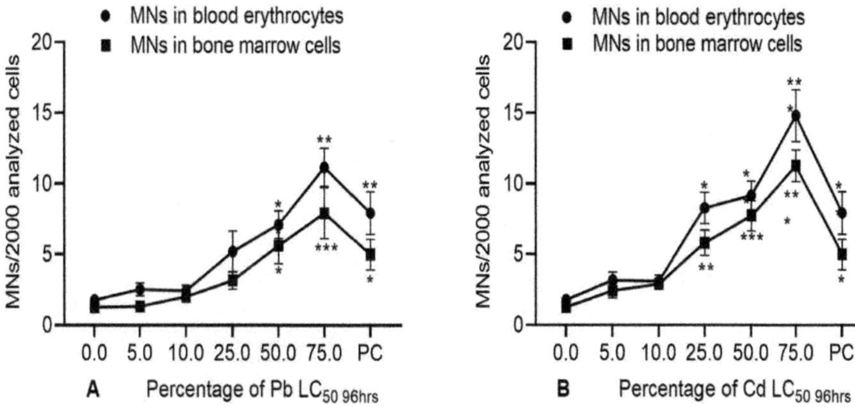

Figure 6.1 Micronuclei induction in bone marrow cells and peripheral blood erythrocytes of the common African toad *Amietophrynus regularis* exposed to sub-lethal concentrations of lead nitrate (Pb) (A) and cadmium chloride (Cd) (B). Results are expressed as mean values ± SE of the mean. Positive control (PC) = 0.01 mL benzene. *, $P < 0.05$; **, $P < 0.01$; ***, $P < 0.001$, significant differences with respect to negative control values.[7]

Furthermore, Cd elicited higher MNs in both peripheral erythrocytes and bone marrow cells compared to those of Pb-exposed toads (see Figure 6.1). This finding is in accordance with the report that Pb is a weaker genotoxic and cytotoxic chemical compared to Cd. It also further agrees with the classification of Pb as a probable human carcinogen (Group 2B)[17] and Cd as a carcinogen (Group A) by the International Agency for Research on Cancer (IARC, WHO).[18] This observation further suggests that the MN test is a sensitive biomarker of genotoxicity in both bone marrow cells and peripheral erythrocytes of toads.[15] The observed NAs include binucleated erythrocytes, nuclear buds, notched nucleus, vacuolated nucleus and lobed nucleus.[15]

Alimba *et al.* also observed a concentration-dependent significant increase in the percentage of apoptotic and necrotic erythrocytes induced by Pb and Cd (see Figures 6.2A and 6.2B).[15] It is very clear that the two most common methods utilized by adult African toads to evade the accumulation of damaged DNA and, hence, prevent possible tumour growth, including apoptosis (see Figure 6.2A) and necrosis (see Figure 6.2B). Therefore, environmental xenobiotics-induced DNA damage in adult African toads may not readily enhance susceptibility to cancer development. However, it is important to state that two amphibian neoplasms, namely renal adenocarcinoma in the Northern leopard frog *Rana pipiens* and lymphosarcoma in the African clawed toad *Xenopus laevis*, have been reported.[19]

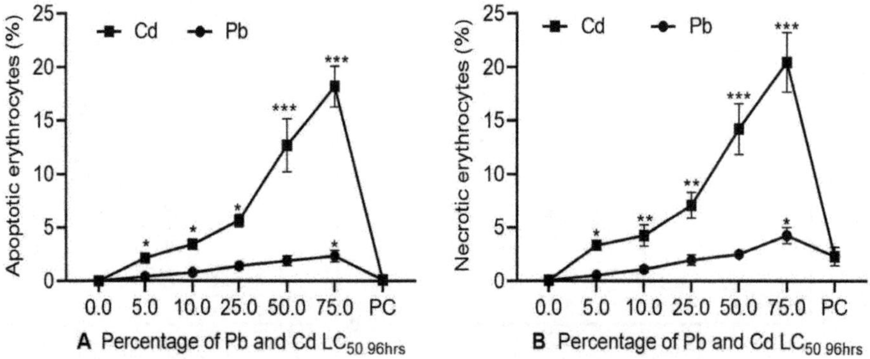

Figure 6.2 Effects of Cd and Pb solutions on the induction of apoptotic (A) and necrotic erythrocytes (B) in the peripheral blood of the common African toad *Amietophrynus regularis* exposed to sub-lethal concentrations of lead nitrate (Pb) and cadmium chloride (Cd). Results are expressed as mean values ± SE of the mean. Positive control (PC) = 0.01 mL benzene. *, $P < 0.05$; **, $P < 0.01$; ***, $P < 0.001$, significant differences with respect to negative control values.[7]

6.3 Adult African Toad as a Bioindicator of Xenobiotic-induced Systemic Toxicity

Environmental xenobiotics induce alterations in the general structure and/or functions of the organ systems of African toads. Ezemonye and Tongo[20] exposed *B. regularis* to sub-lethal concentrations of the organochlorine insecticide/acaricide endosulfan and the organophosphate insecticide diazinon for 28 days and observed a concentration-dependent increase in glutathione-*S*-transferase (GST) expression in the order liver > brain > serum > gastrointestinal tract (GIT) > lungs for both tested pesticides. GST is a major cellular detoxifier of xenobiotics, including pesticides. This observation signals the deleterious effects of pesticides on the health status of the exposed toad, as a decrease in GST suggests suppression of cellular detoxifier mechanisms against xenobiotics. Tongo *et al.* exposed *B. regularis* to sub-lethal concentrations of diazinon for 28 days.[21] At post-exposure, the toads expressed a 96% decrease in acetylcholinesterase (AChE) activity in the brain compared to the unexposed toads. There was also a decrease in AChE in the serum, liver and lungs of the exposed toads. Furthermore, corticosterone and total protein levels were significantly reduced and diazinon accumulated in the organs in the order liver > serum > brain > lungs > GIT. The authors concluded that diazinon caused neurotoxicity and organ damage to the exposed toads via oxidative stress.[21]

Specimens of *Bufo maculatus* (=*S. maculata*) were exposed to sub-lethal concentrations of lead oxide (Pb_2O_3) solutions for 28 days (0.25, 0.50,

1.00, and 2.00 mg Pb_2O_3 L^{-1}). The solutions were administered as a single injection through the dorsal lymph sac of the toads. The sub-lethal concentrations of the Pb_2O_3 solutions caused alterations in their haematological indices. There was a significant decrease in the total erythrocyte count (red blood counts or RBCs), haematocrit (HCT), and haemoglobin (Hb) concentrations compared to unexposed toads (control), while the total leukocyte counts (white blood counts or WBCs) increased.[22] Sayed exposed *B. regularis* to ultraviolet radiation (UVA 320–400 nm) for 3 days in groups of 15, 30, and 60 min day^{-1} compared to control toads (0 min day^{-1}).[23] At post-exposure, blood cell profiles and serum biochemical compositions were measured. UVR 320–400 nm altered the blood cell indices of exposed toads, including RBC count, Hb concentration, HCT and WBC count. It also induced the abnormal formation of red blood cell morphology, including crenated cells, acanthocytes, teardrop-like cells, and sickle cells. Also observed were alterations in serum total protein, glucose, aspartic aminotransferase (AST), alanine aminotransferase (ALT), alkaline phosphatase (ALP), lactate dehydrogenase (LDH), glucose-6-phosphate dehydrogenase (G6PDH) and total bilirubin compared to the controls. This observation suggests that exposure to UVA 340–400 nm increased the biochemical and physiological disturbances in toads, which, in turn, probably increased, amphibian decline *via* morbidity and mortality. Isehunwa *et al.* exposed *B. regularis* to sub-lethal concentrations of 0.14, 0.3, 0.6, and 1.13 mL L^{-1} of a petrol solution for 14 days. At post-exposure, blood glucose significantly increased, while liver and muscle glycogen levels decreased in the exposed toads.[24] The sub-lethal concentrations of petrol caused mortality, hyperglycaemia and low glycogen levels in the liver and muscle of the exposed toads. These observations suggest adverse effects on carbohydrate metabolism, which may lead to death.

Xenobiotic-induced alterations in RBC, Hb, HCT and WBCs observed in adult African toads collected from metal- and pesticide-polluted sites (field studies) and laboratory-simulated exposure to xenobiotics in the environment suggest perturbation of the haematopoietic system. This may also suggest an anaemic condition (poor or failure in the oxygen-carrying capacity of the erythrocytes) and may eventually lead to poor health conditions and mortality of the exposed toads. White blood cells constitute the body's non-specific immune system.[25] Therefore, changes in the quantities and compositions of WBC differentials suggest adverse effects on the body's defence system.[26] Leukocytosis (increase in total WBC) and lymphocytopenia (decrease in the percentage of lymphocytes) along with neutrophilia (increase in the percentage of neutrophils) suggest physiological and inflammatory challenges induced by toxic environmental xenobiotics.[27] These alterations in the WBC composition may develop into immunosuppression.[28] Furthermore, abnormal erythrocyte morphology, as reported by Sayed in *B. regularis* exposed to varying doses of UVR, is in consonance with alterations in haematological indices and suggests adverse health effects due to the exposure to xenobiotics.[23] Abnormal erythrocyte morphologies, including crenated cells, acanthocytes, tear-drop-like, and sickle-shaped cells, are associated with xenobiotic-induced toxicosis via disturbance of osmoregulation[29] and oxidative injury induced by heavy metals on the RBC membrane.[27,30] These findings underpin that changes in haematological indices and abnormal erythrocyte morphology are sensitive biomarkers of xenobiotic-induced toxicological effects.[23,27] Alterations in haematological indices and abnormal erythrocyte morphology should be routinely assessed as biomarkers of xenobiotic-induced health impacts in adult African toads since they signal a risk of deleterious health impacts and mortality.[26,28]

Said *et al.* collected free-ranging *A. regularis* from different locations around the River Nile in Egypt.[31] They examined the toads for histological changes in the liver. They observed dilation of the sinusoids, degenerated hepatocytes, necrotic parenchyma, fibrosis, an inflammatory response, bile secretions, lipofuscin, parenchymal and granular deposits. Heavy metal analysis revealed metal accumulation at different concentrations in the liver and in the surrounding soils from where the toads were collected. Akinsanya *et al.* collected free-ranging *A. regularis* from three different locations with anthropogenic activities in Nigeria.[32] The authors observed severe histological alterations, including glandular hyperplasia, increased connective tissue, severely stunted villi and infiltration of inflammatory cells in the intestine and liver tissues of the toads. The authors also observed a concomitant increase in the concentrations of Cr, Cd, Cu, Ni, and Pb in the examined tissues. Alterations in glutathione peroxidase (GPx), superoxide dismutase (SOD), catalase (CAT) and lipid peroxidation (oxidative stress biomarkers) in the intestine and liver were detected as well. Water and soil samples collected from the three sites also contained high levels of heavy metal contamination. Okeagu *et al.* collected adults of *A. regularis* from three different environments contaminated with pesticides.[7] Histological examinations of the liver and intestine of the collected toads revealed severe stunting of villi and total loss of intestinal glands and lymphoid follicles within the villi. Also, high concentrations of congeners, including aldrin and dieldrin ('classic organochlorine pesticides'), were observed in the intestine and liver of the toads. Tadpoles of the Egyptian anuran *A. regularis* were exposed to individual concentrations of atrazine and $NaNO_3$ at doses of 300 $\mu g\ L^{-1}$ and 200 $mg\ L^{-1}$, respectively. Another group was exposed to a mixture of the doses of herbicide and fertilizer and the last group was neither exposed to herbicide nor to fertilizer; this group served as a control. The atrazine group had altered gonadal development as shown by the increase and decrease in size of the seminiferous tubules in the gonads, but this effect was not observed in the $NaNO_3$-exposed group. The percentage of intersex states detected in toads exposed to the mixture of atrazine and $NaNO_3$ was similar to that of atrazine alone. The authors suggested that at the selected sub-lethal concentrations, there were no interactive effects between atrazine and $NaNO_3$ in the induction of intersex states in the exposed toads.[33] They also argued that the observed gonadal abnormalities may decrease reproductive success and serve as a driver for enhancing the decline in amphibian population.

The various histopathological lesions in the African toads are caused by direct deleterious interactions of the environmental xenobiotics with the exposed tissues of the animals. For instance, toxic metals, pesticides, petrol and fertilizers have been reported to induce dilation of sinusoids, degenerated hepatocytes, parenchymal deposits, necrotic parenchyma, granular deposits, fibrosis, infiltration of inflammatory cells and bile secretion in the liver tissue of toads. Also, lipofuscin deposits, total loss of intestinal glands and lymphoid follicles within the villi and glycogen depletion from the cytoplasm have been observed in the intestinal tissues of toads.[33,34] The pathological lesions observed in the liver and intestines of xenobiotic-exposed toads may also be linked to the induction of oxidative stress and reactive oxygen species induced by the xenobiotics.[32] Some of the reported pathological lesions in the xenobiotic-exposed toads have been associated with health disorders that cause increased morbidity and mortality. For instance, the infiltration of inflammatory cells, a potent mechanism of defence against pathogens and toxic xenobiotics, leads to free radical formation that may cause damage to molecules and cells.[35] Chronic inflammatory reactions or persistent inflammatory cells caused by the presence of xenobiotics may result in triggering autoimmune responses, increased risk of malignancy or neoplastic transformation of susceptible cells and eventual death of the organisms.[36]

6.4 Adult African Toads as Bioindicators of Environmental Xenobiotic Accumulation

Adult African toads have been used as bioindicators of accumulation of pesticides, metals, and other pollutants in the African environment. For instance, specimens of the African common toad *B. gutturalis* (=*S. gutturalis*) collected from a farmland in Zimbabwe sprayed with DDT, accumulated 30–3900 ng g^{-1} ww of DDT in the whole body.[37] Also, *S. garmani*, the Eastern olive toad, collected from the Ndumo Game Reserve and Kruger National Park, South Africa, accumulated 3.9 mg g^{-1} ww of organochlorine pesticide in their body.[38] Male specimens of the spiny toad *B. spinosus* collected from five different sampling sites along the Medjerda basin (Northeastern Algeria) were used as biosentinel species to evaluate the heavy metal pollution status. The following metal (Pb, Fe, Zn and Cu) concentrations were measured in the toad skin (non-invasive skin biopsy) using an atomic absorption flame spectrophotometer. The Pb concentration in the toad skin collected from all the contaminated sample locations exceeded 96 times the standard reference values set by the WHO, while other analysed metals were within the permissible limits.[39] Adult specimens of *A. regularis* and soil samples collected from three major contaminated areas in Lagos, Nigeria, and an uncontaminated site (the control site) were analysed for organochlorine pesticides (OCPs) using GC–MS analysis.[7] The mean concentration of the various congeners of OCPs quantified in the liver and intestine of the toads and soil samples collected from the contaminated site were higher than those from the control site.[7] The effects of the accumulation of xenobiotics on toads include lethal and/or sub-lethal agents that may lead to morbidity and/or mortality, reduced growth rate, high occurrence of developmental abnormalities, alterations in behavioural pattern, reduced fecundity and increased susceptibility to diseases and infections.[39] These adverse effects may increase the decline in the African toad population.

6.5 Final Remarks

There are several possibilities associated with environmental xenobiotics and amphibian decline. Using adult African toads as bioindicator species enables the establishment of a causal linkage between exposure to environmental xenobiotics, both in the field and in laboratory settings, with adverse alterations in toxicity at a systemic and genetic level (see Figure 6.3). Environmental xenobiotics act individually or as a mixture to elicit complex, harmful interactions in amphibians and other organisms.

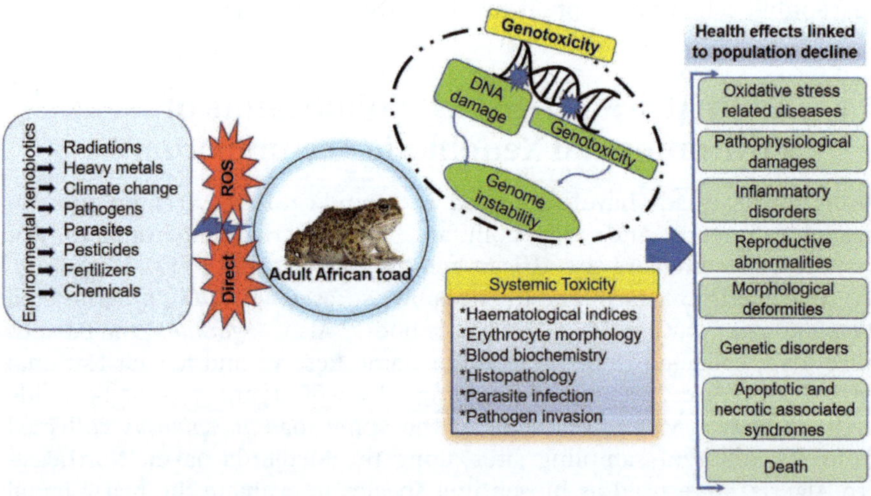

Figure 6.3 Schematic representation of the chain of events induced by environmental xenobiotic effects frequently observed in the common African toad, *Amietophrynus regularis*. Reactive oxygen species (ROS) induce systemic toxicity and genotoxicity. The deleterious health impacts induced may incapacitate the toad and eventually lead to death.

As previously mentioned, the list of deleterious agents includes radiation (such as UVA 340–400 nm), contaminants, parasites/pathogens, toxic metals and climate change, among others, that act as stressors that adversely affect amphibians by causing morbidity or leading to their mortality. Adult African toads were used as bioindicators of xenobiotics to monitor the pattern of damage caused by genomic instability at individual and/or population levels. Xenobiotics accumulate in the viscera of African toads and generate reactive oxygen species (ROS), which are capable of causing systemic toxicity and genotoxicity. These agents alter body biochemistry, haematological indices, body morphology and histopathology and induce micronucleus and nuclear abnormalities. Damage induced may lead to pathophysiological and inflammatory disorders, genetic disturbances and reproductive abnormalities, as well as morphological deformities.

Overall, these factors constitute a threat to amphibian health and are the hallmarks of population decline (see Figure 6.3).

References

1. S. N. Stuart, J. S. Chanson, N. A. Cox, B. E. Young, A. S. Rodrigues, D. L. Fischman and R. W. Waller, Status and trends of amphibian declines and extinctions worldwide, *Science*, 2004, **306**, 1783–1786.
2. I. Horak, S. Horn and R. Pieters, Agrochemicals in freshwater systems and their potential as endocrine disrupting chemicals: A South African context, *Environ. Pollut.*, 2021, **268**, 115718.
3. M. C. Fisher and T. W. J. Garner, Chytrid fungi and global amphibian declines, *Nat. Rev. Microbiol.*, 2020, **18**, 332–343.
4. A. Venturino, E. Rosenbaum, A. C. De Castro, O. L. Anguiano, L. Gauna, T. F. De Schroeder and A. M. P. de D'Angelo, Biomarkers of effect in toads and frogs, *Biomarkers*, 2003, **8**, 167–186.
5. A. S. Said, S. A. L. Saber and B. A. E. El Salkh, Biomarker responses in *Sclerophrys regularis* (*Anura: Bufonidae*) exposed to atrazine and nitrate, *Pollution*, 2022, **8**, 1387–1397.
6. United States Environmental Protection Agency (U.S. EPA), *Methods for evaluating wetland condition: using amphibians in bio-assessments of wetlands*, Office of Water, Washington, DC, 2002, EPA-822-R-022.
7. O. M. Okeagu, B. Akinsanya, P. O. Isibor, J. Daniel-Rugu, B. A. Onadeko, B. Yalwaji and K. O. Adekoya, Bioaccumulation of organochlorine pesticides in the parasite *Cosmocerca* sp. (Nematoda: Cosmocercidae) and the amphibian host *Amietophrynus regularis* (Reuss, 1833) within Lagos metropolis, Nigeria, *Toxicol. Rep.*, 2022, **9**, 136–146.
8. State of the Environment Advisory Council (Australia) (SEAC), *Australia, state of the environment: an independent report presented to the commonwealth minister for the environment by the state of the environment advisory council*, CSIRO Publishing, Collingwood, Vic., 1996.

9. A. Aguilera and T. García-Muse, Causes of genome instability, *Annu. Rev. Genet.*, 2013, **47**, 19–50.
10. B. Kurelec, The genotoxic disease syndrome, *Mar. Environ. Res.*, 1993, **35**, 341–348.
11. R. E. M. Said, S. A. Saber and A. G. M. Osman, Haemotoxic and genotoxic potential of lead on the Egyptian toad *Amietophrynus regularis*, *Inter. J. Ecotoxicol. Ecobiol.*, 2016, **1**, 94–102.
12. N. M. A. Khattab, S. A. Saber, B. A. El-Salkh and R. E. M. Said, Genotoxicity and limbs asymmetry in the Egyptian toad (*Sclerophrys regularis*) as biomarkers for heavy metals toxicity, *Egypt. J. Aqua. Biol. Fish.*, 2021, **25**, 705–717.
13. K. A. Osman, A. Ali, N. S. Ahmed and A. S. El-Seedy, Biochemical and genotoxic effects of some pesticides on the Egyptian Toads, *Sclerophrys regularis* (Reuss, 1833), *Watershed Ecol. Environ.*, 2022, **4**, 125–134.
14. R. M. Khalil and N. K. Elfeky, The blood forming tissues and organs of *Bufo viridis* Laurenti, *Bufo regularis* Reuses and *Rana mascareniensis* Dumerila and Bibron anuran Amphibia, *J. Coll. Sci., King. Saud. Univ.*, 1986, **17**, 87–98.
15. C. G. Alimba, A. M. Aladeyelu, I. A. Nwabisi and A. A. Bakare, Micronucleus cytome assay in the differential assessment of cytotoxicity and genotoxicity of cadmium and lead in *Amietophrynus regularis*, *EXCLI J.*, 2018, **17**, 89–101.
16. International Programme on Chemical Safety (IPCS), Environmental Health Criteria, No. 150, in *Benzene*, WHO, Geneva, 1993.
17. International Agency for Research on Cancer (IARC), Inorganic and organic lead compounds, in *IARC Monographs on the Evaluation of Carcinogenic Risks to Humans*, IARC, Lyon, **vol. 87**, 2006
18. International Agency for Research on Cancer (IARC), Summaries and evaluations: Cadmium and cadmium compounds (Group 1), in *Monographs on the Evaluation of Carcinogenic Risks to Humans*, IARC, Lyon, **vol. 58**, 1993.
19. L. N. Ruben, R. H. Clothier and M. Balls, Cancer resistance in amphibians, *ATLA J.*, 2007, **35**, 463–470.
20. L. Ezemonye and I. Tongo, Sublethal effects of endosulfan and diazinon pesticides on glutathione-S-transferase (GST) in various tissues of adult amphibians (*Bufo regularis*), *Chemosphere*, 2010, **81**, 214–217.
21. I. Tongo, L. Ezemonye and U. Ochei, Diazinon mediated biochemical changes in the African toad (*Bufo regularis*), *J. Xenobiot.*, 2012, **2**, 18–23.
22. L. I. Ezemonye and A. A. Enuneku, Behavioural and haematological changes in adult *Hoplobatrachus occipitalis* and *Bufo maculatus* exposed to sub-lethal concentrations of lead, *Thai J. Toxicol.*, 2012, **27**, 7–15.
23. A. E. H. Sayed, Hematotoxic and biochemical effects of UVA on the Egyptian toad (*Bufo regularis*), *Inter. J. Radiat. Biol.*, 2016, **92**, 35–41.

24. G. O. Isehunwa, I. O. Yusuf and A. R. A. Alada, Effects of petrol exposure on glucose, liver and muscle glycogen levels in the common African toad *Bufo regularis*, *Niger. J. Physiol. Sci.*, 2016, **31**, 139–145.
25. M. R. Romo, D. Perez-Martinez and C. C. Ferrer, Innate immunity in vertebrates: an overview, *Immunology*, 2016, **148**, 125–139.
26. L. Franco-Belussi, D. B. Provete, T. R. F. Leao, M. S. Siqueira, B. S. L. Valverde, O. Martins, C. de Oliveira and C. E. Fernandes, Hematological parameters of a Neotropical wild frog population, with a phylogenetic perspective on blood cell composition in Anura, *Curr. Zool.*, 2022, **68**, 361–369.
27. C. G. Alimba, O. O. Adewumi, O. M. Binuyo and P. G. C. Odeigah, Wild black rats (*Rattus rattus* Linnaeus, 1758) as zoomonitor of genotoxicity and systemic toxicity induced by hazardous emissions from Abule Egba unsanitary landfill, Lagos, Nigeria, *Environ. Sci. Pollut. Res.*, 2021, **28**, 10603–10621.
28. H. L. Anderson, I. E. Brodsky and N. S. Mangalmurti, The evolving erythrocyte: red blood cells as modulators of innate immunity, *J. Immunol.*, 2018, **201**, 1343–1351.
29. B. P. Shaw, S. Dash and A. K. Panigrahi, Effect of methyl mercuric chloride treatment on haematological characteristics and erythrocyte morphology of Swiss mice, *Environ. Pollut.*, 1991, **73**, 43–52.
30. S. H. Jadhav, S. N. Sarkar, M. Aggarwal and H. C. Tripathi, Induction of oxidative stress in erythrocytes of male rats subchronically exposed to a mixture of eight metals found as groundwater contaminants in different parts of India, *Arch. Environ. Contam. Toxicol.*, 2007, **52**, 145–151.
31. R. E. M. Said, S. A. A. Saber, A. G. G. Rab and K. Y. Abouel Fadl, Hepatic alteration of the Egyptian toad *Amietophrynus regularis*, as biomarker to environmental deterioration, *Inter. J. Environ. Monitor. Anal.*, 2015, **3**, 22–29.
32. B. Akinsanya, P. O. Isibor, B. Onadeko and A. Tinuad, Impacts of trace metals on African common toad, *Amietophrynus regularis* (Reuss, 1833) and depuration effects of the toad's enteric parasite, *Amplicaecum africanum* (Taylor, 1924) sampled within Lagos metropolis, Nigeria, *Heliyon*, 2020, **6**, e03570.
33. A. Gadel-Rab, F. A. Mahmoud, R. E. S. Said, S. A. Saber, B. A. El Salkh and A. S. Said, Hermaphroditism induction of sub-lethal dose of atrazine and atrazine-nitrate on the Egyptian toad, *Sclerophrys regularis*, *Inter. J. Ecotoxicol. Ecobiol.*, 2018, **3**, 63–72.
34. F. A. Mahmoud, A. G. Gadel-Rab, R. E. Said, S. A. Saber, B. A. El Salkh, A. S. Said and M. M. Atia, Impact of atrazine and nitrate on liver and kidney of Egyptian toad *Sclerophrys regularis*: bioindicator alarming on ecosystem, *Acta Sci. Biol. Sci*, 2022, **44**, e56386.

35. C. G. Alimba, C. Faggio, S. Sivanesan, A. L. Ogunkanmi and K. Krishnamurthi, Micro(nano)-plastics in the environment and risk of carcinogenesis: Insight into possible mechanisms, *J. Hazard. Mater.*, 2021, **416**, 126143.
36. C. Carey, N. Cohen and L. Rollins-Smith, Amphibian declines: an immunological perspective, *Develop. Comparat. Immunol.*, 1999, **23**, 459–472.
37. M. R. K. Lambert, Residue loads in amphibians used as biomarkers of pesticide levels entering food chains in Sub-Saharan Africa, *Afri. J. Herpetol.*, 2001, **50**, 105–114.
38. N. J. Wolmarans, L. Bervoets, R. Gerber, Y. B. Yohannes, S. M. M. Nakayama, Y. Ikenaka, M. Ishizuka, P. Meire, N. J. Smit and V. Wepener, Bioaccumulation of DDT and other organochlorine pesticides in amphibians from two conservation areas within malaria risk regions of South Africa, *Chemosphere*, 2021, **274**, 129956.
39. N. Guezgouz, C. Parisi, S. Boubsil, G. Grieco, S. A. Hana and G. Guerriero, Heavy metal assessment in the Medjerda River Basin (Northeastern Algeria): a preliminary water analysis and toad skin biopsy, *Proc. Zool. Soc.*, 2021, **74**, 104–113.

CHAPTER 7

Development of Massive Molecular Biomarker Analysis ("Biomarkomics") in Non-model Species: The Transcriptome of Rhinella arenarum (Anura, Bufonidae) Larvae Exposed to Organophosphorus Pesticides as a Case Example

ANDRÉS VENTURINO*[a], DANILO CESCHIN[b] AND CECILIA LASCANO[a]

[a]Centro de Investigaciones en Toxicología Ambiental y Agrobiotecnología del Comahue (CITAAC), Universidad Nacional del Comahue-CONICET, 8300 Buenos Aires 1400, Neuquén, Argentina; [b]Instituto Universitario de Ciencias Biomédicas de Córdoba (IUCBC), Centro de Investigación en Medicina Traslacional "Severo R. Amuchástegui" (CIMETSA); vinculated to: Instituto de Investigación Médica Mercedes y Martín Ferreyra (INIMEC-CONICET- Universidad Nacional de Córdoba), Av. Naciones Unidas 420, 5016 Córdoba, Argentina
*E-mail: a.venturino@conicet.gov.ar

Issues in Toxicology No. 46
Amphibian Species in Environmental Risk Assessment Strategies
Edited by Marcelo L. Larramendy and Guillermo Eli Liwszyc

Published by the Royal Society of Chemistry, www.rsc.org

7.1 Traditional Biomarkers Analysis and Their Pitfalls: Looking for Specific, Sensitive and Early Responses

Environmental risk analysis (ERA) still relies heavily on ecotoxicological parameters, such as lethal or effective fifty concentrations (LC50, EC50), or sometimes the non-observed or lowest-observed effective concentrations in chronic effects (NOEC, LOEC). These endpoints are experimentally established for one or a group of representative species. Currently, the problem with these endpoints is that the effects at the individual, population, or even community levels may be irreversible when the environmental contaminant concentrations reach critical amounts. It has been imperative to find more sensitive endpoints that alert about survival challenges early enough to prevent points of no return. In this sense, regulatory agencies from different countries have been engaged in the drafting and implementation of quality criteria or guidance levels for different substances to protect aquatic life. However, these regulatory bodies still rely on toxicological data (mainly lethal and sometimes behavioural effects) to derive protection limits.[1,2] For long-term effect evaluations, genotoxic or reproductive endpoint parameters are sometimes included in the derivation of quality criteria levels when reliable data are available.

Metabolic, biochemical, molecular targets and responses to chemical exposure are intrinsically involved in the complex net that underlies the effects at the higher individual or population level in the toxicological output. Therefore, the molecular targets and responses to xenobiotics may be interesting biomarkers to be used as early endpoints anticipating later irreversible effects, such as teratogenicity, developmental alterations, physiological-organ failure and finally death. The responses at the biochemical and molecular levels in aquatic biota occur at concentration levels usually far below from those causing morphological and lethal damages.[3,4] A gradient of impact levels is observed in embryos and larvae of the South American common toad *Rhinella arenarum* (Bufonidae) exposed to As. The molecular pathways respond at micromolar concentrations; the underlying antioxidant enzymatic responses, in the sub-millimolar range, anticipate oxidative stress damage at millimolar concentrations; and finally, a compromise of the organisms and their survival occurs at higher exposures (see Figure 7.1). Still, the discussion remains on whether the early-responding and highly sensitive molecular biomarkers should be considered of concern about individuals' survival from an ecotoxicological point of view. Nevertheless, there are several questions that need further elaboration before these biomarkers become useful for ERA. The first is that the metabolite and enzymatic parameters are generally highly variable and fast-responding to homeostatic regulation, as suggested for toad embryos.[5] Currently, research on concentration-effect curves for biochemical or molecular parameters is scarce. Therefore, establishing

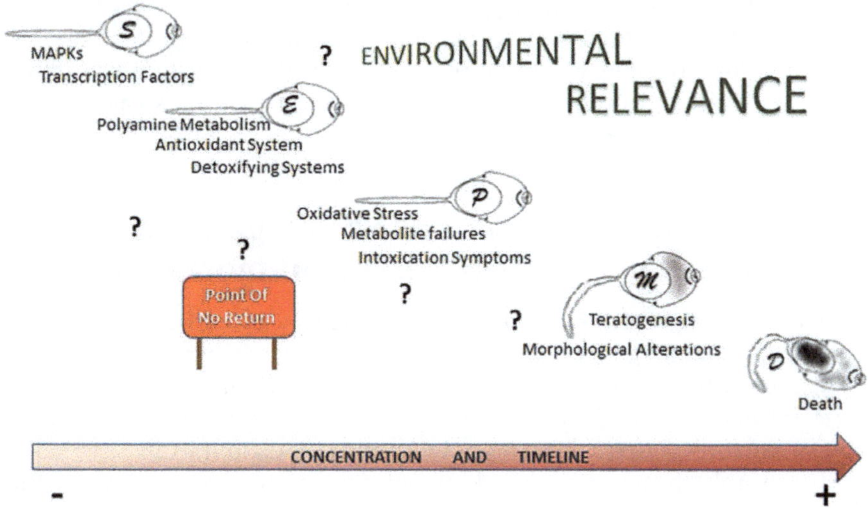

Figure 7.1 The dilemma between environmentally relevant parameters and sensitive plus early-warning biomarkers. In exposed *R. arenarum* larvae, contaminants can trigger early signalling ("S") pathways at very low concentrations, regulating key enzymes ("E") such as those from polyamine metabolism linked to normal development and other protecting enzymes; a higher concentration or time of exposure may generate metabolic alterations, loss of homeostasis and other noticeable physiological ("P") disorders; if the exposure is sustained or intensified, morphological ("M") defects appear and finally the exposed larvae die ("D"). All these steps (not necessarily followed in order in all cases) may be valuable biomarkers, although it would be desirable to set which one is the "point of no return", when the larvae cannot recover from exposure and will be committed to death.

threshold concentration values is either difficult or lacks precision. For these reasons, effective concentrations (EC) related to molecular and biochemical biomarkers are hardly available for ERA in aquatic species. In turn, a battery of biomarkers covering distinct physiological targets is recommended to evaluate the environmental impact in field studies. This has proven to be a good practise for field evaluations of *R. arenarum*.[6]

The advantages of using biomarkers as integrative tools for ERA were envisaged in the 1990s, when there were important advances in studies and proposals for many morphological, behavioural and biochemical parameters as endpoints.[4,7–10] These were incorporated preferentially in autochthonous species and considered valuable bioindicators of contamination. Conventional ERA is based on the chemical assessment of contaminants from a point source at any given time. Biomarker-based assessments would be able to integrate the responses measured on native species in scenarios involving multiple contaminants and their changing concentrations over time. The selection of a battery of biomarkers would serve as a warning system and further chemical analysis or site history searches for possible contamination episodes would be required, as it is not currently possible to assess contaminant type-specific responses, except for a few cases where contaminant metabolites are present.[6,11]

7.2 Molecular Biomarkers as Early Signals in Risk Assessment: Limitations in Non-model Species

The studies on contaminant effects and responses in native species have been accompanied by deepening into the molecular mechanisms, including regulation pathways and the fine crosstalk among them. As stated above, these are early responding and highly sensitive targets, contributing to the assessment of new biomarkers in species of regional interest.[3,12] We have been able to develop these kinds of studies on the exposure of the South American toad *R. arenarum* to organophosphorus pesticides and other contaminants such as As. Antioxidant enzymes, detoxifying enzymes from Phases I and II, polyamine metabolism-related enzymes, lipid metabolism, transcription factors, MAP kinases and other kinases belonging to signalling pathways were studied in *R. arenarum* at the protein and RNA transcript levels.[4,10,13,14] We observed an acceptable correspondence between the progressions from molecular to biochemical to physiological and morphological events with time and concentration factors, leading finally to larval death in *R. arenarum* exposed to contaminants (see Figure 7.1).

Nevertheless, it is not an easy task to develop these analyses for non-model species. There is a limited availability of antibodies with adequate reactivity towards a protein in the range of species that may include the desired one (precisely for being "non-model"). The alternative to trying commercial antibodies with broader species specificity is the development of antibodies using the purified protein or peptide from a particular species. In turn, the study of transcript expression through RT-PCR and qPCR requires primer design based on a known sequence. As the genomes of native, non-model species are largely unknown, the design depends on the genomic sequences of model species, *i.e.*, *Xenopus laevis* and *X. tropicalis* for anuran species. We were able to design primers for selected transcripts in *R. arenarum* using degenerated sequences based on the genome databases of both *Xenopus* species, with 50% of successful amplifications verified by fragment resequencing.[4]

7.3 The Appearance of Next-generation Sequencing and Its Impact on Biomarker Response Analysis

The impressive advance made in the biomarkers area, identifying numerous morphological, behavioural, biological, biochemical and molecular responses in relation to exposure to and interaction with environmental contaminants, could not converge into adequate tools for biomonitoring. The defying complexity and huge variability of environmental exposures derive from the complex and variable responses of biochemical and molecular biomarkers, which are more suitable as early and sensitive predictors. At the beginning of the 2010 decade, the analytical technologies of metabolomics, proteomics, genomics, transcriptomics and other massive studies became potential tools for diagnostics, prediction, or prognosis about the environment. Only when massive information techniques became available and produced large databases of information did the evaluation of multiple (virtually all) biological responses become a realistic task. In principle, the picture of contaminant effects on any organism may be taken at a metabolomic, proteomic, or transcriptomic level. The differential patterns produced at each of these levels can be assumed to be multiple responses, giving a fingerprint from which both generic (specific for a compound or family of compounds) patterns of response and general patterns (related to general stress effects and responses) may be considered in a "biomarkomic" analysis (see Figure 7.2). Therefore, a biomarkomic

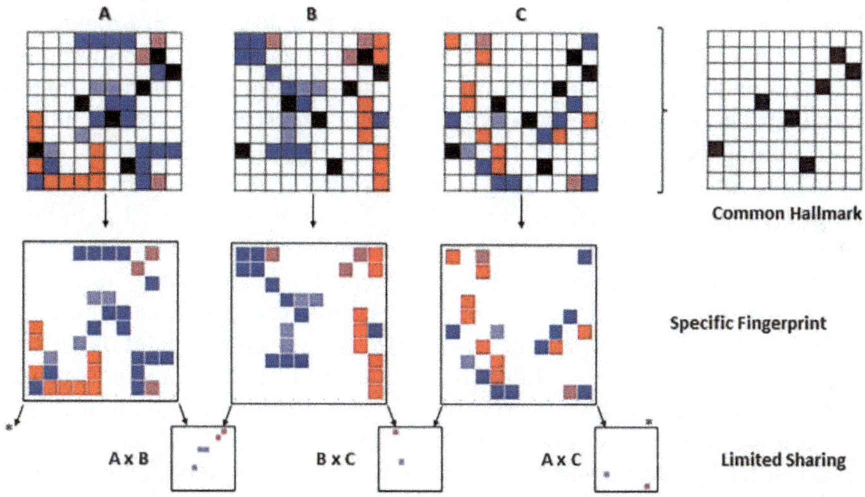

Figure 7.2 Progress of transcriptomic DEG pattern into a biomarkomic tool. The hypothetical transcriptome responses to three contaminants, A, B and C, can be dissected into a hallmark common to all, shown to the right, for those genes either up-exclusively or down-regulated by a general stress signal and specific fingerprints exclusive for a compound or a class of compounds. Note that some responses may be limitedly shared by compounds acting on a common pathway. This diagram may be amplified to multiple dimensions, *i.e.*, to include proteomics and metabolomics into an integrated system biology analysis, different species and time-concentration matrices, among others.

analysis may be defined as the holistic study of pattern responses at different omic layers generated in an organism or tissue by exposure to environmental contaminants under any field conditions. We are particularly interested in transcript biomarkomic responses as a potential tool in environmental impact or risk analysis, with the expectation that these are more stable than proteomic and metabolomic outputs, at least in the first steps of technological development. However, we do hope that a system based on biological biomarkomics may be feasible in the near future. In this direction, the application of big data comparative analysis methods enables the estimation of matrix points as potential biomarker imprints, as the "DIABLO" approach (Data Integration Analysis for Biomarker discovery using Latent cOmponents) applied to transcriptome and lipidome databases from rat pups exposed to bisphenol.[15,16]

Whole RNA sequencing (RNAseq) technologies have been improving rapidly, accompanied by new and highly accessible methodology for the molecular laboratory, increasing precision and notably decreasing costs. At present, it is possible to perform accurate long readings with whole transcript sequencing, identification of transcript isoforms, direct epigenetic modifications and transcript edition detection. These particular changes in transcript qualitative expression are worth mentioning as having a high potential to be used in a biomarkomic approach. The term "epigenome" in this case refers to heritable and sequence-independent genome-wide epigenetic controls, including DNA methylation, histone modification, non-coding RNA expression and transcription factor competition, among others.[17] Epigenetic regulations play a significant role in the development of environmentally induced physiopathological disorders.[18,19] Epigenetic alterations are produced by different environmental pollutants, such as carcinogenic metals, polycyclic aromatic hydrocarbons, endocrine disruptors and pesticides.[20–23] Thus, epigenetic modifications and consequent changes in regulation mechanisms constitute a relatively new field for the study of pollutant impacts on gene expression regulation. In this regard, the epigenome modification connected to pollutant exposure may provide a new dimension to the biomarkomic matrix of genomics, transcriptomics, proteomics and metabolomics.

Considering the degree of advancement in RNAseq technologies, some authors are questioning the need for complementary qPCR validation for the acceptance of DEG results.[24] In a large-scale screening work, it was found that 15–20% of DEG showed non-concordant results with respect to qPCR, but of these, 93% corresponded to fold-changes lower than 2, 80% to fold-changes lower than 1.5, and the rest of non-concordant DEG showed very low levels of expression.[25] As a conclusion, the authors suggest that no more than 2% of non-concordant DEG would show notorious biases.[25]

The direct use of RNAseq as a quantitative technique has been proposed in various applications. There are a good number of publications on transcriptomic studies in autochthonous species evaluating environmental impacts of contaminants, all featuring qPCR confirmations, *e.g.*, for hydrocarbon exploratory works in crabs,[26] for pyrethroids and heavy metals in bivalves[27,28] and endocrine disruptors in fish.[29] The effects of at least three different kinds of contaminants were analysed on the transcriptome of a marine copepod, finding some common pathways such as stress response and development.[30] The concept of using a DEG battery as a biomarkomic warning in ERA shares a similar basis with the "dynamic network biomarker (DNB) method" proposed in clinical diagnosis, which tries to obtain hints in the transcriptomic expression changes in order to establish critical points in disease progression.[31] Very good complementary information about contaminant mixtures has been obtained by combining classical biochemical biomarker responses with transcriptomic analysis of the pathways implied in exposed brown trout and the chemical measurements of contaminants in different sites into an "Aggregated Biomarker Response" (ABR).[32]

7.4 The Advantages of Transcriptomics as a Tool: The Case of *Rhinella arenarum*

The possibility of obtaining whole transcriptome data by RNAseq for *R. arenarum* has opened a range of alternatives while still facing the lack of its genome sequence (a common situation for most local species of interest). First, there are several types of RNA obtained from whole-RNA sequencing that may be of interest in environmental impact studies. From these, the most largely targeted for DEG is mRNA, but long non-coding RNA is also massively sequenced as a potential source for transcript differential expression, as discussed below. Other potential RNAs of environmental interest, such as microRNAs and others, require specific protocols for massive sequencing, differing from common whole RNAseq. Particularly for mRNA massive assembly in *R. arenarum*, we obtained a high-quality mRNA transcriptome, from which a very complete battery of annotated genes has been developed and made available at https://rhinella.uncoma.edu.ar/.[33] This database contains more than 57 thousand annotated sequences, from which primers for conventional and quantitative PCR were designed and produced, with a high rate of successful amplifications further verified by resequencing. This has facilitated, to a great extent, the focused study on selected transcript expression in our contaminant impact studies in *R. arenarum*, *i.e.*, transcripts belonging to the polyamine pathway, detoxifying enzymes, antioxidant enzymes, MAP kinases, and housekeeping genes.[34] At the bioinformatic level, these annotated transcript sequences from *R. arenarum* enabled a phylogenetic analysis within the order Anura, establishing a consensus taxonomic tree. This may be useful to compare and potentially extrapolate expected environmental impacts and responses between close species.

The next step has been the comparison of DEG under different experimental/contaminant conditions. Using bioinformatic tools, a holistic comparison is feasible with no *a priori* hypotheses on possible affected pathways. In our approach, we compared the effects of two organophosphorus pesticides, azinphosmethyl and chlorpyrifos, on the mRNA transcriptome of *R. arenarum* larvae. Results obtained identified a few gene transcripts that are affected in the same way at different times by both pesticides, which could be hallmark biomarkers of organophosphate exposure. It is interesting to note that most of them may be considered novel targets (*i.e.*, not previously described as organophosphate targets), some of them involved in membrane functions, ribosomal proteins, structural or trafficking proteins, as well as gene expression regulators (unpublished results). Those previously described as targets are hardly represented (*i.e.*, serine proteases and other hydrolases). In turn, there are comparatively more gene transcripts responding selectively to either one or the other organophosphate, showing the complexity of effects depending on the specific contaminant molecule and the fingerprint nature of the transcriptomic response. A closer look at the top-ranking DEG in the different exposure conditions leads to relevant candidates to be considered for delimited biomarker studies in traditional approaches. Between the few top-ranking transcripts downregulated in common by azinphosmethyl and chlorpyrifos in exposed *R. arenarum* larvae, we found predominantly those related to DNA and RNA processing, one related to tumour suppression, another regulating proteolysis and a third regulating neurotransmission, all with negatively affected functions. The shared top-upregulated genes included transcripts involved mainly in gene expression regulation and cell proliferation and fate, followed by signalling regulation (unpublished results).

The top-ranking genes which are downregulated in a specific way by azinphosmethyl in *R. arenarum* are mostly related to signalling regulation, connected with GTPase activity, phosphatidylinositol binding and cytoskeletal signalling. In turn, chlorpyrifos-specifically top-downregulated genes include transcripts related to transcription regulation, chromatin remodelling and cell proliferation, while some others are engaged with structural functions, cell fate and proliferation, signalling through protein kinase and phosphatase pathways and protein degradation. On the other hand, in the case of azinphosmethyl, the top-ranking upregulated genes are those related to structural functions, followed by cell fate, signalling and gene expression regulation. In the case of chlorpyrifos, the most frequently affected cellular function is cell proliferation and fate, followed by signalling regulation and gene expression regulation. All these responses in the top-affected transcripts are summarized in Figure 7.3.

On the other hand, the transcriptome of *R. arenarum* has allowed us to perform hypothesis-driven analyses on specific pathways that have been recognized to be impacted by organophosphorus compounds.[34] This has

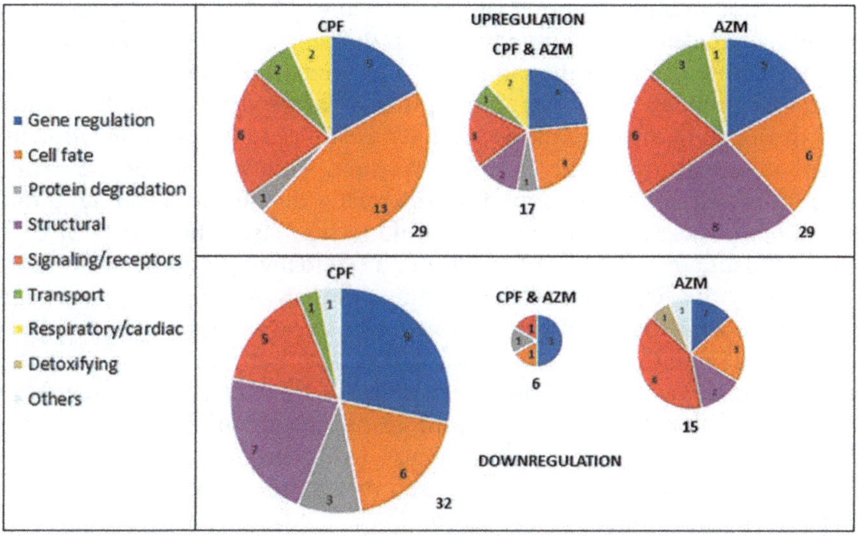

Figure 7.3 Cellular functions associated with the top-ranked DEG in *Rhinella arenarum* larvae. The cellular functions correspond to the top-20 upregulated and the top-20 downregulated transcripts, assessed for genes exclusively affected by chlorpyrifos (CPF), exclusively affected by azinphosmethyl (AZM), or similarly affected by CPF and AZM. Numbers within the portions are the DEG for the corresponding function (note that a transcript may have more than one function associated); numbers in the pie charts are the total of DEG-associated functions, also represented by the areas.

been a manual data-mining process where we looked for all the specific gene transcripts and isoforms belonging to these pathways. The pipeline included both mRNA and protein sequence validation, checking of the relative expression levels and normalization for each experimental condition to the mean expression levels of selected and checked housekeeping genes. After these control steps, all the fragments corresponding to the same transcript and annotated gene, identified as isoforms, were pooled into a unique replicate set, making the statistical analyses more robust. This analysis provided interesting hints, one of which was the greater strength of chlorpyrifos to elicit DEG compared with azinphosmethyl. Interestingly, the fold-change levels (in terms of expression in treated *R. arenarum* larvae *vs.* control specimens) for the hypothesis-selected genes from the pathways expected to be affected by the organophosphates were far below the top-ranked genes ascertained in the whole mRNA transcriptome bioinformatic analysis. Thus, small changes in transcript expression for these genes would be magnified at protein content and activity levels, as shown by biochemical studies.[34] This fact reinforces the interest in deepening the study of highly affected genes at the transcript level for their potential in biomarkomic analysis. Among the genes being affected in a common way by both pesticides, several are involved in the polyamine synthesis pathway and polyamine oxidative degradation. This agrees with the effects of these organophosphorus pesticides on amphibian larvae development. In turn, the expression of antioxidant enzyme genes showed a marked dependence on the compound and the time of exposure. This was evidenced in the different transcript expressions for superoxide dismutase, catalase, glutathione peroxidase isoforms and glutathione synthase. The effects of each organophosphate on the expression of the detoxifying enzymes are more dispersed, with a few genes affected in the same way. Among them, we found cytochrome P450 A1a and glutathione-*S*-transferase isoenzymes mu3, pi1 and 3, tetha3 and the microsomal forms 1 to 3. Other remarkable examples of common targets are the downregulation of the transcription regulator Nrf2, responsible for the antioxidant response and of the AhR nuclear translocator, which is necessary for aryl hydrocarbon receptor-mediated transactivation of detoxifying responses related to the Dioxin/Xenobiotic Response Element in gene regulation.

7.5 Perspectives of Advancement: What Else?

There are some lines of progress in the state-of-the-art research that can be envisaged from the picture presented here. The first comes directly from the RNAseq data we have, *i.e.*, from *R. arenarum* larvae exposed to organophosphorus pesticides, comprising the analysis of lncRNA transcripts detected within the whole transcriptome database. This is a very interesting line of research, considering the regulatory role attributed to lncRNA on DNA function and mRNA transcription, where there is still much to be explored. In contrast with proteomic circuits, this is a highly ordered RNA-based regulatory system through DNA-RNA, RNA-RNA and RNA-protein interactions.[35–37] Next-generation sequencing allows the identification of hundreds of lncRNAs that have been categorized by their function, structure and interacting molecules. This lncRNA regulates gene expression through several mechanisms, such as epigenetic modifications on DNA, alternative splicing, post-transcriptional regulation and mRNA stabilization and translation. Some of the cellular processes involving lncRNA regulation are proliferation, migration, development and apoptosis.[35,37,38] Currently, bioinformatic methods allow the analysis of lncRNA and mRNA co-expression patterns, predicting function regulation.[39] Thus, DEG analysis focused on lncRNA will surely add a functional perspective on contaminant modes of action and effect. There is an increasing recognition of lncRNA as potential biomarkers linked to biological responses in both human and animal model species.[40–45]

In line with the whole transcriptome sequencing for DEG analysis in *R. arenarum* exposed to other contaminants, we are presently engaged in a more comprehensive project including fungicides, oil hydrocarbons, polyaromatic hydrocarbons, toxic metals and nanomaterials of environmental relevance. The leading idea is to develop biomarkomic hallmarks with respect to general stress biomarkers and xenobiotic-specific transcript responses. We are also including other autochthonous anurans, the gammarid *Hyalella curvispina* and the mollusc *Chilina gibbosa*, to obtain transcriptome responses to the mentioned contaminants. With this approach, we expect to obtain a complex network of biomarkomic databases to compare not only the responses of the common South American toad *R. arenarum* but also those of different aquatic species and orders.

Acknowledgements

Our work is supported by the Universidad Nacional del Comahue (Grant 04A134) and the Agencia Nacional de Promoción Científica y Tecnológica (PICT 2017-1529), both from Argentina.

References

1. M. L. Menone, F. G. Iturburu, P. M. Demetrio, A. Venturino, F. L. Pedrozo, P. F. Temporetti, A. Rodriguez, M. V. Amé, K. P. Quaini and P. A. Collins, Calidad del agua y niveles guía para la protección de la biodiversidad acuática. Interacción entre ciencia y gestión, *Ecol. Austral.*, 2022, **32**, 245–257.
2. P. M. Demetrio, F. G. Iturburu, P. A. Collins, M. L. Menone, A. Venturino, P. F. Temporetti, F. L. Pedrozo, M. V. Amé, K. P. Quaini and A. Rodríguez, Metodología para la derivación de niveles guía para la protección de la biodiversidad acuática, *Ecol. Austral.*, 2022, **32**, 258–272.
3. M. Mardirosian, D. Ceschin, C. I. Lascano and A. Venturino, Molecular effectors in the chronic exposure to arsenic as early and sensitive biomarkers in developing *Rhinella arenarum* toads, *Aquat. Toxicol.*, 2017, **186**, 19–27.
4. A. Venturino, Toxicogenomics: New strategies for ecotoxicology studies in autochthonous species I. A glade in the entangled path towards the 'fingerprint' of environmental impact?, *Int. J. Environ. Health*, 2017, **8**, 203–212.
5. A. Ferrari, C. Lascano, L. Anguiano, A. M. Pechen de D'Angelo and A. Venturino, Antioxidant responses to azinphos methyl and carbaryl during embryonic development of the toad *Rhinella (Bufo) arenarum* Hensel, *Aquat. Toxicol.*, 2009, **93**, 37–44.
6. E. A. Rosenbaum, L. Duboscq, J. Soleño, C. M. Montagna, A. Ferrari and A. Venturino, Response of biomarkers in amphibian larvae to in situ

exposures in a fruit producing region at North Patagonia, Argentina, *Environ. Toxicol. Chem.*, 2012, **31**, 2311–2317.

7. D. A. Powers, Fish as model systems, *Science*, 1989, **246**, 352–358.
8. S. M. Adams, J. P. Giesy, L. A. Tremblay and C. T. Eason, The use of biomarkers in ecological risk assessment: Recommendations from the Christchurch conference on biomarkers in ecotoxicology, *Biomarkers*, 2001, **6**, 1–6.
9. U. Borgmann, L. Grapentine, W. P. Norwood, G. Bird, D. G. Dixon and D. Lindeman, Sediment toxicity testing with the freshwater amphipod *Hyalella azteca*: relevance and application, *Chemosphere*, 2005, **61**, 1740–1743.
10. A. Venturino and A. M. Pechen de D'Angelo, Biochemical targets of xenobiotics: Biomarkers in amphibian ecotoxicology, *Appl. Herpetol.*, 2005, **2**, 335–353.
11. Y. Ji, G. Lu, C. Wang, W. Song and H. Wu, Fish transplantation and stress-related biomarkers as useful tools for assessing water quality, *J. Environ. Sci.*, 2010, **22**, 1826–1832.
12. M. Guerreño, M. F. López Armengol, C. M. Luquet and A. Venturino, Comparative study of toxicity and biochemical responses induced by sublethal levels of the pesticide azinphosmethyl in two fish species from North-Patagonia, Argentina, *Aquat. Toxicol.*, 2016, **177**, 365–372.
13. C. I. Lascano, V. Sotomayor, A. Ferrari and A. Venturino, Alteraciones del desarrollo embrionario, poliaminas y estrés oxidativo inducidos por plaguicidas organofosforados en *Rhinella arenarum*, *Acta Toxicol. Argent.*, 2009, **17**, 8–19.
14. M. N. Mardirosian, C. I. Lascano, A. Ferrari, G. A. Bongiovanni and A. Venturino, Acute toxicity of Arsenic and oxidative stress responses in the embryonic development of the common South American toad *Rhinella arenarum*, *Environ. Toxicol. Chem.*, 2015, **34**, 1009–1014.
15. H. T. Nguyen, L. Li, A. Eguchi, K. Kannan, E. Y. Kim and H. Iwata, Effects on the liver lipidome of rat offspring prenatally exposed to bisphenol A, *Sci. Total Environ.*, 2020, **759**, 143466.
16. H. T. Nguyen, L. Li, A. Eguchi, T. Agusa, K. Yamamoto, K. Kannan, E. Y. Kim and H. Iwata, Effects of gestational exposure to bisphenol A on the hepatic transcriptome and lipidome of rat dams: Intergenerational comparison of effects in the offspring, *Sci. Total Environ.*, 2022, **826**, 153990.
17. V. Cavalieri and G. Spinelli, Environmental epigenetics in zebrafish, *Epigenet. Chromatin*, 2017, **10**, 46.
18. M. Kim, M. Bae, H. Na and M. Yang, Environmental toxicants—induced epigenetic alterations and their reversers, *J. Environ. Sci. Health C*, 2012, **30**, 323–367.

19. J. Hu and Y. Yu, Epigenetic response profiles into environmental epigenotoxicant screening and health risk assessment: A critical review, *Chemosphere*, 2019, **226**, 259–272.
20. T. G. Rossman and C. B. Klein, Genetic and epigenetic effects of environmental arsenicals, *Metallomics*, 2011, **3**, 1135–1141.
21. Y. Zhang, S. Dong, H. Wang, S. Tao and R. Kiyama, Biological impact of environmental polycyclic aromatic hydrocarbons (ePAHs) as endocrine disruptors, *Environ. Pollut.*, 2016, **213**, 809–824.
22. S. Rattan and J. A. Flaws, The epigenetic impacts of endocrine disruptors on female reproduction across generations, *Biol. Reprod.*, 2019, **101**, 635–644.
23. H. D. Nicolella and S. de Assis, Epigenetic inheritance: intergenerational effects of pesticides and other endocrine disruptors on cancer development, *Int. J. Mol. Sci.*, 2022, **23**, 4671.
24. T. Coenye, Do results obtained with RNA-sequencing require independent verification?, *Biofilm*, 2021, **3**, 10043.
25. C. Everaert, M. Luypaert, J. L. V. Maag, Q. X. Cheng, M. E. Dinger, J. Hellemans and P. Mestdagh, Benchmarking of RNA-sequencing analysis workflows using whole-transcriptome RT-qPCR expression data, *Sci. Rep.*, 2017, **7**, 1559.
26. J. R. Hall, S. J. Lehnert, E. Gonzalez, S. Kumar, J. M. Hanlon, C. J. Morris and M. L. Rise, Snow crab (*Chionoecetes opilio*) hepatopancreas transcriptome: Identification and testing of candidate molecular biomarkers of seismic survey impact, *Fisheries Res.*, 2021, **234**, 105794.
27. H. Zhang, X. Hong, S. Yan, J. Zha and J. Qin, Environmentally relevant concentrations of bifenthrin induce changes in behaviour, biomarkers, histological characteristics, and the transcriptome in *Corbicula fluminea*, *Sci. Total Environ.*, 2020, **728**, 138821.
28. G. Wang, C. Zhang and B. Huang, Transcriptome analysis and histopathological observations of *Geloinaerosa* gills upon Cr(VI) exposure, *Comp. Biochem. Physiol. C*, 2020, **231**, 108706.
29. F. Yadetie, N. R. Brun, I. Vieweg, J. Nahrgang, O. A. Karlsen and A. Goksøyr, Transcriptome responses in polar cod (*Boreogadus saida*) liver slice culture exposed to benzo[a]pyrene and ethynylestradiol: insights into anti-estrogenic effects, *Toxicol. In Vitro*, 2021, **75**, 105193.
30. A. Barrick, O. Laroche, M. Boundy, J. K. Pearman, T. Wiles, J. Butler, X. Pochon, K. F. Smith and L. A. Tremblay, First transcriptome of the copepod *Gladioferens pectinatus* subjected to chronic contaminant exposures, *Aquat. Toxicol.*, 2022, **243**, 106069.
31. C. Han, J. Zhong, Q. Zhang, J. Hu, R. Liu, H. Liu, Z. Mo, P. Chen and F. Ling, Development of a dynamic network biomarkers method and its application for detecting the tipping point of prior disease development, *Computat. Struct. Biotechnol. J.*, 2022, **20**, 1189–1197.

32. M. Schmitz, B. Deutschmann, N. Markert, T. Backhaus, W. Brack, M. Brauns, M. Brinkmann, T. B. Seiler, P. Fink, S. Tang, S. Beitel, J. A. Doering, M. Hecker, Y. Shao, T. Schulze, M. Weitere, R. Wild, M. Velki and H. Hollert, Demonstration of an aggregated biomarker response approach to assess the impact of point and diffuse contaminant sources in feral fish in a small river case study, *Sci. Total Environ.*, 2022, **804**, 150020.
33. D. G. Ceschin, N. S. Pires, M. N. Mardirosian, C. I. Lascano and A. Venturino, The *Rhinella arenarum* transcriptome: de novo assembly, annotation and gene prediction, *Sci. Rep.*, 2020, **10**, 1053.
34. N. S. Pires, C. I. Lascano, J. Ousset, D. G. Ceschin and A. Venturino, Hypothesis-driven dragging of transcriptomic data to analyze proven targeted pathways in *Rhinella arenarum* larvae exposed to organophosphorus pesticides, *Sci. Rep.*, 2022, **12**, 17712.
35. X. Guo, L. Gao, Y. Wang, D. K. Y. Chiu, T. Wang and Y. Deng, Advances in long noncoding RNAs: identification, structure prediction and function annotation, *Brief. Funct. Genomics*, 2016, **15**, 38–46.
36. J. Jarroux, A. Morillon and M. Pinskaya, in *Long Non Coding RNA Biology. Advances in Experimental Medicine and Biology, History, Discovery, and Classification of lncRNAs, vol. 1008*, ed. M. Rao, Springer, Singapore, 2017, pp. 1–46.
37. J. L. Rinn and H. Y. Chang, Genome regulation by long noncoding RNAs, *Ann. Rev. Biochem.*, 2012, **81**, 145–166.
38. L. Ma, V. B. Bajic and Z. Zhang, On the classification of long non-coding RNAs, *RNA Biol.*, 2013, **10**, 924–933.
39. A. Necsulea, M. Soumillon, M. Warnefors, A. Liechti, T. Daish, U. Zeller, J. C. Baker, F. Grützner and H. Kaessmann, The evolution of lncRNA repertoires and expression patterns in tetrapods, *Nature*, 2014, **505**, 635–640.
40. J. L. Dempsey and J. Y. Cui, Long non-coding RNAs: A novel paradigm for toxicology, *Toxicol. Sci.*, 2017, **155**, 3–21.
41. C. Gao, Z. He, J. Li, X. Li, Q. Bai, Z. Zhang, X. Zhang, S. Wang, X. Xiao, F. Wang, Y. Yan, D. Li, L. Chen, X. Zeng, Y. Xiao, G. Dong, Y. Zheng, Q. Wang and W. Chen, Specific long non-coding RNAs response to occupational PAHs exposure in coke oven workers, *Toxicol. Rep.*, 2016, **3**, 160–166.
42. L. Recio, S. L. Phillips, T. Maynor, M. Waters, A. F. Jackson and C. L. Yauk, Differential expression of long noncoding RNAs in the livers of female B6C3F1 mice exposed to the carcinogen furan, *Toxicol. Sci.*, 2013, **135**, 369–379.
43. L. Wang, H. Yang, Q. Wang, Q. Zhang, Z. Wang, Q. Zhang, S. Wu and H. Li, Paraquat and MPTP induce alteration in the expression profile of

long noncoding RNAs in the substantia nigra of mice: Role of the transcription factor Nrf2, *Toxicol. Lett.*, 2018, **291**, 11–28.

44. K. Xiong, L. Long, X. Zhang, H. Qu, H. Deng, Y. Ding, J. Cai, S. Wang, M. Wang, L. Liao, J. Huang, C. Yi and J. Yan, Overview of long non-coding RNA and mRNA expression in response to methamphetamine treatment *in vitro*, *Toxicol. in Vitro*, 2017, **44**, 1–10.
45. Z. Zhou, H. Liu, C. Wang, Q. Lu, Q. Huang, C. Zheng and Y. Lei, Long non-coding RNAs as novel expression signatures modulate DNA damage and repair in cadmium toxicology, *Sci. Rep.*, 2015, **5**, 15293.

CHAPTER 8

Genotoxic Risk of Solar Ultraviolet Radiation in Amphibians

JAMES EDUARDO LAGO LONDERO[a] AND
ANDRÉ PASSAGLIA SCHUCH*[a]

[a]Department of Biochemistry and Molecular Biology, Post-Graduation Program in Biological Sciences: Toxicological Biochemistry, Federal University of Santa Maria, INPE Building, 1000 Roraima Avenue, Room 3010, P.O. Box 5021, Santa Maria, RS 97105-970, Brazil
*E-mail: andre.schuch@ufsm.br

8.1 Introduction

Biodiversity has suffered intensely due to anthropogenic actions.[1,2] For instance, during the last decades, worldwide declines in amphibian populations have been documented, resulting in relatively high extinction rates in this group.[3] Approximately 41% of the amphibian species are threatened with extinction, according to the International Union for Conservation of Nature (IUCN) Red List.[4] The higher sensitivity of amphibians in comparison to other vertebrate groups is commonly related to their complex life cycle and skin permeability.[4,5] An array of factors and contaminants present in the environment have been linked to these population declines.[3] One environmental factor commonly associated with the amphibian decline phenomenon is solar ultraviolet (UV) radiation.[6,7] This

Issues in Toxicology No. 46
Amphibian Species in Environmental Risk Assessment Strategies
Edited by Marcelo L. Larramendy and Guillermo Eli Liwszyc

Published by the Royal Society of Chemistry, www.rsc.org

can negatively affect the embryonic and larval amphibian developmental stages due to its genotoxicity.[7] Notably, the presence of other stressors in the environment may potentiate the UV-induced effects on amphibians.[6,8]

In this chapter, we review the general aspects of the research field involving the genotoxic risk of solar UV radiation for amphibians and its potential implications for the population decline phenomenon. First, we discuss the environmental relevance of UV radiation for amphibians. Second, we describe the UV-induced DNA lesions. Third, we approach the amphibian defence mechanisms against UV-induced genotoxicity, which include both strategies to avoid and deal with UV damage. Lastly, we discuss the link between the genotoxicity triggered by solar UV exposure and the amphibian decline.

8.2 Environmental Relevance of UV Radiation

Solar UV radiation that reaches the Earth's biosphere is progressively attenuated by an array of factors. Geophysical factors, such as altitude, latitude, solar zenith angle and UV-absorbing compounds, such as the stratospheric ozone layer, atmospheric aerosols and clouds, have an impact on terrestrial UV levels.[9,10] In addition, as early amphibian developmental stages often live in aquatic ecosystems, the water bodies' characteristics should also be considered as a physical barrier to UV radiation. The aquatic vegetation, the depth of the water column and the levels of aquatic solids and particulates are factors that attenuate aquatic penetration of UV radiation.[11,12] Importantly, all these attenuating factors have a major effect on ultraviolet B (UVB; 280–315 nm) relative to ultraviolet A (UVA; 315–400 nm) radiation because UVB wavelengths are more energetic and more strongly absorbed by atmospheric and aquatic molecules.[9,13]

About 95% of the solar UVA radiation and about 5% of the solar UVB radiation reach the Earth's surface.[9,14] Furthermore, aquatic ecosystems receive less UVA and UVB radiation as depth increases.[13] Despite the low percentage of solar UVB radiation that reaches terrestrial and aquatic environments, incident UVB radiation still represents a strong threat to living organisms due to its high genotoxic potential.[12,15–17]

In the 1970s/1980s, seminal investigations indicated that the anthropogenic release of ozone-depleting substances led to decreases in stratospheric ozone concentrations.[18,19] As a counterpart, the Montreal Protocol was established in 1987 to minimize atmospheric contamination by ozone-depleting substances to enable the recovery of the ozone layer. As a consequence of the Protocol, it has been reported that ozone concentrations have been less affected compared to scenarios without applying such a Protocol.[20] Nonetheless, unexpected increases in global emissions of ozone-depleting substances have been reported in the last few years.[21,22] In parallel, events of ozone reduction were also observed at specific localities, such as Southern Brazil, South Africa and New Zealand, as a consequence of the transport of polar air masses with low-ozone content towards these regions.[23–25] Such events are very worrying, as low concentrations of stratospheric ozone lead to increases in ground levels of UVB radiation and possibly other unforeseen environmental consequences.[10,26]

Human actions have changed the climate and land cover. In this scenario, the dynamics of atmospheric compounds, cloudiness, water column depth and aquatic compounds became more complex.[10,27,28] For example, in certain cases, even a high flow of organic matter into aquatic environments (which could be beneficial by blocking exceeding levels of UV radiation) can impose risks to aquatic organisms due to the generation of oxidative stress after photosensitization of these aquatic compounds by UV exposure.[29] The presence of aquatic chemical contaminants can also be energized by UV radiation, leading to DNA damage formation.[30] In addition, deforestation and extreme weather events generate clearings within forests that ultimately lead to the incidence of harmful levels of UV radiation on forest-dwelling amphibians.[8,31] Therefore, climate and land use changes can indirectly affect amphibians by allowing more UV radiation to reach these animals.

8.3 Induction of DNA Damage by Solar UV Radiation

The most relevant deleterious effects of solar UV radiation on living organisms are mediated by DNA damage and the resulting genotoxic effects.[32,33] Both UVB and UVA radiation induce DNA damage in living cells, either *via* direct absorption of UV photons or through the photosensitization of cellular molecules such as molecular oxygen.[9,14]

The direct absorption of UV photons by DNA pyrimidine bases leads to the formation of lesions called pyrimidine dimers (or DNA photoproducts). Cyclobutane pyrimidine dimers (CPDs) represent about 80% of these photoproducts, while pyrimidine (6-4) pyrimidone photoproducts (6-4PPs) represent about 20%.[9,34] Both CPDs and 6-4PPs block DNA replication and transcription, leading to cell death or mutations.[32,33] UVB radiation can cause a higher number of pyrimidine dimers in DNA than UVA and because of this, the toxicity induced by UVB exposure is more detrimental than the toxicity induced by UVA exposure.[9,14]

UV photosensitization of molecular oxygen inside the cells can lead to the formation of oxidatively generated DNA lesions. UV-induced reactive oxygen species can oxidize DNA bases and generate single-strand breaks (SSBs).[14] Figure 8.1 shows structural insights into the general DNA lesions that can be generated in amphibians after UVB/UVA exposure. SSBs and double-strand breaks can be generated after DNA metabolism instability is triggered by unrepaired UV-induced DNA lesions. DNA strand breaks are highly stressful for cells and can result in cell death or mutations at the chromosome level.[33]

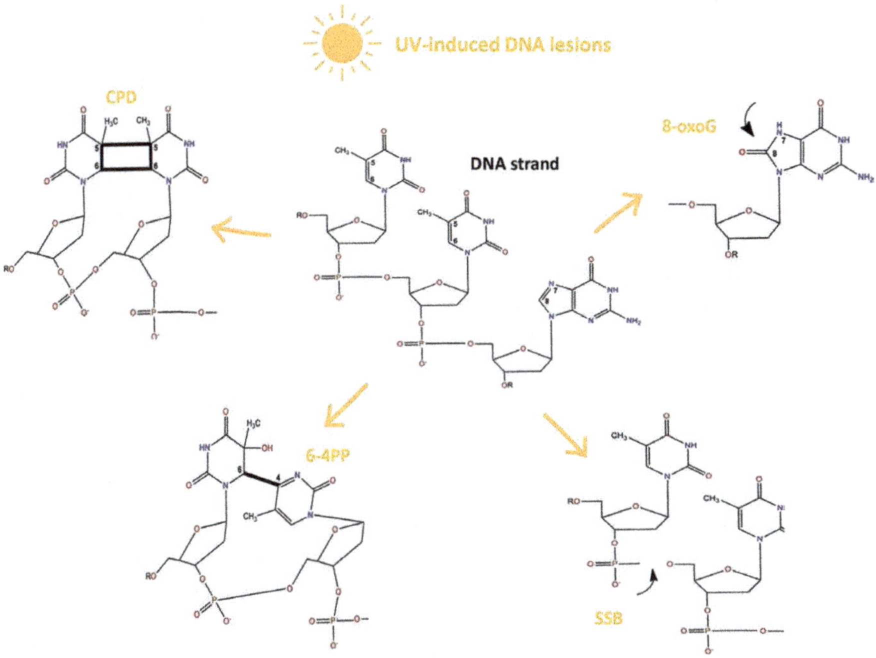

Figure 8.1 UV exposure induces mutagenic and cytotoxic DNA lesions in amphibian cells. Major DNA lesions are illustrated. CPD, cyclobutane pyrimidine dimers; 6-4PP, pyrimidine (6-4) pyrimidone photoproducts; 8-oxoG, 8-oxoguanine (main oxidatively generated lesion); SSB, single-strand breaks.

8.4 Amphibian Defence Mechanisms Against UV-induced Genotoxicity

The persistence of damage in cellular DNA can cause cells to enter programmed cell death (apoptosis), or if cells do not die, DNA damage may lead to mutations.[32,33] At the organismal level, UV-induced DNA damage is strongly associated with the induction of metabolic, immunosuppressive, tumorigenic and lethal effects.[35] Such effects can be significant in terms of population and community ecology (Section 8.5).

UV radiation has been a ubiquitous stressor in life over evolutionary time. This genotoxic stressor probably exerted selection pressure on living organisms, including amphibians, which resulted in the evolution of defence mechanisms.[36] Among these defence mechanisms, behavioural strategies, photoprotection by sunscreen compounds and pigments, and molecular repair stand out in amphibians.[37] However, amphibian species differ in terms of the presence and efficiency of their defence mechanisms against UV radiation.[37]

The choice of suitable oviposition sites by adult amphibians can expose the laid eggs to low levels of UV radiation.[38,39] Amphibian embryos *per se* do not have the option of actively choosing specific microhabitats. Amphibian larvae of some species can perceive and avoid UV photons if shaded microhabitats are available, whereas others cannot.[40–43] However, variations in temperature, pH and aquatic oxygen, competition with others for resources and the presence of predators and pathogens can lead larvae to actively seek out new microhabitats, which can receive high levels of UV radiation.[6] Notably, the presence of certain chemical contaminants in the water bodies, such as the neurotoxic pesticide endosulfan, can affect the UV-avoidance behaviour of tadpoles.[44]

Photoprotective cellular compounds such as sunscreen and pigments present in the jelly of eggs, embryos and larvae can protect the genomes of these animals against UV damage to some extent.[45,46] Interestingly, melanin-containing cells in the amphibian skin and internal organs are a key barrier against UV radiation.[47,48]

As UV-induced damage has been formed in the genome, a series of molecular responses can be activated to deal with the damage. DNA repair, DNA damage tolerance, molecular signalling and cell cycle arrest collectively act to maintain genomic integrity.[35,49] Among these DNA damage responses, the DNA repair mechanisms called photorepair (or photoreactivation), nucleotide excision repair (NER) and base excision repair (BER) stand out in amphibians.[7]

The light-dependent repair by photolyase enzymes (known as photorepair or photoreactivation) uses the energy of visible light and higher UVA photons to directly revert UV-induced pyrimidine dimers to normal bases *via* electron transfer.[34] Amphibians have specific photolyases to repair CPD and 6-4PP lesions: the CPD photolyase, which repairs only CPDs, and the 6-4 photolyase, which repairs only 6-4PPs. Additionally, amphibians have DASH cryptochrome/photolyase, which may repair CPDs in single-stranded DNA.[50] On the other hand, pyrimidine dimers can be removed in a light-independent manner by the NER pathway, a process that requires ATP consumption and the coordinated action of several proteins. The process involves lesion recognition, unwinding of the damaged double helix, cleavage and excision of damaged strand fragments, filling gaps in the molecule by DNA synthesis and finally ligation.[51] In addition, UV-induced DNA oxidation can be repaired by the BER pathway. Initially, DNA glycosylases recognize and remove oxidized bases, generating abasic sites (AP). Then AP endonucleases cleave abasic sites, forming single-strand breaks, which can then be processed by the replacement of a single nucleotide or 2–10 nucleotides.[52]

Although solar UV radiation has been defined as a driver of amphibian decline worldwide, especially due to its genotoxicity, the responses to sunlight-induced DNA damage in amphibian species are still poorly understood.[7] To date, UV-induced DNA damage has been directly measured in amphibians in seven environmentally relevant studies. One of these studies showed a relationship between CPDs, teratogenesis and growth inhibition in tadpoles of the African clawed frog *Xenopus laevis* (Anura: Pipidae) exposed to UVB radiation.[53] Another study that assessed CPD and 6-4PP lesions showed that UV attenuation caused by aquatic organic matter significantly reduced the effects of UV radiation in embryos of the spotted salamander *Ambystoma maculatum* (Caudata: Ambystomatidae).[54]

The five other studies assessed DNA repair events. Elevated levels of pyrimidine dimers were detected after UVB-pesticide co-exposures (*i.e.*, endosulfan/UVB and α-cypermethrin/UVB) compared to UVB exposure alone in *X. laevis* embryos, related to altered transcription levels of NER genes.[55] Another study revealed the persistence rates of CPDs and 6-4PP lesions over time and biological effects in tadpoles of the Montevideo tree frog *Boana pulchella* (Anura: Hylidae) subjected to UVB or UVA radiation and subsequently to treatments in which tadpoles recovered under the light (for photorepair) or in the dark (for NER). By revealing the key role of photorepair in maintaining tadpole survival, this study provides significant insights into the UV-sensitivity hypothesis.[43] This hypothesis correlated photolyase activity measured with *in vitro* egg protein assays from 10 amphibian species (the African clawed frog *X. laevis*, the Cascade frog *Rana cascadae*, the Pacific tree frog *Hyla regilla*, the Western toad *Bufo boreas*, the rough-skinned newt *Taricha granulosa* and the salamanders, namely, the Northwestern salamander *Ambystoma gracile*, the long-toed salamander *Ambystoma macrodactylum*, the Dunn's salamander *Plethodon dunni*, the Western red-backed salamander *Plethodon vehiculum* and Southern torrent salamander *Rhyacotriton variegatus*) and the survival of eggs from three species (*H. regilla*, *B. boreas* and *R. cascadae*) exposed to UV radiation in controlled field experiments.[56] Two other studies revealed an effect of temperature on DNA damage induction and repair in UV-exposed tadpoles and that amphibian species respond differently to temperature after UV exposure.[57,58] Morison *et al.* used the striped marsh frog *Limnodynastes peronii* (Anura: Limnodynastidae) as the study species,[57] whereas Hird *et al.* used the spotted grass frog *Limnodynastes tasmaniensis* (Anura: Limnodynastidae) and the ornate burrowing frog *Platyplectrum ornatum* (Anura: Limnodynastidae).[58] In addition, a study with tadpoles of the yellow-spotted tree frog *B. curupi* (Anura: Hylidae) revealed that UVB-trichlorfon co-exposures reduce pyrimidine dimer levels induced by UVB exposure alone, suggesting that trichlorfon stimulates DNA repair pathways.[59]

8.5 Implications of UV-induced Genotoxicity for the Amphibian Decline

Exposure level to sunlight and the ability to limit the induction or repair UV-induced DNA lesions determine amphibian fitness.[7,37] Physiologically debilitated individuals in stressful environments may become even more susceptible to UV exposure and its genotoxic effects.[6] Although mortality is the most extreme effect of UV radiation, the sublethal effects of solar UV radiation have negative implications for amphibian biology. Controlled experiments using tadpoles of the South American tree frogs *B. pulchella* and *B. curupi* (Anura: Hylidae) (Figure 8.2) indicated that mortality and reduced performances of development, locomotion and feeding, as well as skin alterations, are induced mainly in tadpoles exposed to UVB rather than UVA radiation and are primarily associated with genotoxic effects.[25,43,59–61] In addition, carry-over effects of UV exposure from one life stage to subsequent stages may be a source of fitness variation, as demonstrated with the European common frog *Rana temporaria* (Anura: Ranidae).[62] The induction of telomere shortening observed in the juvenile stage of the Australian green tree frog *Litoria caerulea* (Anura: Pelodryadidae) as a consequence of the tadpole's early exposure to UV radiation may be a related mechanism.[63]

The physiological costs of UV exposure may alter amphibian ecology. Individuals can become more disadvantaged in the competition for resources and more susceptible to predation or pathogens.[6,7,64] Some amphibian species are more susceptible to decline than others.[65] These differences may result in changes in the structure of the ecological community and in the characteristics of ecosystems. Continued mortality after sublethal damage

Figure 8.2 Tree frog species commonly used by our group to investigate the effects of UV radiation on amphibian larvae. Top: representative adult examples of the Montevideo tree frog *Boana pulchella* (left) and the yellow-spotted tree frog *Boana curupi* (right). Bottom: representative examples of the habitats occupied by *B. pulchella* (pond in an open area) and *B. curupi* (stream in a forested area).

induction may ultimately contribute to population decline.[65] Nonetheless, there is still a lack of concrete evidence about the link between the genotoxicity induced by solar UV radiation and amphibian population decline. Molecular mechanistic investigations can reveal new insights into ecological patterns and processes potentially linked to population extirpation.

References

1. J. S. Albert, G. Destouni, S. M. Duke-Sylvester, A. E. Magurran, T. Oberdorff, R. E. Reis, K. O. Winemiller and W. J. Ripple, Scientists' warning to humanity on the freshwater biodiversity crisis, *Ambio.*, 2021, **50**, 85–94.
2. R. H. Cowie, P. Bouchet and B. Fontaine, The sixth mass extinction: Fact, fiction or speculation?, *Biol. Rev.*, 2022, **97**, 640–663.
3. J. P. Collins and M. L. Crump, in *Extinction in our times: global amphibian decline*, Oxford University Press, New York NY, 1st edn, 2009.
4. IUCN Red List of Threatened Species, Available from: https://www.iucnredlist.org/resources/summary-statistics. Accessed 23 April 2023.
5. M. L. Crump, in *Amphibian ecology and conservation. A handbook of techniques,1st edn, Amphibian diversity and life history*, ed. C. K. Dodd Jr, Oxford University Press, New York NY, 2009, pp. 3–19.
6. L. A. Alton and C. E. Franklin, Drivers of amphibian declines: effects of ultraviolet radiation and interactions with other environmental factors, *Clim. Chang. Resp.*, 2017, **4**, 6.
7. J. E. L. Londero, M. B. dos Santos and A. P. Schuch, Impact of solar UV radiation on amphibians: Focus on genotoxic stress, *Mutat. Res. Genet. Toxicol. Environ. Mutagen.*, 2019, **842**, 14–21.
8. M. C. Rocha, M. B. dos Santos, R. Zanella, O. D. Prestes, A. S. Gonçalves and A. P. Schuch, Preserved riparian forest protects endangered forest-specialists amphibian species against the genotoxic impact of sunlight and agrochemicals, *Biol. Conserv.*, 2020, **249**, 108746.
9. A. P. Schuch, C. C. M. Garcia, K. Makita and C. F. M. Menck, DNA damage as a biological sensor for environmental sunlight, *Photochem. Photobiol. Sci.*, 2013, **12**, 1259–1272.
10. A. F. Bais, G. Bernhard, R. L. McKenzie, P. J. Aucamp, P. J. Young, M. Ilyas, P. Jöckel and M. Deushi, Ozone-climate interactions and effects on solar ultraviolet radiation, *Photochem. Photobiol. Sci.*, 2019, **18**, 602–640.
11. G. S. Peterson, L. B. Johnson, R. P. Axler and S. A. Diamond, Assessment of the risk of solar ultraviolet radiation to amphibians. II. *In situ* characterization of exposure in amphibian habitats, *Environ. Sci. Technol.*, 2003, **37**, 5080.

12. M. B.dos Santos, J. E. L. Londero, Á. A. Mainardi, S. Z. Cechin and A. P. Schuch, Genotoxic risk assessment of solar UV radiation in tadpoles from Brazilian wetlands, *Mutat. Res. Genet. Toxicol. Environ. Mutagen.*, 2023, **885**, 503578.
13. C. E. Williamson and K. C. Rose, When UV meets fresh water, *Science*, 2010, **329**, 637–639.
14. A. P. Schuch, N. C. Moreno, N. J. Schuch, C. F. M. Menck and C. C. M. Garcia, Sunlight damage to cellular DNA: Focus on oxidatively generated lesions, *Free. Radic. Biol. Med.*, 2017, **107**, 110–124.
15. D. Karentz and L. H. Lutze, Evaluation of biologically harmful ultraviolet radiation in Antarctica with a biological dosimeter designed for aquatic environments, *Limnol. Oceanogr.*, 1990, **35**, 549–561.
16. A. P. Schuch, T. Yagura, K. Makita, H. Yamamoto, N. J. Schuch, L. F. Agnez-Lima, R. M. MacMahon and C. F. M. Menck, DNA damage profiles induced by sunlight at different latitudes, *Environ. Mol. Mutagen.*, 2012, **53**, 198–206.
17. F. Fuentes-León, A. Peres de Oliveira, N. Quintero-Ruiz, V. Munford, G. Satoru Kajitani, A. Coimbra Brum, A. P. Schuch, P. Colepicolo, A. Sánchez-Lamar and C. F. M. Menck, DNA damage induced by late spring sunlight in Antarctica, *Photochem. Photobiol.*, 2020, **96**, 1215–1220.
18. M. J. Molina and F. S. Rowland, Stratospheric sink for chlorofluoromethanes: chlorine atom-catalysed destruction of ozone, *Nature*, 1974, **249**, 810–812.
19. J. C. Farman, B. G. Gardiner and J. D. Shanklin, Large losses of total ozone in Antarctica reveal seasonal ClOx/NOx interaction, *Nature*, 1985, **315**, 207–210.
20. P. A. Newman and R. McKenzie, UV impacts avoided by the Montreal Protocol, *Photochem. Photobiol. Sci.*, 2011, **10**, 1152–1160.
21. S. A. Montzka, G. S. Dutton, P. Yu, E. Ray, R. W. Portmann, J. S. Daniel, L. Kuijpers, B. D. Hall, D. Mondeel, C. Siso, J. D. Nance, M. Rigby, A. J. Manning, L. Hu, F. Moore, B. R. Miller and J. W. Elkins, An unexpected and persistent increase in global emissions of ozone-depleting CFC-11, *Nature*, 2018, **557**, 413–417.
22. X. Fang, J. A. Pyle, M. P. Chipperfield, J. S. Daniel, S. Park and R. G. Prinn, Challenges for the recovery of the ozone layer, *Nat. Geosci.*, 2019, **12**, 592–596.
23. E. J. Brinksma, Y. J. Meijer, B. J. Connor, G. L. Manney, J. B. Bergwerff, G. E. Bodeker, I. S. Boyd, J. B. Liley, W. Hogervorst, J. W. Hovenier, N. J. Livesey and D. P. J. Swart, Analysis of record-low ozone values during the 1997 winter over Lauder, New Zealand, *Geophys. Res. Lett.*, 1998, **25**, 2785–2788.

24. N. Semane, H. Bencherif, B. Morel, A. Hauchecorne and R. D. Diab, An unusual stratospheric ozone decrease in the Southern Hemisphere subtropics linked to isentropic air-mass transport as observed over Irene (25.5° S, 28.1° E) in mid-May 2002, *Atmos. Chem. Phys.*, 2006, **6**, 1927–1936.

25. A. P. Schuch, M. B. dos Santos, V. M. Lipinski, L. V. Peres, C. P. dos Santos, S. Z. Cechin, N. J. Schuch, D. K. Pinheiro and E. L. da S. Loreto, Identification of influential events concerning the Antarctic ozone hole over southern Brazil and the biological effects induced by UVB and UVA radiation in an endemic tree frog species, *Ecotoxicol. Environ. Saf.*, 2015, **118**, 190–198.

26. J. B. Kerr and C. T. McElroy, Evidence for large upward trends of ultraviolet-B radiation linked to ozone depletion, *Science*, 1993, **262**, 1032–1034.

27. J. M. Kiesecker, A. R. Blaustein and L. K. Belden, Complex causes of amphibian population declines, *Nature*, 2001, **410**, 681–684.

28. C. E. Williamson, R. G. Zepp, R. M. Lucas, S. Madronich, A. T. Austin, C. L. Ballaré, M. Norval, B. Sulzberger, A. F. Bais, R. L. McKenzie, S. A. Robinson, D. P. Häder, N. D. Paul and J. F. Bornman, Solar ultraviolet radiation in a changing climate, *Nat. Clim. Change*, 2014, **4**, 434–441.

29. R. Wolf, T. Andersen, D. O. Hessen and K. Hylland, The influence of dissolved organic carbon and ultraviolet radiation on the genomic integrity of *Daphnia magna*, *Funct. Ecol.*, 2017, **31**, 848–855.

30. A. P. Roberts, M. M. Alloy and J. T. Oris, Review of the photo-induced toxicity of environmental contaminants, *Comp. Biochem. Physiol. C, Toxicol. Pharmacol.*, 2017, **191**, 160–167.

31. M. A. Clark, W. M. Ota, S. J. Smith, B. K. Muramoto, S. Ngo, G. E. Chan, M. A. Kenyon, M. C. Sturtevant, M. G. Diamond, G. M. Bucciarelli and L. B. Kats, A natural experiment identifies an impending ecological trap for a neotropical amphibian in response to extreme weather events, *Ecol. Evol.*, 2022, **12**, e8848.

32. R. P. Sinha and D. P. Häder, UV-induced DNA damage and repair: A review, *Photochem. Photobiol. Sci.*, 2002, **1**, 225–236.

33. R. P. Rastogi, A. Kumar, M. B. Tyagi and R. P. Sinha, Molecular mechanisms of ultraviolet radiation-induced DNA damage and repair, *J. Nucleic Acids.*, **2010**, 592980.

34. A. Sancar, Structure and function of photolyase and *in vivo* enzymology: 50th anniversary, *J. Biol. Chem.*, 2008, **283**, 32153–32157.

35. N. Chatterjee and G. C. Walker, Mechanisms of DNA damage, repair, and mutagenesis, *Environ. Mol. Mutagen.*, 2017, **58**, 235–263.

36. A. B. Blaustein, L. K. Belden, A. C. Hatch, L. B. Kats, P. D. Hoffman, J. B. Hays, A. Marco, D. P. Chivers and J. M. Kiesecker, in *Ecosystems, Evolution, and Ultraviolet Radiation, 1stedn, Ultraviolet radiation and*

amphibians, ed. C. S. Cockell and A. R. Blaustein, Springer, New York, 2001, pp. 63–79.

37. A. R. Blaustein and L. K. Belden, Amphibian defenses against ultraviolet-B radiation, *Evol. Dev.*, 2003, **5**, 89–97.

38. W. J. Palen and D. E. Schindler, Water clarity, maternal behavior, and physiology combine to eliminate UV radiation risk to amphibians in a montane landscape, *Proc. Natl. Acad. Sci. USA.*, 2010, **107**, 9701–9706.

39. L. B. Kats, G. M. Bucciarelli, D. E. Schlais, A. R. Blaustein and B. A. Han, Ultraviolet radiation influences perch selection by a Neotropical poison-dart frog, *PLoS One*, 2012, **7**, e51364.

40. T. F. Van De Mortel and W. A. Buttemer, Avoidance of ultraviolet-B radiation in frogs and tadpoles of the species *Litoriaaurea, L. dentata* and *L. peronii*, *Proc. Linnean Soc. New South Wales*, 1998, **119**, 173–179.

41. L. K. Belden, I. T. Moore, R. T. Mason, J. C. Wingfield and A. R. Blaustein, Survival, the hormonal stress response and UV-B avoidance in Cascades Frog tadpoles (*Rana cascadae*) exposed to UV-B radiation, *Funct. Ecol.*, 2003, **17**, 409–416.

42. M. Pahkala, A. Laurila and J. Merilä, Effects of ultraviolet-B radiation on behaviour and growth of three species of amphibian larvae, *Chemosphere*, 2003, **51**, 197–204.

43. A. P. Schuch, V. M. Lipinski, M. B. Santos, C. P. Santos, S. S. Jardim, S. Z. Cechin and E. L. S. Loreto, Molecular and sensory mechanisms to mitigate sunlight-induced DNA damage in treefrog tadpoles, *J. Exp. Biol.*, 2015, **218**, 3059–3067.

44. S. Yu, S. M. Weir, G. P. Cobb and J. D. Maul, The effects of pesticide exposure on ultraviolet-B radiation avoidance behavior in tadpoles, *Sci. Total Environ.*, 2014, **481**, 75–80.

45. L. J. Hansen, D. L. Fabacher and R. Calfee, The role of the egg jelly coat in protecting *Hyla regilla* and *Bufo canorus* embryos from ultraviolet B radiation during development, *Environ. Sci. Pollut. Res. Int.*, 2002, **9**, 412–416.

46. M. G. Perotti and M. del C. Diéguez, Effect of UV-B exposure on eggs and embryos of patagonian anurans and evidence of photoprotection, *Chemosphere*, 2006, **65**, 2063–2070.

47. L. Franco-Belussi, H. N. Sköld and C. de Oliveira, Internal pigment cells respond to external UV radiation in frogs, *J. Exp. Biol.*, 2016, **219**, 1378–1383.

48. L. Franco-Belussi, L. Z. Fanali and C. de Oliveira, UV-B affects the immune system and promotes nuclear abnormalities in pigmented and non-pigmented bullfrog tadpoles, *J. Photochem. Photobiol. B*, 2018, **180**, 109–117.

49. G. Giglia-Mari, A. Zotter and W. Vermeulen, DNA damage response, *Cold Spring Harb. Perspect. Biol.*, 2011, **3**, 1–19.

50. Q. Mei and V. Dvornyk, Evolutionary history of the potolyase/cryptochrome superfamily in Eukaryotes, *PLoS One*, 2015, **10**, e0135940.
51. R. dos S. Feltrin, A. L. A. Segatto, T. A. de Souza and A. P. Schuch, Open gaps in the evolution of the eukaryotic nucleotide excision repair, *DNA Repair (Amst)*, 2020, **95**, 102955.
52. H. E. Krokan and M. Bjørås, Base excision repair, *Cold Spring Harb. Perspect. Biol.*, 2013, **5**, a012583.
53. D. J. Bruggeman, J. A. Bantle and C. Goad, Linking teratogenesis, growth, and DNA photodamage to artificial ultraviolet b radiation in *Xenopus laevis* larvae, *Environ. Toxicol. Chem.*, 1998, **17**, 2114–2121.
54. M. P. Lesser, S. L. Turtle, J. H. Farrell and C. W. Walker, Exposure to ultraviolet radiation (290-400 nm) causes oxidative stress, DNA damage, and expression of p53/p73 in laboratory experiments on embryos of the spotted salamander, *Ambystoma maculatum*, *Physiol. Biochem. Zool.*, 2001, **74**, 733–741.
55. S. Yu, S. Tang, G. D. Mayer, G. P. Cobb and J. D. Maul, Interactive effects of ultraviolet-B radiation and pesticide exposure on DNA photo-adduct accumulation and expression of DNA damage and repair genes in *Xenopus laevis* embryos, *Aquat. Toxicol.*, 2015, **159**, 256–266.
56. A. R. Blaustein, P. D. Hoffman, D. G. Hokit, J. M. Kiesecker, S. C. Walls and J. B. Hays, UV repair and resistance to solar UV-B in amphibian eggs: a link to population declines?, *Proc. Natl. Acad. Sci. USA.*, 1994, **91**, 1791–1795.
57. S. A. Morison, R. L. Cramp, L. A. Alton and C. E. Franklin, Cooler temperatures slow the repair of DNA damage in tadpoles exposed to ultraviolet radiation: Implications for amphibian declines at high altitude, *Glob. Change Biol.*, 2020, **26**, 1225–1234.
58. C. Hird, C. E. Franklin and R. L. Cramp, Temperature causes species-specific responses to UV-induced DNA damage in amphibian larvae, *Biol. Lett.*, 2022, **18**, 20220358.
59. C. R. Schavinski, M. B. dos Santos, J. E. L. Londero, M. C. da Rocha, A. M. B. do Amaral, N. Q. Ruiz, G. da S. Leandro, V. L. Loro and A. P. Schuch, Effects of isolated and combined exposures of *Boana curupi* (Anura: Hylidae) tadpoles to environmental doses of trichlorfon and ultraviolet radiation, *Mutat. Res. Genet. Toxicol. Environ. Mutagen.*, 2022, **883–884**, 503549.
60. J. E. L. Londero, C. P. dos Santos, A. L. A. Segatto and A. P. Schuch, Impacts of UVB radiation on food consumption of forest specialist tadpoles, *Ecotoxicol. Environ. Saf.*, 2017, **143**, 12–18.
61. C. P. Santos, J. E. L. Londero, M. B. dos Santos, R. dos S. Feltrin, L. Loebens, L. B. Moura, S. Z. Cechin and A. P. Schuch, Sunlight-induced genotoxicity and damage in keratin structures decrease tadpole performance, *J. Photochem. Photobiol. B*, 2018, **181**, 134–142.

62. M. Pahkala, A. Laurila and J. Merilä, Carry-over effects of ultraviolet-B radiation on larval fitness in *Rana temporaria*, *Proc. R. Soc. Lond., B, Biol, Sci.*, 2001, **268**, 1699–1706.
63. N. U. Lundsgaard, R. L. Cramp and C. E. Franklin, Early exposure to UV radiation causes telomere shortening and poorer condition later in life, *J. Exp. Biol.*, 2022, **225**, jeb243924.
64. L. A. Alton, C. R. White, R. S. Wilson and C. E. Franklin, The energetic cost of exposure to UV radiation for tadpoles is greater when they live with predators, *Funct. Ecol.*, 2012, **26**, 94–103.
65. A. R. Blaustein and J. M. Kiesecker, Complexity in conservation: lessons from the global decline of amphibian populations, *Ecol. Lett.*, 2002, **5**, 597–608.

CHAPTER 9

Mortality Induced by Mixtures of Pesticides in Tadpoles from the Pampas Region of Argentina

CELESTE RUIZ DE ARCAUTE*[a,b], MILAGROS LABORDE[a,b] AND SONIA SOLONESKI[a,b]

[a]Cátedra de Citología, Facultad de Ciencias Naturales y Museo, Universidad Nacional de La Plata, Calle 64 N° 3, B1904AMA La Plata, Argentina; [b]Consejo Nacional de Investigaciones Científicas y Técnicas (CONICET), Argentina
*E-mail: cruizdearcaute@fcnym.unlp.edu.ar

9.1 Introduction

In recent decades, freshwater environments have been threatened by different anthropogenic activities, such as waste disposal, burning fossil fuels, deforestation and pollution from agrarian, industrial and urban runoffs, which have adverse impacts on aquatic species.[1–3] In order to protect the aquatic environment and its biota, it is imperative to identify the main pollutants, their sources and their fates to promote realistic and sustainable land use.[4] Since the early 1990s, the expansion of agricultural land in South America has brought large-scale pesticide use, as evidenced by an average amount in the use of these compounds from 2.06 to 2.42

Issues in Toxicology No. 46
Amphibian Species in Environmental Risk Assessment Strategies
Edited by Marcelo L. Larramendy and Guillermo Eli Liwszyc

Published by the Royal Society of Chemistry, www.rsc.org

kg ha^{-1} between 2000 and 2018, respectively.[5] In Argentina, in particular, agricultural land has expanded in the last 20 years, increasing habitat degradation and contamination, with the actual cultivated area reaching approx. 36 million ha.[6–8] The Pampas region, one of the most fertile arable lands in the world, with its mild weather and fertile soil, is the site of extensive crops such as grains and oilseeds, including soybean and corn, to name a few.[6,9] In these circumstances, pesticides have been massively employed without rigorous governmental regulation and proper information to prevent misuse, accidents and inadequate package disposal, among other issues.[10]

Commercial formulations of pesticides are constituted by the active ingredient(s) and several types of additives, such as solvents, coadjuvants, carrier substances and other components, in order to improve the active ingredient´s absorption, emulsion, spread, stickiness and wetting. In addition, these chemicals also reduce evaporation, drift and volatilization.[11] This fact calls for the evaluation of the detrimental effects that formulated products could induce and the identification of the individual source (or sources) responsible for such effects extremely difficult.[12] Moreover, it is widely known that some commercial formulations can exert higher toxicity than the active ingredients themselves, with the excipients able to cause toxicity and cellular damage, among other side effects.[13–19]

Field studies have detected multiple agrochemicals in surface waters from Argentina, mainly in Pampasic Mesopotamia[20–23] and Buenos Aires Province waterways.[24–34] The widely employed herbicide glyphosate (GLY) was detected at values ranging from 0.001 to 1.80 mg L^{-1}.[20,21,31–33] Moreover, its major degradation product, aminomethylphosphonic acid, was detected within a concentration range of 0.5 to 2.3 μg L^{-1}.[20,32] The pesticides atrazine (0.64 to 1.4 μg L^{-1}),[27,30,34] diethyltoluamide (0.043 to 0.701 μg L^{-1}),[34] tebuconazole (0.030 to 0.035 μg L^{-1}),[34] endosulfan (0.01 to 4.26 μg L^{-1}),[22] cypermethrin (CYP, 0.01 to 194 μg L^{-1}),[22,24,25] chlorpyrifos (CPF, 0.01 to 10.8 μg L^{-1}),[22,24–26] azoxystrobin (0.53 to 4.26 μg L^{-1}),[23] cyproconazole (0.51 to 3.23 μg L^{-1}),[23] deltamethrin (0.18 to 2.73 μg L^{-1}),[23,24] epoxiconazole (0.54 to 2.74 μg L^{-1}),[23] and penconazole (0.52 to 2.45 μg L^{-1})[23] were also detected. A concentration of 0.010 μg L^{-1} was determined for λ-cyhalothrin by Mac Laughlin and collaborators.[24]

At present, the practise of applying a mixture of pesticides is a commonly used strategy to control pests and minimize the emergence of resistant species.[35,36] In addition, when multiple crops are cultivated simultaneously, different types of agrochemicals are required. Consequently, different kinds of substances converge and form complex mixtures in the environment. These, in turn, represent a new group of environmental pollutants,[37–39] thus increasing the need to include mixtures in risk assessment evaluations.[40–42] In this type of scenario, chemicals interact with each other and with biological matrices, concomitantly modifying the toxicity of the individual substances, inducing lethality,[43–46] oxidative stress[47] and genotoxicity,[3] among other related effects.

Amphibians have certain characteristics, such as their biphasic life cycle, permeable skin, sensitivity and unprotected eggs, among other factors, that make them suitable organisms to evaluate and monitor pollution effects.[48] For the last three decades, there has been increasing evidence of a worldwide decline in amphibian populations.[3,45,49] Other factors that can lead to this situation include climate change, habitat degradation because of chemical and biological pollution and the introduction of exotic species and infectious diseases.[8,43,50–53] According to the International Union for Conservation of Nature (IUCN),[54] almost 41% of amphibian species are under threat and regrettably, 32 species became extinct during the last century.[53,54] The Neotropical Region is not exempt from this situation, and several studies have evaluated the impact of pollution on Neotropical anuran species, stressing the need to employ native species to adequately select an indicator species.[55–61]

To the best of our knowledge, in Argentina, there are approximately 60 species of amphibians reported for the whole country. Over 23 anuran species inhabiting the Pampas region belong to the Bufonidae, Leptodactylidae, Ceratophryidae, Odontophrynidae, Microhylidae and Hylidae families. The South American common frog *Rhinella arenarum* (Bufonidae) (see Figure 9.1A) is listed in the IUCN Red List as one of the Pampas amphibian species of least concern in light of its tolerance, presumed large population, wide distribution and range of habitats.[54] This species has been selected for laboratory assays evaluating mixtures of agrochemicals in the aforementioned region, as it is considered an appropriate *in vivo* model. *Rhinella arenarum* has been widely employed in recent decades for detecting either lethal or sublethal effects exerted by agrochemicals, such as toxicity,[7,62–70] genotoxicity,[3,62,64,71–73] biochemical changes,[74] oxidative stress[65,69,70,72,74–78] and behavioural and morphological alterations.[7,66,67,69,78] Other commonly employed species for evaluation of the toxicity induced by xenobiotics are the Montevideo tree frog *Boana pulchella* (Hylidae) (see Figure 9.1B), the lesser foam frog *Leptodactylus latrans* (Leptodactylidae), the oven frog *Leptodactylus latinasus* (Leptodactylidae) and the Bella Vista toad *R. fernandezae* (Bufonidae), among others.

9.2 Evaluation of Agrochemical Mixtures

To predict the toxicity of a mixture of chemicals, two approaches can be employed: either to test the whole mixture (top-down) or the individual components of the mixture (bottom-up).[79] The first option is mostly applied to environmental samples, with the caveat that the substances responsible for the response evaluated often remain unidentified. In contrast, the bottom-up approach is more commonly employed but requires further information about toxicity, concentrations and the

Figure 9.1 Anuran adult specimens of the South American common frog *Rhinella arenarum* (Bufonidae) (A) and the Montevideo tree frog *Boana pulchella* (Hylidae) (B).

acknowledgement of the mode of action of the constituents off the mixture.[79]

It is a well-known fact that for a mixture of two different substances, three types of interactions can be defined: first, concentration addition (CA) that requires a similar mode of action of the components of the mixture; second, independent action (IA) that requires a dissimilar mode of action of the components of the mixture; and third, interactions among components of the mixture (synergism or antagonism).[79] Moreover, three different types of interaction are acknowledged in order to assess simultaneous exposure to multiple chemicals: additivity, synergism and antagonism.[80,81] Additivity (or additive joint action) refers to the toxic effect of a mixture equal to that expected from adding the toxicity of the single compounds of the mixture. Synergism (a greater than additive effect or potentiation) exists when the toxic effect induced by the mixture is greater than expected if the mixture were additive and lastly, antagonism (less than an additive effect) exists when the toxicity is lower than estimated if the mixture were additive. Moreover, when the components of a mixture exert their effect alone, the non-interaction hypothesis must be detailed in terms of a proper additivity model, such as IA, dose addition (DA), CA or a mixed model. In this case, synergy is referred to as "more-than-additive" and antagonism as "less-than-additive".[82]

When only one component in a binary mixture induces toxicity, the term "non-interaction" implies that the mixture is not more toxic than that active ingredient by itself. Thus, synergy is equal to the term potentiation and refers to an increased effect caused by the second compound, whereas the lack of an effect is indicated as "inertism". Similarly, antagonism refers to a diminished effect observed when the second component is added to form a mixture. The term potentiation is employed when the toxic effects of a chemical are enhanced by one or more other substances without inducing that effect on its own. For binary combinations, where the effect is induced by one of the components, and the mixture is more toxic than the active compound itself, the terms potentiation and synergism are commonly used as synonyms.[82] Finally, the term interaction is used when deviations of non-interaction or additivity in mixtures occur, like synergism or antagonism. This term is descriptive, as it does not imply any underlying mechanism(s).[82]

The assessment of the effects of pesticide mixtures can be performed using different methodologies. One possibility is first to test each pesticide separately and then test them in combinations. Studies employing mixtures of pesticides evaluate mostly equitoxic mixtures, defined as such when each component in the mixture is present in the same proportion of its individual toxicity. According to Warne, for testing of mixtures, the laboratory assays for equitoxic mixtures must be performed using the same protocols proposed for a single toxicant. Instead of expressing concentrations in mg L^{-1}, this assay employs an arbitrary unit (S factor) that equals the toxic effect of each component of the mixture.[81] For these studies, it has been suggested to use at least five different experimental treatments: two treatments with an S value greater than 1, one treatment with an S value equal to 1, and two treatments with an S value less than 1. When analysing the results obtained after the treatment, if S is equal to 1, the equitoxic mixture should be considered additive. On the other hand, if S is less than or greater than 1, the mixture should be considered synergistic or antagonistic, respectively.[81] Following Warne,[81] mixtures can be evaluated in non-equitoxic concentrations, a much more realistic scenario since pollutants in the environment converge and possibly interact, potentially changing their toxicological effects.[79,83] Most tests conducted employing these mixtures have been performed with binary mixtures.[7,43,45,84] Results are usually presented in isobolograms, which summarize the results obtained for the series of non-equitoxic mixtures. The diagonal line or isobole

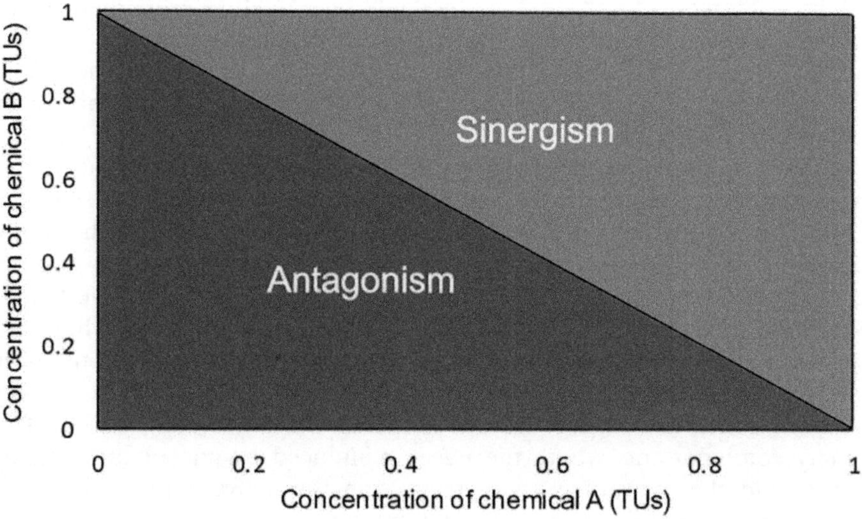

Figure 9.2 Isobologram showing the diagonal line or isobole indicating an additive effect of the components of the mixture. The area below and to the left of the isobole represents a synergistic effect, whereas the area above and to the right of the isobole represents an antagonistic effect.

connecting the values of 1 TU on the X- and Y-axes corresponds to the effect of concentration addition. If combinations that cause 50% mortality are located above and to the right of the additivity line, the effect of the mixture is antagonistic. On the contrary, if the combination results are located below and to the left of the additivity line, the effect produced by the interaction is considered synergistic (see Figure 9.2).[81]

9.3 Studies Evaluating Mixtures

In Argentina, several studies have evaluated the toxic effects of pesticide mixtures and the resulting interaction. All but one of them were performed in laboratory conditions employing *R. arenarum* tadpoles as a test organism. A list of these studies is presented in Table 9.1.

Lajmanovich and coworkers evaluated the toxicological interactions between different herbicides on *R. arenarum* tadpoles.[7] They analysed the following herbicides: Ultra-Max® (GLY 74.7%, Monsanto Co.), Metsulfuron 60® (metsulfuron-methyl, MET 60%, Ciagro S.A.), Ectran® (bispyribac-sodium, BIS 40%, Ipesa S.A.) and Tordon 24-K® (picloram, PIC 27.7%, Dow Agrosciences). To determine the effect induced by the mixtures, they employed a 50 : 50 combination according to the equation proposed by Marking:[88]

Table 9.1 Toxicity studies evaluating mixtures of pesticides in anuran embryos and tadpoles of the South American common frog *Rhinella arenarum* (Bufonidae) performed under laboratory conditions.

Mixtures (a.i.)	Developmental stage	Types of chemical	Type of exposure	Time of exposure	Resulting interaction	Ref.
GLY + PIC/GLY + BIS/ GLY + MET	Tadpoles	Commercial formulations	50 : 50 combination	48 h	Synergism	7
GLY + CYP	Tadpoles	Commercial formulations	Equitoxic mixtures	96 h	Synergism	43
			Non-equitoxic mixtures		Synergism	
DZN + NP	Embryos Tadpoles	Pure compound	Equitoxic mixtures Non-equitoxic mixtures	96 and 168 h	Synergism Synergism	85
END + CYP	Embryos Tadpoles	Pure compound (END) and commercial formulation (CYP)	Equitoxic mixtures	96, 168 and 240 h	Antagonism Additive	86
GLY + 2,4-D	Tadpoles	Commercial formulations	Equitoxic mixtures Non-equitoxic mixtures	48, 96, 168 and 240 h	Antagonism Synergism	87
GLY + DIC/GLY + FLC	Tadpoles	Commercial formulations	Equitoxic mixtures	96 h	Weak antagonist interaction	45
			Non-equitoxic mixtures		Mostly additive, one slightly antagonist	

BIS, Bispyribac-sodium; CYP, Cypermethrin; 2,4-D, 2,4-Dichlorophenoxiacetic acid; DIC, Dicamba; DZN, Diazinon; END, Endosulfan; FLC, Flurochloridone; GLY, Glyphosate; MET, Metsulfuron; NP, Nonylphenol; PIC, Picloram; a.i., Active ingredient.

$$\left(\frac{Am}{Ai}\right) + \left(\frac{Bm}{Bi}\right) = S$$

where A and B correspond to the herbicides analysed alone, and i and m correspond to the LC_{50} values of the individual herbicides and their mixtures, respectively, while S represents the sum of the observed activity. As stated previously for Warne's method, the effect of the mixture was considered synergistic if $S > 1$, antagonistic if $S < 1$ or additive if $S = 1$. Moreover, it allowed the calculation of an additivity index (AI), which takes a value of $AI = 1/S - q$ if $S \leq 1.0$, whereas for $S \geq 1.0$, $AI = -S + 1$. Accordingly, negative and positive values of AI indicate less and more additivity, respectively, whereas AI = 0 refers to an additive effect. Results demonstrated that the toxicity gradient for the combinations assay was GLY + PIC, GLY + BIS and GLY + MET, from more to less toxic binary combinations. According to the additivity index employed, all mixtures induced a synergistic effect. Finally, the negative results of AI indicated low additivity.[7]

Brodeur and collaborators analysed the results of interactions between the herbicide GLY and the insecticide cypermethrin (CYP)-based commercial formulations in tadpoles of *R. arenarum*.[43] To evaluate the mixtures, they employed the protocol proposed by Warne,[81] with two different combinations of commercial formulations: Glifoglex® (GLY 48%, Gleba) + Glextrin® (CYP 25%, Gleba) and GLY Atanor® (48% GLY, Atanor) + Xiper® (CYP 25%, Icona). The $LC_{50\ 96h}$ values (or toxic unit, TU) of the individual compounds were 72.8, 30.2, 19.4 and 6.8 mg L^{-1} for Glifoglex®, Glexttin®, GLY Atanor® and Xiper®, respectively.[43] Results demonstrated that the four series of equitoxic and non-equitoxic mixtures were synergic in all evaluated combinations. Moreover, according to the results, they could establish that the synergism was 1.6–3.9 times more relevant in the mixture of GLY Atanor® + Xiper® than in the combination of Glifoglex® + Xiper®. This information could allow for opting for the least toxic combination when applying these pesticides in the field. Moreover, it points out the role of the excipients and adjuvants included in commercial formulations, which, in turn, alter the toxicity of the active principles. Moreover, Aronzon and collaborators evaluated the toxic effects of mixtures of diazinon (DZN, a.i. purity 99%, Supelco Analytical) and nonylphenol (NP, technical grade, 96.9%, Sigma Aldrich) on *R. arenarum* embryos and tadpoles.[85] They combined the compounds, employing different proportions based on the corresponding $LC_{50\ 168h}$ values. The isobologram showed synergistic interactions for embryos and tadpoles after 96 and 168 h of exposure.[85] These results were obtained from mixtures of the active ingredients, and they provide preliminary information about the toxicity that the commercial formulations, which are the compounds applied in the field, could induce in *R. arenarum*. Svartz and collaborators studied the effects induced by endosulfan (a.i. PS81, Supelco) and Glextrin (CYP 25%, Gleba S.A.) on *R. arenarum* embryos and tadpoles.[86] The assay was performed employing environmentally relevant equitoxic concentrations of both compounds. The interactions resulting after acute and chronic exposure were antagonistic for embryos and additive for tadpoles, demonstrating that mixtures of these insecticides induce toxicity that is affected by both the exposure time and the developmental stage of exposed individuals.[86] Finally, Peluso and collaborators performed two studies employing *R. arenarum*.[52,87] In the first one, they evaluated the quality of water and sediment samples from three sites of streams adjacent to crop regions on embryos and tadpoles after acute (96 h) and short-term chronic exposure (504 h). For all sampling sites, CPF, endosulfan and GLY were detected in maximum concentrations of 2.5, 0.9 and 2 μg L^{-1}, respectively. These concentrations caused lethal effects on tadpoles after chronic exposure, while samples from only one site were toxic for embryos after only acute exposure.[52] These results highlight the importance of evaluating the toxicity of mixtures over chronic periods of exposure. Finally, Peluso and collaborators evaluated the toxicity induced by Atanor® (GLY 43.8%, Atanor S.A.) and Asi Max 50® (2,4-D 50% Chemotecnica S.A.) and their mixtures in tadpoles of the same species.[87] The equitoxic mixture values for each compound were calculated from the $LC_{50\ 168h}$, while the non-equitoxic mixture was prepared following the recommendations from the supplier of GLY when combined with the amine salt form of 2,4-D.[87] The results showed mainly antagonist interactions for equitoxic mixtures. In contrast, the non-equitoxic mixture showed a synergistic response for all exposure times.[87] The results obtained show different kinds of interactions, depending upon the proportion of agrochemicals present in the mixture and the time of exposure.

Years ago, our research group evaluated the toxicity of the binary mixtures of the herbicides GLY + dicamba (DIC) and GLY + flurochloridone (FLC) on *R. arenarum* tadpoles following the protocol detailed by Warne.[45,81] Accordingly, the mixtures were performed with the commercial formulations Credit® (GLY 48%, Dow AgroSciences S.A.), Banvel® (DIC 57.71%, Syngenta Agro S.A., Argentina) and Twin Pack Gold (FLC 25%, Syngenta Agro S.A., Argentina). Equitoxic and non-equitoxic mixtures were performed relative to the $LC_{50\ 96h}$ value obtained for each individual compound. Results for equitoxic mixtures revealed that both mixtures, GLY + DIC and GLY + FLC, showed a weak antagonistic pattern of interaction, as both values were slightly above 1 (1.74 and 1.54 for GLY + DIC and GLY + FLC, respectively), the value expected for additivity. Finally, for non-equitoxic mixtures, most mixtures tested were additive, while only when DIC and FLC were fixed at 0.33 TUs, the interaction was slightly antagonistic.[45]

As for *in situ* studies, Agostini and collaborators evaluated the effects of commercial formulations of CYP, CPF, endosulfan, GLY and 2,4-D, alone and in mixtures, on tadpoles of four amphibian species, namely *B. pulchella*, *L. latrans*, *R. fernandezae* and *R. arenarum*.[49] They evaluated the lethality and mobility impairment induced in tadpoles by mixtures of GLY + CYP, GLY + CPF, GLY + CPF + endosulfan and GLY + 2,4-D.[49] The exposure was performed by placing the tadpoles in enclosures situated in either ponds exposed to pesticides *vis-à-vis* reference ponds. Mortality was evaluated at three different intervals: 24 h before mixture application and 24 h and 48 h after mixture application. In their results, Agostini and co-workers found significant decreases in survival after 13 of the 15 pesticide mixture applications.[49] All mixtures containing endosulfan caused a significant decrease in survival at 24 and 48 h after exposure. On the other hand, no tadpoles survived after the application of CYP + GLY + endosulfan. All the applications of CPF+ GLY reduced survival either at 24 or 48 h after exposure. Moreover, all mixtures containing CYP + GLY, except for one case where the mixture did not reach the pond, produced a significant decrease in survival at both evaluated exposure times. Out of three applications of GLY + 2,4-D, only one induced a significant decrease in tadpole survival. Finally, results showed that survival at 48 h after exposure was significantly lower compared with that obtained 24 h earlier in 8 out of the 13 applications. The average survival rates detected in control ponds were over 99% for all evaluated times.[49] The results obtained after 48 h of exposure show that mixtures containing endosulfan or CPF reduced tadpole survival to <1%, while a reduced survival rate of 10% was observed for mixtures containing CYP.[49]

9.4 Final Remarks

Nowadays, it is a fact that pesticides are often applied in the form of mixtures of at least two different agrochemicals, which makes it difficult to estimate the effect that each chemical component present within the mixture may exert on both target and non-target species. Moreover, sometimes there is a lack of information on the single substances involved in mixtures, such as their toxicity, induction of sublethal effects, or mode of action, among other side effects.[37] Priority in enlarging this information is a *sine qua non*-requisite to properly identify and characterize mixtures that need to be faced first.[37] For the above-mentioned reasons, it becomes evident that before studying the potential toxic effect of a mixture, it is necessary to determine the individual toxic effects of all the tested compounds. As we have previously mentioned, commercial agrochemical products are composed of not only active ingredients, but also adjuvants and other additives. In other words, they are intrinsically mixed. To further support this line of reasoning, most (if not all) results indicate the existence of variations in the toxicity induced by different commercial formulations of the same active principle.[43,89] In Argentina, adjuvants and other additives present in commercial formulations are protected trade secrets. Therefore, they are not considered in the discussion of the effects induced by a commercial product on target as well as non-target species. Several studies have clearly stated that these additives can produce cellular damage by themselves,[15,61,62,90–94] toxicity,[12,63] and genotoxic effects[15,61,62,90] in different biotic matrices.

In the literature, we find many studies regarding the deleterious effects of contaminated waters (with mixtures of chemicals, some of which have not been determined), but there is a lack of information on how these chemicals interact and if this interaction affects their toxic capability. Moreover, all the studies performed employing different anuran species we presented used binary equitoxic and/or non-equitoxic mixtures. The advantage of evaluating non-equitoxic mixtures according to the methodology proposed by Warne[81] is that the toxicological interaction is drawn from four independent experimental observations, so its relevant conclusions should be considered more robust and representative of the real conditions to which organisms are exposed. The bottom-up approach employed to predict the effect of mixtures is well established and currently employed in several European Union countries and by the World Health Organization legislations.[95,96] Nevertheless, this methodology can lead to underestimating the hazard the mixture represents, as the "real" composition may not be fully known.[97] In addition, in nature, the components of the mixture under research may interact not only with each other but also with other components or factors that can alter their effect.[49,98] The challenge is to employ the information generated from mixture studies to implement proper risk management strategies.

Most of the studies performed in the laboratory found synergistic or antagonistic interactions (see Table 9.1). Nevertheless, this type of interaction is believed to be less probable to be found, at least with a significant standard deviation from additivity.[43,80,81] As for the results observed, it would be of high importance to determine whether the synergistic interaction observed between the commercial formulations evaluated after acute exposure is also present after chronic exposure, especially to low but environmentally relevant concentrations of the active principles. In addition, while most studies are carried out in a laboratory and try to extrapolate the observed effects to natural populations, with the difficulty of comparing results, others are performed under field conditions. To the best of our knowledge, only one study evaluated the effects of several pesticides, such as GLY, CPF, CYP and endosulfan, applied in different combinations on four native amphibians inhabiting the Argentinean Pampas region, such as *B. pulchellus*, *L. latrans*, *R. fernandezae* and *R. arenarum*.[49] The selected pesticides and the application techniques evaluated are commonly employed to control pests in the Pampas region. In addition, the four amphibian species studied inhabit the same areas, so the results obtained are relevant at a regional level. Their results indicate that mixtures containing insecticides reached ponds in high concentrations, causing deleterious effects on the studied species.[49]

From the previously mentioned results, we can conclude that contamination by mixtures of agrochemicals negatively affects adult anuran survival and therefore, their diversity in the Pampas region. These lethal and sublethal effects may negatively affect anuran fitness, a factor that may lead to a decline in the local population and/or changes in their distribution.[8] Therefore, it is necessary to assess the toxicity of a mixture at different proportions of each component (non-equitoxic mixtures), and different biomarkers could give more information. It is also necessary to perform the evaluation employing environmentally relevant concentrations and at different times of exposure (either both acute and chronic effects), since this reveals a more environmentally truthful situation.[87] Determining the toxic effects that pesticides induce is a challenging but necessary task that both laboratory and field scenario studies contribute to elucidate.

Acknowledgements

Funding was provided by the National University of La Plata (Grant 11/N926) and the National Council for Scientific and Technological Research (CONICET, PIP N° 3086) from Argentina.

References

1. M. P. Verhougstraete, S. L. Martin, A. D. Kendall, D. W. Hyndman and J. B. Rose, Linking fecal bacteria in rivers to landscape, geochemical, and hydrologic factors and sources at the basin scale, *Proc. Natl. Acad. Sci.*, 2015, **112**, 10419–10424.
2. R. Beiras, in *Marine Pollution, Sources, Fate and Effects of Pollutants in Coastal Ecosystems*, Elsevier, Amsterdam, 2018.
3. W. F. Carvalho, C. Ruiz de Arcaute, J. M. Pérez-Iglesias, M. R. R. Laborde, S. Soloneski and M. L. Larramendy, DNA damage exerted by mixtures of commercial formulations of glyphosate and imazethapyr herbicides in *Rhinella arenarum* (Anura, Bufonidae) tadpoles, *Ecotoxicology*, 2019, **28**, 367–377.
4. P. Amoatey and M. S. Baawain, Effects of pollution on freshwater aquatic organisms, *Water Environ. Res.*, 2019, **91**, 1272–1287.
5. FAO, FAOSTAT agri-environmental indicators/pesticides, *Food and Agriculture Organization*, 2020, Available from: http://www.fao.org/faostat/en/#data/EP. Accessed February 2023.
6. J. C. Brodeur, M. J. D'amonts, D. E. Rojas, D. Cristos, C. Vargas, M. B. Poliserpi and A. E. Andriulo, Concentration of current-use pesticides in frogs from the Pampa region and correlation of a mixture toxicity index with biological effects, *Environ. Res.*, 2022, **204**, 112354.
7. R. C. Lajmanovich, C. M. Junges, A. M. Attademo, P. M. Peltzer, M. C. Cabagna-Zenklusen and A. Basso, Individual and mixture toxicity of commercial formulations containing glyphosate, metsulfuron-methyl, bispyribac-sodium, and picloram on *Rhinella arenarum* tadpoles, *Water Air Soil Pollut.*, 2013, **224**, 1404.

8. R. P. Suárez, A. P. Goijman, S. Cappelletti, L. M. Solari, D. Cristos, D. Rojas, P. Krug, K. J. Babbitt and G. I. Gavier-Pizarro, Combined effects of agrochemical contamination and forest loss on anuran diversity in agroecosystems of east-central Argentina, *Sci. Total Environ.*, 2021, **759**, 143435.
9. F. G. Iturburu, G. Calderon, M. V. Amé and M. L. Menone, Ecological risk assessment (ERA) of pesticides from freshwater ecosystems in the Pampas region of Argentina: legacy and current use chemicals contribution, *Sci. Total Environ.*, 2019, **691**, 476–482.
10. S. López, D. Aiassa, S. Benítez-Leite, R. Lajmanovich, F. Mañas, G. Poletta, N. Sánchez, M. Simonello and A. Carrasco, Pesticides used in South American GMO-based agriculture: a review of their effects on humans and animal models, *Adv. Mol. Toxicol.*, 2012, **6**, 41–75.
11. WHO, Public health impacts of pesticides used in agriculture (WHO in collaboration with the United Nations Environment Programme, Geneva, 1990), *World Health Organization & United Nations Environment Programme*, 1990.
12. R. M. Mann and J. R. Bidwell, The toxicity of glyphosate and several glyphosate formulations to four species of Southwestern Australian frogs, *Arch. Environ. Contam. Toxicol.*, 1999, **36**, 193–199.
13. N. V. González, S. Soloneski and M. L. Larramendy, The chlorophenoxy herbicide dicamba and its commercial formulation Banvel® induce genotoxicity and cytotoxicity in Chinese hamster ovary (CHO) cells, *Mutat. Res.*, 2007, **634**, 60–68.
14. K. Nagy, R. C. Duca, S. Lovas, M. Creta, P. T. J. Scheepers, L. Godderis and B. Ádám, Systematic review of comparative studies assessing the toxicity of pesticide active ingredients and their product formulation, *Environ. Res.*, 2019, **181**, 108926.
15. N. Nikoloff, M. L. Larramendy and S. Soloneski, Assessment of DNA damage, cytotoxicity, and apoptosis in human hepatoma (HepG2) cells after flurochloridone herbicide exposure, *Food Chem. Toxicol.*, 2014, **65**, 233–241.
16. S. Soloneski, M. Kujawski, A. Scuto and M. L. Larramendy, Carbamates: A study on genotoxic, cytotoxic, and apoptotic effects induced in Chinese hamster ovary (CHO-K1) cells, *Toxicol. In Vitro*, 2015, **29**, 834–844.
17. S. Soloneski and M. L. Larramendy, Sister chromatid exchanges and chromosomal aberrations in Chinese hamster ovary (CHO-K1) cells treated with the insecticide pirimicarb, *J. Hazard. Mater.*, 2010, **174**, 410–415.
18. S. Soloneski, N. Nikoloff and M. L. Larramendy, Analysis of possible genotoxicity of the herbicide flurochloridone and its commercial formulations: Endo III and Fpg alkaline comet assays in Chinese hamster ovary (CHO-K1) cells, *Mutat. Res.*, 2016, **797**, 46–52.

19. D. Zeljezic, V. Garaj-Vrhovac and P. Perkovic, Evaluation of DNA damage induced by atrazine and atrazine-based herbicide in human lymphocytes *in vitro* using a comet and DNA diffusion assay, *Toxicol. In Vitro*, 2006, **20**, 923–935.
20. J. E. Primost, D. J. G. Marino, V. C. Aparicio, J. L. Costa and P. Carriquiriborde, Glyphosate and AMPA, "pseudo-persistent" pollutants under real-world agricultural management practices in the Mesopotamic Pampas agroecosystem, Argentina, *Environ. Pollut.*, 2017, **229**, 771–779.
21. A. E. Ronco, D. J. G. Marino, M. Abelando, P. Almada and C. D. Apartin, Water quality of the main tributaries of the Paraná Basin: glyphosate and AMPA in surface water and bottom sediments, *Environ. Monit. Assess.*, 2016, **188**, 458.
22. M. A. Etchegoyen, A. E. Ronco, P. Almada, M. Abelando and D. Marino, Occurrence and fate of pesticides in the argentine stretch of the Paraguay-Paraná basin, *Environ. Monit. Assess.*, 2017, **189**, 63.
23. M. B. Medina, M. S. Munitz and S. L. Resnik, Validation and expanded uncertainty determination of pesticides in water and their survey on paddy rice irrigation water from Argentina, *J. Environ. Sci. Health. B*, 2020, **55**, 983–989.
24. T. M. Mac Loughlin, M. L. Peluso and D. J. G. Marino, Multiple pesticides occurrence, fate, and environmental risk assessment in a small horticultural stream of Argentina, *Sci. Total Environ.*, 2022, **802**, 149893.
25. D. J. G. Marino and A. E. Ronco, Cypermethrin and chlorpyrifos concentration levels in surface water bodies of the Pampa Ondulada, Argentina, *Bull. Environ. Contam. Toxicol.*, 2005, **75**, 820–826.
26. M. G. Agostini, F. Kacoliris, P. Demetrio, G. S. Natale, C. Bonetto and A. E. Ronco, Abnormalities in amphibian populations inhabiting agroecosystems in Northeastern Buenos Aires Province, Argentina, *Dis. Aquat. Org.*, 2013, **104**, 163–171.
27. S. Corcoran, C. D. Metcalfe, T. Sultana, M. V. Amé and M. L. Menone, Pesticides in surface waters in Argentina monitored using polar organic chemical integrative samplers, *Bull. Environ. Contam. Toxicol.*, 2020, **104**, 21–26.
28. M. Menone, J. Aizpún de Moreno, V. J. Moreno, A. L. Lafranchi, T. Metcalfe and C. D. Metcalfe, Organochlorine pesticides and PCBs in a Southern Atlantic coastal lagoon watershed, Argentina, *Arch. Environ. Contam. Toxicol.*, 2001, **40**, 355–362.
29. D. J. Pérez, E. Okada, E. De Gerónimo, M. L. Menone, V. C. Aparicio and J. L. Costa, Spatial and temporal trends and flow dynamics of glyphosate and other pesticides within an agricultural watershed in Argentina, *Environ. Toxicol. Chem.*, 2017, **36**, 3206–3216.
30. N. Urseler, R. Bachetti, V. Morgante, E. Agostini and C. Morgante, Groundwater quality and vulnerability in farms from agricultural-dairy

basin of the Argentine Pampas, *Environ. Sci. Pollut. Res.*, 2022, **29**, 63655–63673.

31. P. J. Peruzzo, A. Porta and A. E. Ronco, Levels of glyphosate in surface waters, sediments and soils associated with direct sowing soybean cultivation in North pampasic region of Argentina, *Environ. Pollut.*, 2008, **156**, 61–66.
32. V. C. Aparicio, E. De Gerónimo, D. Marino, J. Primost, P. Carriquiriborde and J. L. Costa, Environmental fate of glyphosate and aminomethylphosphonic acid in surface waters and soil of agricultural basins, *Chemosphere*, 2013, **93**, 1866–1873.
33. L. Lupi, K. S. B. Miglioranza, V. C. Aparicio, D. Marino, F. Bedmar and D. A. Wunderlin, Occurrence of glyphosate and AMPA in an agricultural watershed from the Southeastern region of Argentina, *Sci. Total Environ.*, 2015, **536**, 687–694.
34. E. De Gerónimo, V. C. Aparicio, S. Bárbaro, R. Portocarrero, S. Jaime and J. L. Costa, Presence of pesticides in surface water from four subbasins in Argentina, *Chemosphere*, 2014, **107**, 423–431.
35. N. J. Hawkins, C. Bass, A. Dixon and P. Neve, The evolutionary origins of pesticide resistance, *Biol. Rev.*, 2019, **94**, 135–155.
36. H. Ge, Z. Lin, Z. Yao, Y. Gao, Y. Cong and H. Yu, Balance between herbicidal activity and toxicity effect: a case study of the joint effects of triazine and phenylurea herbicides on *Selenastrum capricornutum* and *Photobacterium phosphoreum*, *Aquat. Toxicol.*, 2014, **150**, 165–174.
37. S. K. Bopp, A. Kienzler, A.-N. Richarz, S. C. Van der Linden, A. Paini, N. Parissis and A. P. Worth, Regulatory assessment and risk management of chemical mixtures: challenges and ways forward, *Crit. Rev. Toxicol.*, 2019, **49**, 174–189.
38. K. A. Heys, R. F. Shore, M. G. Pereira, K. C. Jones and F. L. Martin, Risk assessment of environmental mixture effects, *RSC Adv.*, 2016, **6**, 47844–47857.
39. M. Lydy, J. Belden, C. Wheelock, B. Hammock and D. Denton, Challenges in regulating pesticide mixtures, *Ecol. Soc.*, 2004, **9**, 1–13.
40. E. Drakvik, R. Altenburger, Y. Aoki, T. Backhaus, T. Bahadori, R. Barouki, W. Brack, M. T. D. Cronin, B. Demeneix, S. Hougaard Bennekou, J. van Klaveren, C. Kneuer, M. Kolossa-Gehring, E. Lebret, L. Posthuma, L. Reiber, C. Rider, J. Rüegg, G. Testa, B. van der Burg, H. van der Voet, A. M. Warhurst, B. van de Water, K. Yamazaki, M. Öberg and Å. Bergman, Statement on advancing the assessment of chemical mixtures and their risks for human health and the environment, *Environ. Int.*, 2020, **134**, 105267.
41. A. Kortenkamp and M. Faust, Regulate to reduce chemical mixture risk, *Science*, 2018, **361**, 224–226.

42. A. Kortenkamp, M. Faust, T. Backhaus, R. Altenburger, M. Scholze, C. Müller, S. Ermler, L. Posthuma and W. Brack, Mixture risks threaten water quality: the European Collaborative Project SOLUTIONS recommends changes to the WFD and better coordination across all pieces of European chemicals legislation to improve protection from exposure of the aquatic environment to multiple pollutants, *Environ. Sci. Eur.*, 2019, **31**, 69.
43. C. M. J. Brodeur, M. B. Poliserpi, M. F. DAndrea and M. Sánchez, Synergy between glyphosate- and cypermethrin-based pesticides during acute exposures in tadpoles of the common South American toad *Rhinella arenarum*, *Chemosphere*, 2014, **112**, 70–75.
44. J. C. Brodeur, S. Melpel, A. B. Anglesio, D. Cristos, M. F. D'Andrea and M. B. Poliserpi, Toxicities of glyphosate- and cypermethrin-based pesticides are antagonic in the tenspotted livebearer fish (*Cnesterodon decemmaculatus*), *Chemosphere*, 2016, **155**, 429–435.
45. C. Ruiz de Arcaute, J. C. Brodeur, S. Soloneski and M. L. Larramendy, Toxicity to *Rhinella arenarum* tadpoles (Anura, Bufonidae) of herbicide mixtures commonly used to treat fallow containing resistant weeds: glyphosate–dicamba and glyphosate–flurochloridone, *Chemosphere*, 2020, 125623.
46. C. Ruiz de Arcaute, S. Soloneski and M. L. Larramendy, Synergism of mixtures of dicamba and 2,4-dichlorophenoxyacetic acid herbicide formulations on the neotropical fish *Cnesterodon decemmaculatus* (Pisces, Poeciliidae), *Environ. Pollut.*, 2018, **236**, 33–39.
47. A. P. Cuzziol Boccioni, G. Lener, J. Peluso, P. M. Peltzer, A. M. Attademo, C. Aronzon, M. F. Simoniello, L. D. Demonte, M. R. Repetti and R. C. Lajmanovich, Comparative assessment of individual and mixture chronic toxicity of glyphosate and glufosinate ammonium on amphibian tadpoles: A multibiomarker approach, *Chemosphere*, 2022, **309**, 136554.
48. W. Hopkins and C. L. Rowe, in *Ecotoxicology of Reptiles and Amphibians*, SETAC Press, Pensacola, FL, 2010, p. 325.
49. M. G. Agostini, I. Roesler, C. Bonetto, A. E. Ronco and D. Bilenca, Pesticides in the real world: The consequences of GMO-based intensive agriculture on native amphibians, *Biol. Conserv.*, 2020, **241**, 108355.
50. A. Egea-Serrano, R. A. Relyea, M. Tejedo and M. Torralva, Understanding of the impact of chemicals on amphibians: A meta-analytic review, *Ecol. Evol.*, 2012, **2**, 1382–1397.
51. L. Kehoe, T. Kuemmerle, C. Meyer, C. Levers, T. Václavík and H. Kreft, Global patterns of agricultural land-use intensity and vertebrate diversity, *Divers. Distrib.*, 2015, **21**, 1308–1318.
52. J. Peluso, C. M. Aronzon and C. S. Pérez Coll, Assessment of environmental quality of water bodies next to agricultural areas of Buenos

Aires province (Argentina) by means of ecotoxicological studies with *Rhinella arenarum*, *J. Environ. Sci. Heal. B*, 2019, **54**, 655–664.

53. M. Usal, S. Veyrenc, M. Darracq-Ghitalla-Ciock, S. Sroda, J. B. Fini, M. Raveton and S. Reynaud, Exposure to a mixture of benzo[a]pyrene and triclosan induces multi-and transgenerational metabolic disorders associated with decreased female investment in reproduction in *Silurana (Xenopus) tropicalis*, *Environ. Pollut.*, 2022, **292**, 118418.
54. IUCN, International Union for Conservation of Nature, Red list of threatened species, 2023, Available from: https://www.iucnredlist.org/. Accessed 7 February 2023.
55. M. G. Agostini, G. S. Natale and A. E. Ronco, Impact of endosulphan and cypermethrin mixture on amphibians under field use for biotech soya bean production, *Int. J. Environ. Health Res.*, 2009, **3**, 379–389.
56. M. G. Agostini, G. S. Natale and A. E. Ronco, Lethal and sublethal effects of cypermethrin to *Hypsiboas pulchellus* tadpoles, *Ecotoxicology*, 2010, **19**, 1545–1550.
57. G. Berger, F. Graef and H. Pfeffer, Temporal coincidence of migrating amphibians with mineral fertiliser applications on arable fields, *Agric. Ecosys. Environ.*, 2012, **155**, 62–69.
58. R. C. Lajmanovich, C. M. Junges, M. C. Cabagna-Zenklusen, A. M. Attademo, P. M. Peltzer and M. Maglianese, Toxicity of *Bacillus thuringiensis var. israelensis* in aqueous suspension on the South American common frog *Leptodactylus latrans* (Anura: Leptodactylidae) tadpoles, *Environ. Res.*, 2015, **136**, 205–212.
59. R. C. Lajmanovich, M. T. Sandoval and P. M. Peltzer, Induction of mortality and malformation in *Scinax nasicus* tadpoles exposed to glyphosate formulations, *Bull. Environ. Contam. Toxicol.*, 2003, **70**, 612–618.
60. J. M. Pérez-Iglesias, L. Franco-Belussi, G. S. Natale and C. de Oliveira, Biomarkers at different levels of organisation after atrazine formulation (SIPTRAN 500SC®) exposure in *Rhinella schineideri* (Anura: Bufonidae) Neotropical tadpoles, *Environ. Pollut.*, 2019, **244**, 733–746.
61. J. M. Pérez-Iglesias, C. Ruiz de Arcaute, N. Nikoloff, L. Dury, S. Soloneski, G. S. Natale and M. L. Larramendy, The genotoxic effects of the imidacloprid-based insecticide formulation Glacoxan Imida on Montevideo tree frog *Hypsiboas pulchellus* tadpoles (Anura, Hylidae), *Ecotoxicol. Environ. Saf.*, 2014, **104**, 120–126.
62. J. Vera-Candioti, G. Natale, S. Soloneski, A. E. Ronco and M. L. Larramendy, Sublethal and lethal effects on *Rhinella arenarum* (Anura, Bufonidae) tadpoles exerted by the pirimicarb-containing technical formulation insecticide Aficida®, *Chemosphere*, 2010, **78**, 249–255.

63. G. V. Svartz and C. S. Pérez-Coll, Comparative toxicity of cypermethrin and a commercial formulation on *Rhinella arenarum* larval development (Anura: Bufonidae), *Environ. Health*, 2013, **6**, 320–329.

64. M. Acquaroni, J. Peluso, G. Svartz, C. Aronzon and C. Pérez Coll, Characterization of acute toxicity, genotoxicity, and oxidative stress of dimethoate in *Rhinella arenarum* larvae, *Environ. Sci. Pollut. Res.*, 2021, **28**, 41772–41779.

65. M. Acquaroni, G. Svartz and C. Pérez Coll, Acute, chronic and neurotoxic effects of dimethoate pesticide on *Rhinella arenarum* throughout the development, *J. Environ. Sci. Health, B*, 2022, **57**, 142–152.

66. C. M. Aronzon, M. T. Sandoval, J. Herkovits and C. Pérez-Coll, Stage-dependent toxicity of 2,4-dichlorophenoxyacetic on the embryonic development of a South American toad, *Rhinella arenarum*, *Environ. Toxicol. Chem.*, 2011, **26**, 373–381.

67. C. M. J. Brodeur, G. V. Svartz, C. S. Perez-Coll, D. J. G. Marino and J. Herkovits, Comparative susceptibility to atrazine of three developmental stages of *Rhinella arenarum* and influence on metamorphosis: Non-monotonous acceleration of the time to climax and delayed tail resorption, *Aquat. Toxicol.*, 2009, **91**, 161–170.

68. R. C. Lajmanovich, A. M. Attademo, P. M. Peltzer, C. M. Junges and M. C. Cabagna, Toxicity of four herbicide formulations with glyphosate on *Rhinella arenarum* (Anura: Bufonidae) tadpoles: B-esterases and glutathione S-transferase inhibitors, *Arch. Environ. Contam. Toxicol.*, 2011, **60**, 681–689.

69. N. Liendro, A. Ferrari, M. Mardirosian, C. I. Lascano and A. Venturino, Toxicity of the insecticide chlorpyrifos to the South American toad *Rhinella arenarum* at larval developmental stage, *Environ. Toxicol. Pharmacol.*, 2015, **39**, 525–535.

70. C. S. Martinuzzi, A. M. Attademo, P. M. Peltzer, T. M. Mac Loughlin, D. J. G. Marino and R. C. Lajmanovich, Comparative toxicity of two different dimethoate formulations in the common toad (*Rhinella arenarum*) tadpoles, *Bull. Environ. Contam. Toxicol.*, 2020, **104**, 35–40.

71. B. Bosch, F. Mañas, N. Gorla and D. Aiassa, Micronucleus test in post metamorphic *Odontophrynus cordobae* and *Rhinella arenarum* (Amphibia: Anura) for environmental monitoring, *J. Toxicol. Environ. Health Sci.*, 2011, **3**, 155–163.

72. R. C. Lajmanovich, A. M. Attademo, M. F. Simoniello, G. L. Poletta, C. M. Junges, P. M. Peltzer, P. Grenón and M. C. Cabagna-Zenklusen, Harmful effects of the dermal intake of commercial formulations containing chlorpyrifos, 2,4-D, and glyphosate on the common toad *Rhinella arenarum* (Anura: Bufonidae), *Water Air Soil Pollut.*, 2015, **226**, 427.

73. R. C. Lajmanovich, M. C. Cabagna-Zenklusen, A. M. Attademo, C. M. Junges, P. M. Peltzer, A. Bassó and E. Lorenzatti, Induction of

micronuclei and nuclear abnormalities in tadpoles of the common toad (*Rhinella arenarum*) treated with the herbicides Liberty® and glufosinate-ammonium, *Mutat. Res.*, 2014, **769**, 7–12.

74. V. Sotomayor, T. S. Chiriotto, A. M. Pechen and A. Venturino, Biochemical biomarkers of sublethal effects in *Rhinella arenarum* late gastrula exposed to the organophosphate chlorpyrifos, *Pestic. Biochem. Phys.*, 2015, **219**, 48–53.

75. A. Ferrari, C. Lascano, A. M. Pechen de D'Angelo and A. Venturino, Antioxidant responses to azinphos methyl and carbaryl during the embryonic development of the toad *Rhinella* (Bufo) *arenarum* Hensel, *Comp. Biochem. Physiol., C, Toxicol. Pharmacol.*, 2011, **153**, 34–39.

76. J. A. Guzman and T. Guardia, Effects of an organophosphorous insecticide on the cholinesteratic activities of *Bufo arenarum* (H), *Bull. Environ. Contam. Toxicol.*, 1978, **20**, 52–58.

77. A. Ferrari, C. I. Lascano, O. L. Anguiano, A. M. Pechen de D'Angelo and A. Venturino, Antioxidant responses to azinphos methyl and carbaryl during the embryonic development of the toad *Rhinella* (Bufo) *arenarum* Hensel, *Aquat. Toxicol.*, 2009, **93**, 37–44.

78. C. I. Lascano, A. Ferrari and A. Venturino, Sublethal concentrations of azinphos-methyl induce biochemical and morphological alterations in *Rhinella arenarum* embryos, *Chem. Ecol.*, 2011, **27**, 557–568.

79. A. F. Hernández, F. Gil and M. Lacasaña, Toxicological interactions of pesticide mixtures: an update, *Arch. Toxicol.*, 2017, **91**, 3211–3223.

80. B. I. Escher, H. M. Stapleton and E. L. Schymanski, Tracking complex mixtures of chemicals in our changing environment, *Sci. Total Environ.*, 2020, **367**, 388–392.

81. M. J. Warne, in *A review of the ecotoxicity of mixtures, approaches to, and recommendations for their management*, Environment Protection & Health Council, Adelaide, 2003, pp. 253–276.

82. O. Martin, M. Scholze, S. Ermler, J. McPhie, S. K. Bopp, A. Kienzler, N. Parissis and A. Kortenkamp, Ten years of research on synergisms and antagonisms in chemical mixtures: A systematic review and quantitative reappraisal of mixture studies, *Environ. Int.*, 2021, **146**, 106206.

83. T. K. Reffstrup, J. C. Larsen and O. Meyer, Risk assessment of mixtures of pesticides. Current approaches and future strategies, *Regul. Toxicol. Pharmacol.*, 2010, **56**, 174–192.

84. F. A. Pavan, C. G. Samojeden, C. F. Rutkoski, A. Folador, S. P. Da Fré, C. Müller, P. A. Hartmann and M. T. Hartmann, Morphological, behavioral and genotoxic effects of glyphosate and 2,4-D mixture in tadpoles of two native species of South American amphibians, *Environ. Toxicol. Pharmacol.*, 2021, **85**, 103637.

85. C. M. Aronzon, G. V. Svartz and C. S. Pérez Coll, Synergy between diazinon and nonylphenol in toxicity during the early development of the *Rhinella arenarum* toad, *Water Air Soil Pollut.*, 2016, **227**, 139.

86. G. V. Svartz, C. M. Aronzon and C. S. Pérez Coll, Combined endosulfan and cypermethrin-induced toxicity to embryo-larval development of *Rhinella arenarum*, *J. Toxicol. Environ. Health A*, 2016, **79**, 197–209.

87. J. Peluso, A. F. Lanuza, C. S. Pérez Coll and C. M. Aronzon, Synergistic effects of glyphosate- and 2,4-D-based pesticide mixtures on *Rhinella arenarum* larvae, *Environ. Sci. Pollut. Res.*, 2021, **29**, 14443–14452.

88. L. L. Marking, in *Fundamentals on aquatic toxicology, methods and applications, Toxicity of chemical mixtures*, G. M. Corporation, 1985, pp. 164–176.

89. USEPA, *Hazard evaluation: wildlife and aquatic organisms*, United States Environmental Protection Agency, Washington DC, 1982, EPA-540/9-82-024.

90. C. Ruiz de Arcaute, J. M. Pérez-Iglesias, N. Nikoloff, G. S. Natale, S. Soloneski and M. L. Larramendy, Genotoxicity evaluation of the insecticide imidacloprid on circulating blood cells of Montevideo tree frog *Hypsiboas pulchellus* tadpoles (Anura, Hylidae) by comet and micronucleus bioassays, *Ecol. Indic.*, 2014, **45**, 632–639.

91. N. C. Bach, D. J. G. Marino, G. S. Natale and G. M. Somoza, Effects of glyphosate and its commercial formulation, Roundup® Ultramax, on liver histology of tadpoles of the neotropical frog, *Leptodactylus latrans* (amphibia: Anura), *Chemosphere*, 2018, **202**, 289–297.

92. G. Chaufan, I. Coalova and M. C. Ríos de Molina, Glyphosate commercial formulation causes cytotoxicity, oxidative effects, and apoptosis on human cells: differences with its active ingredient, *Int. J. Toxicol.*, 2014, **33**, 29–38.

93. S. D. Gomez, P. S. Bustos, V. G. Sánchez, M. G. Ortega and N. Guiñazú, Trophoblast toxicity of the neonicotinoid insecticide acetamiprid and an acetamiprid-based formulation Author links open overlay panel, *Toxicology*, 2020, **431**, 152363.

94. M. C. Sobrero, F. Rimoldi and A. E. Ronco, Effects of the glyphosate active ingredient and a formulation on *Lemna gibba* L. at different exposure levels and assessment end-points, *Bull. Environ. Contam. Toxicol.*, 2007, **79**, 537–543.

95. M. E. Meek, A. R. Boobis, K. M. Crofton, G. Heinemeyer, M. Van Raaij and C. Vickers, Risk assessment of combined exposure to multiple chemicals: A WHO/IPCS framework, *Regul. Toxicol. Pharmacol.*, 2011, **60**, 1–14.

96. P. Prince, E. Dhein, M. Hamer, X. Han, M. Heneweer, M. Junghans, P. Kunz, C. Magyar, H. Penning and C. Rodriguez, A decision tree for

assessing effects from exposures to multiple substances, *Environ. Sci. Eur.*, 2012, **24**, 26.

97. P. A. Neale, S. Ait-Aissa, W. Brack, N. Creusot, M. S. Denison, B. Deutschmann, K. Hilscherová, H. Hollert, M. Krauss, J. Novák, T. Schulze, T. B. Seiler, H. Serra, Y. Shao and B. I. Scher, Linking *in vitro* effects and detected organic micropollutants in surface water using mixture-toxicity modeling, *Environ. Sci. Technol.*, 2015, **49**, 14614–14624.
98. A. C. Johnson and J. P. Sumpter, Are we going about chemical risk assessment for the aquatic environment the wrong way?, *Environ. Toxicol. Chem.*, 2016, **35**, 1609–1616.

CHAPTER 10

The Direct-developing Frog Eleutherodactylus johnstonei: A Promising but Underused Amphibian Model for Ecological Risk Assessment

FABIO LEONARDO MEZA-JOYA*[a], LUIS FELIPE ATUESTA-PALOMINO[b] AND JORGE LUIS FUENTES*[b]

[a]Wildlife & Ecology Group, Massey University, School of Natural Sciences, Private Bag 11-222, Palmerston North, New Zealand; [b]Laboratorio de Microbiología y Mutagénesis Ambiental, Grupo de Investigación en Microbiología y Genética, Universidad Industrial de Santander, Escuela de Biología, 680001, Bucaramanga, Colombia
*E-mail: f.l.mezajoya@massey.ac.nz, jfuentes@uis.edu.co, jlfl2014@live.com

10.1 Introduction

Although considered a globally declining taxon,[1,2] amphibians (Anura: frogs and toads; Caudata: salamanders and newts; and Gymnophiona: caecilians) are not routinely incorporated in ecological risk assessment, especially in

Issues in Toxicology No. 46
Amphibian Species in Environmental Risk Assessment Strategies
Edited by Marcelo L. Larramendy and Guillermo Eli Liwszyc

Published by the Royal Society of Chemistry, www.rsc.org

a regulatory context. Rather, amphibian 'typical' life stages (*i.e.*, aquatic larvae) and terrestrial forms (*i.e.*, post-metamorphic phases and direct-developing species) are assumed to be covered by surrogate toxicity data from fish or mammals and birds, respectively.[3–7] While recent studies point out the possibility of using fish as surrogates for aquatic-phase amphibians, data from birds and mammals provide little evidence that these organisms can effectively reflect toxicity in terrestrial amphibians.[6,7] This mismatched surrogacy may introduce biases and lead to spurious conclusions regarding the sensibility of land-based amphibians to environmental pollutants.

Research on amphibian sensitivity to environmental pollutants is on the rise, yet a large data gap persists and many substances await testing.[6–10] Although researchers are increasingly using alternate endpoints that report potential sublethal effects (*e.g.*, behavioral, metabolic, cellular, and molecular), most of the existing ecotoxicological literature for this group is based on acute toxicity endpoints on anurans (*e.g.*, Howe *et al.*,[11] Relyea,[12] Relyea and Jones,[13] Linger *et al.*,[14] Moore *et al.*[15]). Exposing embryos or larval stages of metamorphic species to waterborne pollutants under controlled conditions is a common practice, as these early stages are assumed to be more sensitive to toxicants than adults,[8,9] which impedes accurately predicting the relative sensitivity of terrestrial forms to environmental toxicants.[6] A great deal of variation in amphibian life history and morphology further challenges efforts to identify a candidate amphibian model species to represent the whole group in risk assessment.[5,6]

Despite long-standing and well-recognized data gaps, the ecotoxicological literature on amphibians is vast, and a systematic review of the role of this group in ecological risk assessment is outside the scope of this chapter. Interested readers can find comprehensive reviews compiling toxicity data from amphibians in other recent publications (*e.g.*, Ortiz-Santaliestra *et al.*,[6] Glaberman *et al.*[7]). Therefore, we focused on outlining the relative sensitivity of the direct-developing frog *Eleutherodactylus johnstonei* (see Figure 10.1), commonly known as the Lesser Antillean whistling frog, to chemical and physical stressors in the context of amphibian risk assessment. To achieve this goal, this chapter is divided into three sections evaluating the role of this species in ecotoxicology—pesticide toxicity, chemical cyto-genotoxicity, and ultraviolet (UV) radiation-induced cyto-genotoxicity—based on existing and novel data. In each section, we summarized some commonly used methods and compared the sensitivity of relevant ecotoxicological endpoints between *E. johnstonei* and other amphibian species.

10.2 The Lesser Antillean Whistling Frog in Environmental Risk Assessment

Eleutherodactylus johnstonei, a Caribbean frog native of the Lesser Antilles, has a long history of human-mediated range expansion (reviewed in Kaiser,[16] Meza-Joya *et al.*[17]). Today, this species' invasive range greatly

Figure 10.1 Adult male (top) and female (bottom) of *Eleutherodactylus johnstonei* from Parque de las Orquídeas, Bucaramanga, Colombia.

surpasses its limited native distribution, making it considered a very adaptable and highly invasive taxon.[18] These features are germane for chemical screening as this frog's plasticity allows housing under laboratory conditions at low maintenance costs,[17] while its invasive nature and low-risk status ease mandatory research permits when required. Besides this, the direct-developing life cycle of *E. johnstonei* is radically different from the ancestral biphasic mode observed in most living frogs, including *Xenopus laevis*, the most popular amphibian model in standard toxicological studies.[5–7] Unlike metamorphosing frogs, direct developers are emancipated from aquatic breeding and hatch as terrestrial froglets.[19] Therefore, *E. johnstonei* might be relevant for risk assessment in the terrestrial environment.[17] The sensitivity of different ecotoxicity test endpoints to some widely used pesticides studied on this frog lends weight to this idea (*e.g.*, Linzey *et al.*,[3] Meza-Joya *et al.*[20]).

10.2.1 Pesticide Toxicity

Pesticide pollution has been pointed out as a primary ecological issue contributing to the global amphibian decline in both natural and agricultural areas.[21–23] To provide quantitative data about the direct impacts of pesticides on non-target organisms, the common toxicological practice is to conduct single-species acute toxicity tests under laboratory conditions—usually exposing larval stages of metamorphosing frogs to a given pesticide for 96 h—to generate dose-response curves and estimate median lethal concentrations (LC_{50}), the pesticide concentration that kills half the members of the exposed population.[10,24,25] For risk assessment, toxicity data obtained from these acute tests are compared to existing data across multiple species to assess the relative toxicity of a given pesticide and set a safe threshold for its concentrations in nature.[10,25] Although critical for risk monitoring, toxicity data represent only a part of the equation as risk; in its simplest form, it is a function of a toxicant's potential effects by the likelihood of being exposed to it.[26]

Considering their distinctive terrestrially, direct-developing frogs—including *E. johnstonei*—are prone to direct exposure to airborne agrochemicals in field situations, as are terrestrial post-metamorphic amphibians.[17,20] Although under certain field spraying conditions, pesticide interception by vegetation canopy would reduce dermal exposure risk (*e.g.*, Lynch and Arroyo,[27] Rodríguez-Gil *et al.*[28]), which would not always be the case as the less dense the canopy, the more exposure risk (*e.g.*, Lynch and Arroyo[27]). Not to mention that some canopy dwellers might experience enhanced exposure in such situations. Hence, the direct uptake route across the skin caused by overspraying is highly relevant,[4] especially for terrestrial forms in open habitats or aquatic forms in shallow wetlands (*e.g.*, Relyea,[25] Solomon *et al.*[29]). Exposure through other routes (*e.g.*, *via* food intake) is also germane, as it might result in enhanced chemical uptake[4] and bioaccumulation.[3] Sensitivity is also a critical but often an overlooked part of the risk equation, as toxicity risk is also determined by the interaction between exposure and susceptibility to a given toxicant.

Over the last decade or so, interest in *E. johnstonei* as a model for ecological risk assessment of amphibians has been encouraged,[17] given its sensitivity to pesticides in laboratory[20] and outdoor studies.[3] Field monitoring of introduced wild populations of *E. johnstonei* revealed residues of organochlorine pesticides (up to 1.2 μg g^{-1}) in the livers and fat bodies of this frog, likely resulting from food intake and posterior bioaccumulation.[3] More recently, laboratory bioassays investigating the toxicity of a glyphosate (GLY)-based formulation oversprayed directly on adults and neonates of this species[20] found that nominal LC_{50}-96 h (4.9 and 5.4 kg a.e. ha^{-1} for males and females, respectively) were in range with those reported for native terrestrial frogs,[30] while that for the neonates was lower (1.2 kg a.e. ha^{-1}). Despite these promising results, there is still reticence toward utilizing this species for amphibian chemical screening; the reasons for this are unknown, but they could be related to the dearth of toxicity data for terrestrial amphibians as a whole.

Today, pesticide toxicity data for *E. johnstonei* is restricted to the GLY-based formulation analyzed by Meza-Joya *et al.*,[20] which has been used for the eradication of illicit coca crops in Colombia. This formulation is a mixture (55 : 44 : 1) of water, Roundup® SL—containing 360 g L^{-1} of acid equivalent as isopropylamine (IPA) salt of GLY and polyoxyethylene amine (POEA) surfactant—and the agricultural adjuvant Cosmoflux®411F—containing light petroleum oil. Toxicity data for equivalent GLY-Cosmoflux mixtures are only available for a few Colombian anurans,[30,31] but allow for direct comparisons in order to assess variation in species' sensitivity to this chemical cocktail (see Table 10.1). For this purpose, we generated species sensitivity distributions (SSDs) for aquatic-phased (*i.e.*, Gosner's[32] stage 25 tadpoles) and terrestrial-phased anurans (*i.e.*, neonates, juveniles, and adults) by fitting toxicity data (*i.e.*, LC_{50}-96 h) *via* maximum likelihood estimation to six possible distributions (log-normal, log-logistic, Weibull, Pareto, gamma, or exponential) using the R package 'SSD tools' v1.0.2.[34] Throughout this chapter, we will follow the updated amphibian taxonomy proposed by Frost.[33]

These SSDs indicate that *E. johnstonei* is, in general, equally or more sensitive to the tested chemical cocktail than other adult and juvenile terrestrial anurans (see Figure 10.2). Toxicity values for this species are very much in line with those of other native terrestrial frogs, although neonates are highly sensitive at concentrations that might be considered environmentally relevant (*i.e.*, HC_5). Toxicity estimates for adult *E. johnstonei* closely relate to those of the direct-developing frog *Pristimantis taeniatus* (Anura, Strabomantidae). Terrestrial juveniles of the centrolenid frog *Espadarana prosoblepon* (Anura, Centrolenidae) are also highly sensitive, although aquatic tadpoles of this species show the opposite trend. In aquatic assays, tadpoles of the hylid frog

Table 10.1 Summary of median lethal concentrations at 96 h (LC_{50}-96 h) for *Eleutherodactylus johnstonei* and other anurans after *in vivo* exposure to equivalent GLY formulations-Cosmoflux®411F. Tadpole life stage refers to Gosner's[32] stage 25 larvae. Confidence limit values are shown in parenthesis (when reported). GLY-based formulations compared here (*i.e.*, Glyphos® and Roundup® SL) contain POEA surfactant and similar concentrations of active ingredient: 354 and 360 grams of GLY acid equivalent (as IPA salt) per liter (g a.e. L^{-1}), respectively. Species names follow Frost.[33]

Species	Family	Life cycle	Life stage	Formulation	Assay type	Exposure pathway	LC_{50}-96 h	Source
(a) Terrestrial-phased anurans: LC_{50} (kg a.e. ha^{-1}, kilograms of glyphosate acid equivalent per hectare)								
Eleutherodactylus johnstonei	Eleutherodactylidae	Direct	Neonate	Roundup SL	Bioassay	Overspray	1.2 (1.1–1.4)	20
Espadarana prosoblepon	Centrolenidae	Biphasic	Juvenile	Glyphos	Microcosm	Overspray	4.5 (n.a.)	30
Rhinella humboldti	Bufonidae	Biphasic	Juvenile	Glyphos	Microcosm	Overspray	6.5 (n.a.)	30
Scinax ruber	Hylidae	Biphasic	Juvenile	Glyphos	Microcosm	Overspray	7.3 (n.a.)	30
Rhinella sterno-signata	Bufonidae	Biphasic	Juvenile	Glyphos	Microcosm	Overspray	14.8 (n.a.)	30
Engystomops pustulosus	Leptodactylidae	Biphasic	Juvenile	Glyphos	Microcosm	Overspray	19.6 (n.a.)	30
Rhinella horribilis	Bufonidae	Biphasic	Juvenile	Glyphos	Microcosm	Overspray	22.8 (n.a.)	30
Eleutherodactylus johnstonei	Eleutherodactylidae	Direct	Adult	Roundup SL	Bioassay	Overspray	5.1 (4.5–5.5)	20
Pristimantis taeniatus	Strabomantidae	Direct	Adult	Glyphos	Microcosm	Overspray	5.6 (n.a.)	30
(b) Aquatic-phased anurans: LC_{50} (mg a.e L^{-1}, milligrams of glyphosate acid equivalent per litre)								
Dendropsophus microcephalus	Hylidae	Biphasic	Tadpole	Glyphos	Microcosm	Waterborne	1.2 (n.a.)	31
Rhinella sterno-signata	Bufonidae	Biphasic	Tadpole	Glyphos	Microcosm	Waterborne	1.5 (n.a.)	31
Scinax ruber	Hylidae	Biphasic	Tadpole	Glyphos	Microcosm	Waterborne	1.6 (1.5–1.8)	31

(*continued*)

Table 10.1 (*continued*)

Species	Family	Life cycle	Life stage	Formulation	Assay type	Exposure pathway	LC_{50}-96 h	Source
Boana platanera	Hylidae	Biphasic	Tadpole	Glyphos	Microcosm	Waterborne	2.1 (1.8–2.3)	31
Rhinella humboldti	Bufonidae	Biphasic	Tadpole	Glyphos	Microcosm	Waterborne	2.3 (2.0–2.6)	31
Espadarana prosoblepon	Centrolenidae	Biphasic	Tadpole	Glyphos	Microcosm	Waterborne	2.4 (n.a.)	31
Rhinella horribilis	Bufonidae	Biphasic	Tadpole	Glyphos	Microcosm	Waterborne	2.7 (2.5–3.0)	31
Engystomops pustulosus	Leptodactylidae	Biphasic	Tadpole	Glyphos	Microcosm	Waterborne	2.8 (2.5–3.1)	31

n.a., Not available.

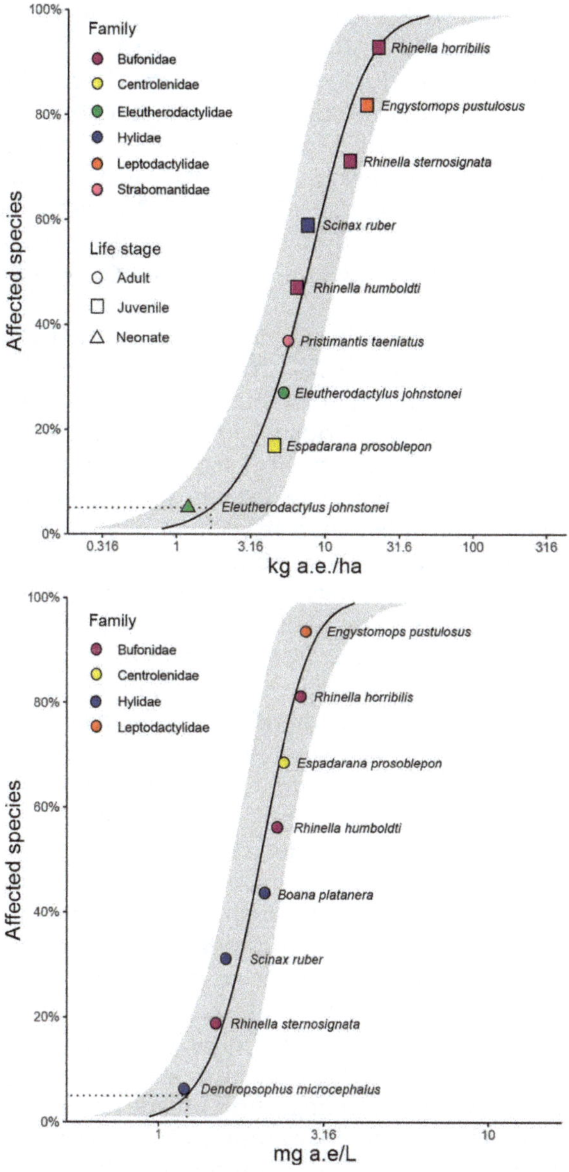

Figure 10.2 Species sensitivity distributions (SSDs) for the acute toxicity of the GLY active ingredient to terrestrial- (top) and aquatic-phased amphibians (bottom). All data were normalized to acid equivalents (a.e.) to allow direct comparison (Table 10.1). Confidence intervals (95%) for the distribution parameters and distribution estimates were calculated *via* 10 000 parametric bootstrapping iterations. The dashed line represents the model-averaged hazard concentration for 5% of species (HC_5), the typically used threshold for risk assessment.

Dendropsophus microcephalus (Anura, Hylidae) are more sensitive than other tested frogs (see Table 10.1 and Figure 10.2). This result is not surprising, as the congeneric species *D. minutus* is considered to be a good bioindicator for the assessment of pollutants in Neotropical aquatic environments.[35–38] It is worth noting that we aimed to compare species' sensitivity to an equivalent chemical cocktail rather than draw conclusions about the potential effects of GLY-based herbicides on amphibians. This topic has been extensively debated and reviewed elsewhere (*e.g.*, Brühl *et al.*,[4] Relyea,[25] Rodríguez-Gil *et al.*,[28] Wagner *et al.*[39]).

While these comparisons help us identify general patterns of sensitivity across species and life stages, inferences must be made with caution as other factors than just actual differences in species sensitivity may lead to spurious conclusions. Critically, most toxicity data from Bernal *et al.*[30,31] lack estimates of uncertainty (see Table 10.1), which are crucial for strengthening confidence in toxicity predictions and preventing actions that might raise rather than lower risk inferences. Differences in experimental setup among studies could represent an additional source of variation. Unlike bioassays (see Meza-Joya *et al.*[20]), laboratory-based microcosms were supplemented with soil and leaf litter (see Bernal *et al.*[30]), yet we expect variation related to this fact to be minimal, as it has been shown that adding soil to a microcosm has no apparent influence on the measured effects of GLY-based herbicides (*e.g.*, Relyea[12]). Despite these caveats, we consider our comparisons informative enough to describe the wide variation in acute responses among species. It is important to note that laboratory-based assays—whether soil is added or not—are simplifications of reality and, as such, are not representative of field conditions as they fail to incorporate additive and/or synergistic effects of multiple biotic and abiotic stress factors.[10,40] Much more research on this issue is urgently needed, as relying only on toxicity data from acute exposures could mischaracterize the risk some xenobiotics may pose to terrestrial anurans.

The high sensitivity of *E. johnstonei* to a chemical mixture dominated by a GLY-based herbicide, the most widely applied broad-spectrum herbicide in the world, supports its use for ecotoxicological screening in terms of toxicant susceptibility. Further testing with a broader spectrum of xenobiotic contaminants will help validate the routine use of this species for chemical screening. The scarcity of toxicity data for terrestrial anurans hampers our ability to reliably identify candidate families for selecting suitable model species, yet it is worth mentioning that Neotropical direct developers with terrestrial breeding within the anuran clade Terrarana might represent promising candidates (see Figure 10.2). This highlights the need for further research with a greater diversity of terrestrial amphibians across different life stages, especially earlier terrestrial forms (*e.g.*, neonates and juveniles) that seem more sensitive than adults. Beyond acute toxicity data, accumulating evidence points out that chronic exposures to environmentally relevant pollutant concentrations have detrimental effects on aquatic amphibian growth, development, reproduction, and behavior (*e.g.*, Howe et al.,[11] Williams and Semlitsch,[41] Weltjeet *et al.*,[42] Boccioni *et al.*[43]). While the importance of long-term (*i.e.*, chronic) toxicity tests in risk assessment is unquestionable, to our knowledge, no study has rigorously assessed such effects on terrestrial amphibians.

10.2.2 Chemical Cyto-genotoxicity

Standard screening on amphibians—acute and chronic—is often limited to gross toxicity endpoints (*e.g.*, survival and growth) that fail to capture the wide array of sublethal effects induced by pollutants (*e.g.*, genotoxicity, immunotoxicity, and neurotoxicity). Out of the many techniques used to detect DNA damage, the single-cell gel electrophoresis bioassay (now known as the comet assay) has become one of the standards in amphibian ecotoxicology. The popularity of this test relies on its ability to detect the accumulation of DNA strand breaks and its repair in amphibian models, as well as its versatility, sensitivity, reproducibility, and low cost.[38,44,45] Since the first use in amphibian ecotoxicology by Ralph *et al.*,[46] the comet assay has been mostly conducted on aquatic larvae of metamorphosing frogs to assess the genotoxicity of pesticides and other clastogenic substances (*e.g.*, Carvalho *et al.*,[23] Lopes *et al.*,[38] Ralph *et al.*,[46] Mouchet *et al.*,[47] Banner *et al.*[48]). Recently, an enzymatic version employing proteinase K on this assay was developed to study *in vitro* genetic damage in *E. johnstonei*,[45] thus opening new possibilities for ecotoxicology and environmental monitoring of terrestrial amphibians.

Using *E. johnstonei* as a study system, the enzymatic comet assay has proven to be a highly reproducible and sensitive method for detecting DNA strand breakages induced by reference mutagens such as bleomycin, 4-nitroquinoline-1-oxide,[45] and GLY-based formulations both *in vitro* and *in vivo*.[20] The efficiency of repair mechanisms is critical in genetic toxicology research, and the latter study indicates that repair capacity in this frog decreased as exposure time increased.[20] These studies are good examples of how promising frogs of this species are in this research field. To date, *E. johnstonei* remains the only direct-developing frog in which chemical-induced DNA damage and repair have been researched using the comet assay. Although estimates of genetic damage induced by identical compounds (*i.e.*, bleomycin and GLY-based herbicides) are scarce in amphibian literature, similar substances have been investigated in a number of metamorphic anurans using the comet assay, allowing for gross comparisons of damage induction (see Table 10.2).

Common mutagens such as bleomycin have been tested using the comet assay after *in vitro* exposure of spleen lymphocytes in two *Xenopus* frogs, including *X. laevis*.[48] Genetic damage estimates from this study

Table 10.2 Overview of amphibian species used in the comet assay to assess DNA damage induced by (a) bleomycin and (b) GLY-based herbicides. For comparative purposes, sublethal concentrations of GLY-based herbicides are expressed as the percentage of the mean lethal concentration (LC_{50}) at 96 h for each species. Data from field-based studies were not considered as they could be neither related to specific pesticide treatments nor single exposure pathways. Species names follow Frost.[33]

Species	Study objective	Exposure pathway and target tissue	Main results and conclusions	Source
(a) Bleomycin (reference mutagen)				
Xenopus laevis *Xenopus tropicalis*	Examine the bleomycin-induced genetic damage and repair to test cancer predictability.	Acute; 0.5 h *in vitro* exposure of adult lymphocytes.	Species-dependent genotoxicity and repair; *X. tropicalis* (>20 µg mL^{-1}) is more susceptible than *X. laevis* (>60 µg mL^{-1}).	48
Eleutherodactylus johnstonei	Assess the genotoxic effects of reference mutagens, including bleomycin.	Acute; >12 h *in vitro* exposure of adult erythrocytes.	Increased genotoxicity from 4.7 µg mL^{-1} of bleomycin; successful modification of the comet assay (enzymatic version).	45
(b) GLY-based herbicides				
Aquarana catesbeiana	Examine the genotoxicity of five herbicides, including a GLY-based herbicide.	Acute; 24 h *in vivo* waterborne exposure of tadpole erythrocytes.	Genotoxic effects from concentrations higher than the LC_{50} of 2.8 mg a.e L^{-1} estimated by Moore *et al.*[15]	49
Eleutherodactylus johnstonei	Assess the cyto-genotoxic effects of a GLY-based herbicide plus a Cosmo-flux adjuvant.	Acute; <24 h *in vivo* overspray of frogs and *in vitro* exposure of adult erythrocytes.	Increased cyto-genotoxic effects after sublethal exposure (3–15% LC_{50}); DNA repair is likely to occur.	20
Rhinella arenarum	Investigate the genotoxicity of dicamba- and GLY-based herbicides.	Acute; 96 h *in vivo* waterborne exposure of tadpole erythrocytes.	Genotoxic effects after sublethal exposure (5–10% LC_{50}); synergistic effect when mixing both herbicides.	50

(*continued*)

Table 10.2 (*continued*)

Species	Study objective	Exposure pathway and target tissue	Main results and conclusions	Source
Dendropsophus minutus	Evaluate the genotoxic and mutagenic effects of a GLY-based herbicide.	Acute; 96 h *in vivo* waterborne exposure of tadpole erythrocytes.	Increased genetic damage but no micronucleus induction after sublethal exposure (11% LC_{50}).	35
Rhinella arenarum	Examine the genotoxic effects of imazethapyr- and GLY-based herbicides.	Acute; 96 h *in vivo* waterborne exposure of tadpole erythrocytes.	Genotoxic after sublethal exposure (5–10% LC_{50}); increased damage when mixing both herbicides.	23
Rhinella arenarum	Assess the genotoxicity of a GLY-based herbicide and an arsenite.	Chronic; 22 day *in vivo* waterborne exposure of tadpole erythrocytes.	Genotoxic after sublethal exposure (1.2% LC_{50} by Carvalho *et al.*);[23] synergistic genotoxicity when mixed.	51
Dendropsophus minutus	Investigate the genotoxic effects of a GLY-based herbicide and a POEA surfactant.	Acute; 96 h *in vivo* waterborne exposure of tadpole erythrocytes.	Genotoxic after sublethal exposure (8–16% LC_{50}); potential mutagenic and hepatotoxic effects.	38
Rhinella arenarum	Evaluate the genotoxicity of glufosinate ammonium and GLY-based herbicides.	Chronic; 45 day *in vivo* waterborne exposure of tadpole erythrocytes.	Genotoxic after sublethal exposure (3.2% LC_{50} by Carvalho *et al.*);[23] teratogenic and neurotoxic effects.	43

suggest that *Xenopus* lymphocytes are five times more resistant—damage induction at concentrations ≥20 μg mL^{-1}—than *E. johnstonei* erythrocytes, in which damage can be induced at concentrations ≥4.7 μg mL^{-1}.[45] Although comet-based estimates from these studies are not directly comparable, the minimal concentration required for inducing genetic damage suggests that *E. johnstonei* erythrocytes are sensitive enough to be used as a target tissue for genotoxicity assessment. The comet assay on erythrocytes has also been used to measure genetic damage in several different anuran species after exposure to GLY-based herbicides. Most of these studies were conducted on tadpoles of metamorphic frogs after *in vivo* acute waterborne exposure (≤96 h) to sublethal concentrations of herbicide, that is, concentrations equivalent to less than 20% of the LC_{50}-96 h reported for each tested species (see Table 10.2).

At sub-lethal concentrations, GLY-based herbicides can induce significant increases in DNA strand breakage in all but one species (see Clements *et al.*,[49] Table 10.2). While damage estimates from these studies are not directly comparable because of differences in assay conditions and comet scoring methods, comparisons of the minimal concentration required for inducing genetic impairment, estimated as a percentage of the LC_{50}-96 h for each species, suggest *E. johnstonei* is as, or more sensitive than other anuran species. Genetic damage induction after acute exposure in this species (<24 h) can be observed at concentrations as low as 3% of its LC_{50}-96 h.[20] This minimal concentration is similar but lower than that derived from some popular models for ecotoxicological research, such as *Rhinella arenarum* (5% of its LC_{50}-96 h)[23,50] and *Dendropsophus minutes* (8% of its LC_{50}-96 h).[35,38] Not surprisingly, recent studies indicate damage induction after prolonged exposures to such herbicides at even lower concentrations. Furthermore, it has been demonstrated that chronic *in vivo* exposure (>20 days) of *R. arenarum* tadpoles may lead to damage induction at concentrations as low as 1.2% of its LC_{50}-96 h.[43,51]

Cytotoxicity is another critical endpoint in toxicology research[52] that has rarely been studied in amphibians. To the best of our knowledge, this endpoint has been assessed in only three anuran species exposed to pesticides and UV radiation—alone or in combination—by estimating (i) the median hemolytic application rate in adult *E. johnstonei*,[20] (ii) the rate of erythrocytic nuclear abnormalities in *Aquarana catesbeiana* tadpoles,[53] and (iii) the frequency of apoptotic cells in *Boana curupi* tadpoles.[54] The sensitivity of this endpoint in *E. johnstonei* has been verified by a marked decrease in erythrocyte viability after *in vivo* exposure to sublethal concentrations (≥10.5% of its LC_{50}-96 h) of a GLY-based mixture.[20] Lack of cytotoxicity data for other amphibians exposed to GLY-based formulations precluded any comparison (*e.g.*, Bernal *et al.*[30]).

Overall, the sensitivity of the cytotoxic and genotoxic endpoints assessed in *E. johnstonei* underpins the usefulness of this species as a tool for assessing the toxicological effects of environmental contaminants. These endpoints have great potential for contributing to ecological risk assessment, given their simplicity, versatility, repeatability, and cost-efficiency (discussed above). Alternative endpoints assessing different clastogenic and genotoxic effects (*e.g.*, erythrocyte nuclear abnormality tests) may nicely complement the above-mentioned biomarkers and provide additional evidence on the potential genotoxic and mutagenic properties of chemical xenobiotics (*e.g.*, Lopes *et al.*[38]). These biomarkers are particularly promising when combined, for instance, with gross toxicity (*e.g.*, survival), behavioral (*e.g.*, alterations in locomotor performance), and histopathological endpoints (*e.g.*, tissue and organ changes), as all of these together may lead to a more holistic evaluation of the potential effects of xenobiotics on amphibians.

10.2.3 UV Radiation Cyto-genotoxicity

Research on the mechanisms underlying amphibian population declines has revealed intricate linkages between abiotic factors and human-mediated environmental changes.[55] Among abiotic factors, solar UV radiation is one of the most important and ubiquitous environmental stressors affecting aquatic and terrestrial ecosystems.[56,57] Although the two components of sunlight that reach the earth's surface—ultraviolet B (UVB, 280–315 nm) and ultraviolet A (UVA, 315–400 nm)—are capable of inducing damage in biomolecules (*e.g.*, DNA, proteins, and lipids), lesions induced by UVB are more detrimental because it has enough energy to cause a higher number of distortive DNA lesions.[58–60] The energy generated by UVB radiation can be directly absorbed by DNA pyrimidine bases, resulting in structural lesions such as cis-syn cyclobutane pyrimidine dimers (CPDs) and pyrimidine (6,4) primidone photoproducts (6,4PPs).[58,60] Depending on the extent of the DNA damage and repair capacity, these lesions may affect DNA metabolism, disrupt DNA replication and transcription, and ultimately lead to mutations and cell demise.[58,61]

There is an extensive body of literature dealing with the adverse effects of UV radiation on amphibians, and comprehensive revisions are available elsewhere (*e.g.*, Blaustein and Kiesecker,[55] Londero *et al.*,[60] Blaustein and Belden,[62] Alton and Franklin[63]). Notably, most harmful effects of this type of radiation are mediated through genetic damage—DNA is the primary target of UVB radiation in living cells[59,60]—and can be expressed *via* survival, phenotypic, physiological, developmental, and behavioral responses, among others.[55,60,62,63] The intensity of these effects may vary among species and life stages; early stages are much more sensitive than adults, but they also depend on ecological parameters determining UVB exposure in nature.[62–64] Amphibians have also evolved mechanisms to circumvent the damaging effects of UVB radiation *via* behavioral (*e.g.*, microhabitat choices), physiological (*e.g.*, heat shock proteins), and molecular (*e.g.*, DNA repair) defenses (reviewed in Londero *et al.*,[60] Blaustein and Belden,[62] Alton and Franklin[63]).

Genetic damage is a significant factor in the induction of lethal and sublethal effects after UVB exposure that remains largely understudied in amphibian ecotoxicology.[60] The comet assay is a simple tool for detecting DNA damage and repair that has been extensively utilized for chemical screening in amphibians (discussed in the previous section), but no attempt has been made to adapt this bioassay for detecting UVB-induced genotoxicity in amphibian models. Genetic damage induced by UVB radiation is generally estimated by quantifying CPDs, and the T4 endonuclease V (EndoV–T4) has proven high efficiency at performing this task.[65] With the inclusion of this enzyme in the methodological protocol, the enzymatic comet assay developed by Valencia *et al.*[45] was successfully adapted to assess DNA breaks on amphibian erythrocytes after *in vitro* exposure to artificial UVB radiation. *Eleutherodactylus johnstonei* was chosen as a model because of the well-known advantages of using this species as a study system (reviewed in Meza-Joya *et al.*)[17] and its routine use in our ecotoxicity bioassays.

10.2.3.1 Experimental Procedures

Adult *E. johnstonei* frogs were captured at several sites in the Bucaramanga metropolitan area (Santander, Colombia) and kept in glass terrariums simulating natural conditions until used in bioassays (see Meza-Joya *et al.*[17]). Blood samples were obtained by direct cardiac puncture of cold-anesthetized frogs and kept in the fridge (4 °C) until use. Blood cells were diluted in a 0.75% (w/v) NaCl solution and then counted in a Neubauer chamber, ensuring a cell density of 3.55×10^6 cells mL^{-1}. Cell viability was computed as the percentage of healthy cells—based on cell membrane integrity—in the sample using the Trypan Blue (TB) dye exclusion assay.[66] Then, cells were distributed into 5 cm diameter Petri dishes for irradiation using a UV irradiation chamber BS02 (Opsytec Dr. Grobel, Germany) equipped with a UV-MAT radiation controller from the same company. Initially, range-finding tests were performed using doses ranging from 0 to 10 000 J m^{-2}, at increasing intervals from 200 to 1000 J m^{-2}. Lethal doses of LD_{20}, LD_{30}, and LD_{50} were determined by interpolation after regressing cell viability (%) as a linear function of UVB doses.[67] All assays included at least three independent experiments with duplicate doses, including negative controls (non-irradiated cells).

Sublethal doses below the estimated LD_{30} were chosen for UVB genotoxicity assessment, ranging from 0 to 2000 J m^{-2} at intervals of 500 J m^{-2}. The effect of UVB radiation on DNA integrity was assessed using the enzymatic version of the comet assay,[45] with a specific adaptation for detecting CPDs comprising a second enzymatic lysis step after the first lysis with proteinase K, using EndoV–T4 enzyme as implemented by Collins *et al.*[68] in rodent mutant cell lines. A step-by-step protocol for this new comet variant is presented in Box 10.1. DNA damage was expressed in arbitrary units based on the classification of comets into five categories (0–4) as previously proposed,[69] and a genetic damage index (GDI) was calculated for each treatment as follows: GDI = (N0 × 0 + N1 × 1 + N2 × 2 + N3 × 3 + N4 × 4)/*n*, where N*i* is the number of nuclei scored in each category and *n* is the number of scored cells per slide.[70] Two hundred cells per slide and two slides per blood sample were analyzed using a Nikon Eclipse E200 microscope, and the results from at least three independent experiments—including non-irradiated negative controls—were averaged to obtain the GDI values for each treatment.

Box 10.1 Amphibian double-enzymatic (proteinase K + EndoV–T4) comet assay protocol.

Step 1 Count cells in a Neubauer chamber and dilute, if required, in a 0.75% (w/v) NaCl solution to a cell density of 3.55×10^6 cells mL^{-1}. **ΔCRITICAL:** once reached the desired density, keep the cell solution at 4 °C in the dark until use; all protocols must be performed under dark/red light conditions. **CAUTION!:** use cell solution as soon as possible to ensure high cell viability.
Step 2 Spread cells into Petri dishes with a 5 cm diameter for irradiation at the chosen doses. **ΔCRITICAL:** use subtoxic doses (*e.g.*, $\leq DL_{30}$) to avoid large numbers of apoptotic cells. **CAUTION!:** omit this step for non-irradiated negative controls.
Step 3 Centrifugate the irradiated cell solution at 1800 rpm for 3 min, resuspend the pellets in 20 µL of proteinase K solution prepared in buffer PK (50 mM Tris–HCl, 10 mM $CaCl_2$, pH 8) at 0.19 mg mL^{-1}, and incubate for 1 h at 37 °C in a Thermomixer (Eppendorf, Sao Paulo, Brazil) for cell membrane proteolysis. **ΔCRITICAL:** keep the cells in the dark between lysis steps.
Step 4 Centrifugate the cell solution at 2200 rpm for 3 min, resuspend the pellet in 40 µL (20 units) of EndoV–T4 enzyme prepared in enzymatic buffer (25 mM Na_2PO4, 100 mM NaCl, 1 mM EDTA, 1 mM dithiothreitol, pH 7.2), and incubate for 30 min at 37 °C in a Thermomixer for CPD hydrolyses. **TROUBLESHOOTING:** unexpected DNA breaks in controls: pipette tips with wider openings help to reduce mechanical damage.
Step 5 Spread 400 µL of 1.3% molecular-grade agarose prepared in water on clean microscopic slides and incubate for 30 min at 37 °C to ensure adherence.
Step 6 Centrifugate the cell solution at 2200 rpm for 3 min, resuspend the pellet in 40 µL of 0.75% NaCl solution, mix the cell suspension with 75 µL of 1% low-melting-point agarose at 37 °C, spread on slides containing a layer of 1.3% molecular grade agarose (Step 5), and cover with coverslips at 8 °C for 15 min for the agarose to solidify. **ΔCRITICAL:** prepare at least two slides for treatment.
Step 7 Remove coverslips after agarose gelification, place the slides at 4 °C in a comet assay tank (Cleaver Scientific Ltd, UK) containing cold (4 °C) alkaline electrophoresis buffer (0.3 N NaOH, 1 mM EDTA, pH 13) for 30 min to enable nuclei decompaction, and run electrophoresis at 25 V and 300 mA for 30 min. **ΔCRITICAL:** spread the slides over the whole electrophoretic area and cover them with no more than 5 mm of alkaline buffer to ensure uniformity in the electric field. **CAUTION!:** the whole step must be performed in the dark.

Step 8 Wash the slides with neutralizing buffer (0.4 M Tris, pH 7.5) in the dark for 5 min, rinse in distilled water for 1 min, and dry at 37 °C for 1 h. **CAUTION!:** once dried, keep the slides in the fridge until used for staining (the sooner, the better).

Step 9 Stain slides using the silver-staining protocol,[45] which consists of ten consecutive steps: (1) distilled water for 1 min; (2) fixing solution for 10 min (15% trichloroacetic acid, 5% zinc sulfate heptahydrate, 5% glycerol); (3) distilled water for 1 min; (4) dry at 37 °C for 30 min; (5) freshly prepared stain solution for 10–15 min [34 mL of vigorously mixed stock solution B (0.1% ammonium nitrate, 0.1% silver nitrate, 0.25% tungstosilicic acid, 0.15% formaldehyde, v/v), and 66 mL of stock solution A (5% sodium carbonate)]; (6) repeat previous step (5) for 4–5 min if needed; (7) distilled water for 1 min; (8) stop solution for 1 min (acetic acid 1%); (9) distilled water for 1 min; and (10) dry at room temperature before microscope analyzes. **CAUTION!:** all specified staining solutions must be pre-set as any delay between steps may affect staining reproducibility; slides must be placed back-to-back in horizontal or vertical staining jars.

Reagents: All reagents were obtained from New England Biolabs (Beverly, MA), J.T. Baker (Phillipsburg, NJ), and Merck (Kenilworth, NJ).

10.2.3.2 UVB-induced Cyto-genotoxicity Estimates

The cytotoxicity of UVB radiation was assayed based on the cell viability (%CV) obtained by the TB exclusion assay. The dose-response curve based on viability data showed a significant and negative correlation between radiation dose and cell viability (R^2 = 0.934, P < 0.001). Based on the regression slope, lethal doses were calculated as follows: LD_{50} = 4814 J m^{-2}, LD_{30} = 2036 J m^{-2}, and LD_{20} = 647 J m^{-2} (see Figure 10.3). Cell viability at doses lower than the LD_{30} was higher than 70%. The genotoxic effect of UVB on erythrocytes was measured using a modified comet assay, which used a double enzymatic lysis procedure (proteinase K + EndoV–T4). In the absence of EndoV–T4, estimates of DNA breaks were consistently low and likely promoted by UVB-induced oxidative stress (R^2 = 0.850, P < 0.017, see Figure 10.3). By combining both enzymes in two consecutive lysis steps, we obtained bare nuclei with very low levels of DNA breaks in non-irradiated cells and significant levels of UVB-induced breaks in all irradiated treatments. The dose-response curve based on GDI values showed a significant and positive correlation between radiation dose and UVB-induced genotoxicity (R^2 = 0.971, P < 0.001; see Figure 10.3).

10.2.3.3 Estimates of UVB-induced Genetic Damage in Amphibians

Despite growing knowledge regarding the influence of UV-induced genotoxic stress in amphibians (reviewed in Londero *et al.*[60]), only six studies have directly evaluated the formation of CPDs and 6,4PPs in tadpoles of different anuran species. Briefly, two laboratory-based studies have assessed the induction of DNA photoproducts and repair activity *via*

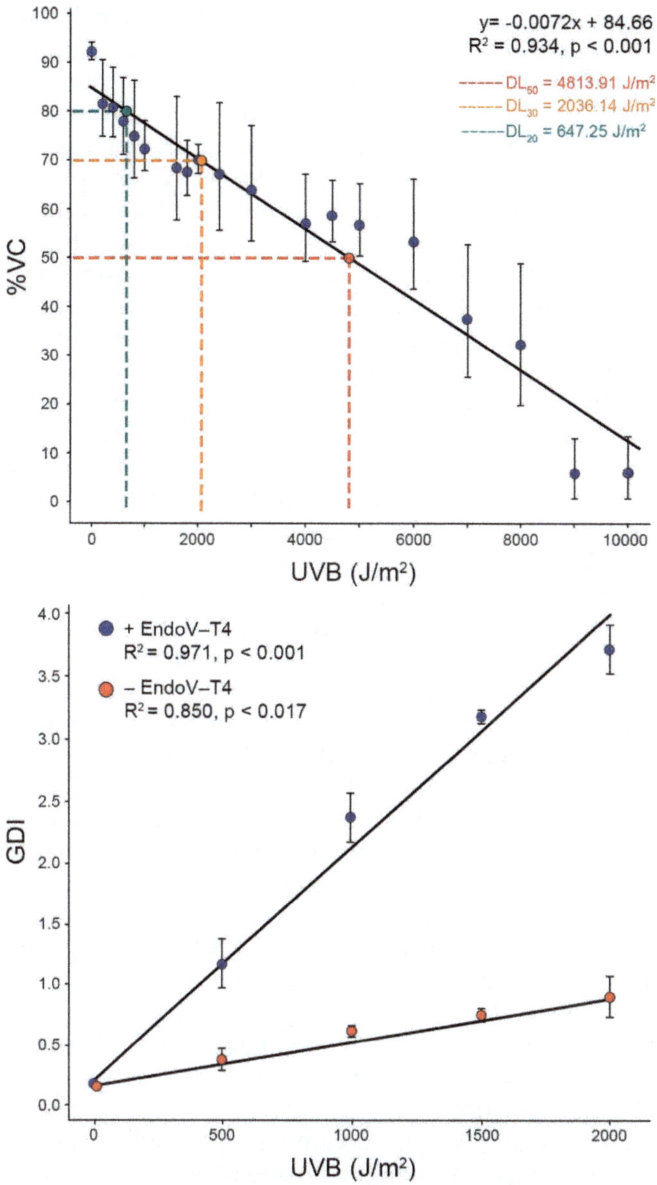

Figure 10.3 UVB-induced cytotoxic (top) and genotoxic (bottom) effects in *Eleutherodactylus johnstonei* erythrocytes after exposure *in vitro* to UVB radiation. Standard errors calculated from three independent experiments with two replicates each are presented. Cytotoxic effects based on cell viability using the TB dye exclusion assay. The dashed lines represent the lethal doses of LD_{20}, LD_{30}, and LD_{50}. Genotoxic effects based on the comet assay in the presence (+) and absence (–) of the enzyme EndoV–T4.

immunological detection in tadpoles of two Neotropical hylid frogs: *Boana curupi*[54] and *B. pulchella*.[58] Similarly, two indoor investigations evaluated the induction of photoproducts with UV-specific endonucleases[71] and the expression of DNA photo-adducts and DNA damage and repair genes[72] in embryos of the well-known anuran model *X. laevis*. One additional study quantified the thermal sensitivity of DNA repair rates and photolyase gene expression in tadpoles of *Limnodynastes peronii* acutely exposed to artificial UVB at both 20 °C and 30 °C.[73] Lastly, one study assessed the induction of photoproducts *via* antibodies binding and repair activity through gene expression on salamander embryos of *Ambystoma maculatum* (Urodela, Ambystomidae) exposed under laboratory and field conditions.[74] General information on these studies' aims, methodological approaches, results, and conclusions is presented in Table 10.3. A number of additional studies have explored the influence of UV genotoxic stress using indirect biomarkers of DNA damage in early anuran developmental stages (*e.g.*, nuclear abnormalities, cell death, photoreactivation post-UV treatments) expressed *via* phenotypic, physiological, or behavioral responses (see Londero *et al.*[60]).

To our knowledge, this work presents the first use of the comet assay for directly detecting UVB-induced DNA breaks (*i.e.*, CPDs) in an amphibian model after *in vitro* exposure to artificial UVB radiation. Direct estimates of genetic damage may be seen as early indicators of potential UVB-related sublethal effects on affected organisms, especially in individuals that lack efficient mechanisms to protect themselves from damage. The results from our assays reinforce the usefulness of *E. johnstonei* as a study system for UV radiation screening. While we focused here on damage, the efficiency of repair mechanisms can be explored using the double-enzymatic comet protocol, as this frog repair capacity has been explored using a related comet setup.[20] Although amphibians have evolved some molecular pathways to either limit the induction of DNA lesions or repair them once the damage has occurred, little is known about the efficacy of such repair mechanisms and/or their interplay with behavioral (*e.g.*, nocturnality) and physiological (*e.g.*, skin darkening) strategies.[58,60] This knowledge is critical for identifying integrally amphibian species that may be particularly vulnerable to solar radiation.

In a rapidly changing world, developing rapid and sensitive screening methods for evaluating DNA damage is an urgent need, especially for globally declining taxa such as amphibians. The double-enzymatic (proteinase K + EndoV–T4) comet assay was highly sensitive for detecting *in vitro* UVB-induced DNA damage on blood cells of *E. johnstonei*. This comet protocol used the proteinase K enzyme to lyse membrane proteins under neutral pH conditions since lysis using detergent and salts under alkaline treatments is particularly aggressive to nuclear stability in *E. johnstonei* blood cells.[45] This enzyme also digests proteins associated with nuclei and eliminates DNA-protein bonds, facilitating DNA unwinding and electrophoretic migration of

Table 10.3 Overview of studies directly estimating UVB-induced genetic stress (*e.g.*, CPDs) on amphibians after UV radiation exposure. Species names follow Frost.[33]

Species	Study objective	Methodological approach	Main results and conclusions	Source
Boana curupi	Assess the effects of UV radiation and trichlorfon using lethal and sublethal endpoints.	*In vivo* exposure of tadpoles to UVA/B and trichlorfon assessing survival, morphology, apoptosis, DNA integrity, and biochemical responses.	UVB induces high mortality, apoptosis, DNA, and protein damage; co-exposure (UV + trichlorfon) buffered UVB-induced negative effects.	54
Boana pulchella	Investigate UV-induced DNA damage/repair and explore behavioral responses to avoid UV exposure.	*In vivo* exposure of tadpoles to UVA/B quantifying DNA photo-adducts and repair and assessing behavioral avoidance to UV radiation.	UVB induces high rates of apoptosis and DNA photo-adducts, likely by slowing DNA repair; sensorial escape is key to maintaining genomic integrity.	58
Xenopus laevis	Identify the sensitive sublethal endpoints indicative of UV radiation exposure.	*In vivo* exposure of embryos to UVA/B assessing mortality, malformations, growth, and DNA damage.	UVB-induced DNA damage correlates with teratogenesis and growth inhibition. DNA damage varies between egg clutches.	71
Xenopus laevis	Evaluate the effects of UVB radiation and two pesticides (endosulfan/cypermethrin) on DNA stability and repair.	*In vivo* exposure of embryos to UVB and pesticides quantifying DNA photo-adducts and nucleotide excision repair *via* gene expression.	Pesticides enhance the accumulation of UVB-induced DNA photo-adducts, likely by altering the expression of key nucleotide excision repair genes.	72

(*continued*)

Table 10.3 (*continued*)

Species	Study objective	Methodological approach	Main results and conclusions	Source
Limnodynastes peronii	Examine the thermal sensitivity of DNA repair mechanisms against UVB-induced genetic damage.	*In vivo* exposure of tadpoles to UVB radiation at 20 or 30 °C examining UVB-induced damage and photolyase repair.	Contrary to photoproduct induction, DNA repair rates are thermally sensitive; photolyase expression is upregulated by UVB exposure time.	73
Ambystoma maculatum	Assess UV-related oxidative stress, genetic damage, and repair under laboratory and field experiments.	*In vivo* exposure of embryos to UVA/B assessing mortality, photo-adduct induction, p53/p73 expression, and pigment concentration.	Laboratory exposures to UV induce mortality, DNA photo-adducts, and melanin synthesis, p53/p73 expression leads to repair or apoptosis; field results suggest little effect to UV radiation.	74

DNA breaks, respectively.[45,75,76] The second enzymatic lysis step with EndoV–T4 recognizes DNA breaks associated with CPD enzymatic hydrolysis,[68] thus providing a direct evaluation of UV-induced DNA damage. The ability of this comet variant to detect *in vitro* UVB-induced DNA breaks related to CPDs may have interesting applications for *in vivo* solar radiation screening, especially when combined with routine screening of pollutant genotoxicity, as previously recommended.[55,63]

10.3 Final Remarks

Terrestrial amphibian ecotoxicology is a developing research field that requires sensitive and versatile model organisms. *Eleutherodactylus johnstonei* has served as a model for studies in diverse research fields such as zoonotic diseases,[77] reproductive ecology,[78] biological invasions,[79] development, and morphology;[80] so it is somewhat surprising that its potential for ecotoxicology has not received the attention that it merits. Beyond indoor experimentation, city-dwelling populations of this frog might help explore the effects of urban pollutants on amphibians (*e.g.*, light, noise, and air emissions). Researchers may take advantage of the invasive nature of this species to avoid, at least to some extent, the paradox of generating ecotoxicity data for numerous substances while reducing animal testing on indigenous species. While reducing animal testing on native amphibians is desirable, it cannot prevail over ensuring the long-term viability of wild populations; hence, choosing model organisms must be properly supported. We hope the overview presented in this chapter will encourage other researchers to capitalize on the enormous potential of this species in ecotoxicological research and ecological risk assessment.

Acknowledgements

The authors thank funding from the Ministerio de Ciencia, Tecnología e Innovación, Ministerio de Educación Nacional, Ministerio de Comercio, Industria y Turismo, and ICETEX-Colombia Científica, Fund Francisco José de Caldas (Grant No. RC-FP44842-212-2018). The 'Corporación para la Defensa de la Meseta de Bucaramanga (CDMB)' and the 'Autoridad Nacional de Licencias Ambientales (ANLA)' granted to the Universidad Industrial de Santander collect permits of wild species of biological diversity for noncommercial scientific research purposes (Resolution CDMB No. 001368, Resolution ANLA No. 0047). The authors declare that they have no conflicts of interest.

References

1. S. N. Stuart, J. S. Chanson, N. A. Cox, B. E. Young, A. S. L. Rodrigues, D. L. Fischman and R. W. Waller, Status and trends of amphibian declines and extinctions worldwide, *Science*, 2004, **306**, 1783–1786.
2. D. B. Wake and V. T. Vredenburg, Colloquium paper: are we in the midst of the sixth mass extinction? A view from the world of amphibians, *Proc. Natl. Acad.Sci. USA*, 2008, **105**, 11466–11473.
3. D. Linzey, J. Burroughs, L. Hudson, M. Marini, J. Robertson, J. Bacon, M. Nagarkatti and P. Nagarkatti, Role of environmental pollutants on immune functions, parasitic infections and limb malformations in marine toads and whistling frogs from Bermuda, *Int. J. Environ. Health. Res.*, 2003, **13**, 125–148.
4. C. A. Brühl, S. Pieper and B. Weber, Amphibians at risk? Susceptibility of terrestrial amphibian life stages to pesticides, *Environ.Toxicol. Chem.*, 2011, **30**, 2465–2472.
5. M. S. Johnson, C. Aubee, C. J. Salice, K. B. Leigh, E. Liu, U. Pott and D. Pillard, A review of ecological risk assessment methods for amphibians: Comparative assessment of testing methodologies and available data, *Integr. Environ. Assess Manag.*, 2017, **13**, 601–613.
6. M. E. Ortiz-Santaliestra, J. P. Maia, A. Egea-Serrano and I. Lopes, Validity of fish, birds and mammals as surrogates for amphibians and reptiles in pesticide toxicity assessment, *Ecotoxicology*, 2018, **27**, 819–833.
7. S. Glaberman, J. Kiwiet and C. B. Aubee, Evaluating the role of fish as surrogates for amphibians in pesticide ecological risk assessment, *Chemosphere*, 2019, **235**, 952–958.
8. D. W. Sparling, G. Linder, C. A. Bishop and S. K. Krest, in *Ecotoxicology of Amphibians and Reptiles*, CRC Press, Taylor and Francis Group, New York NY, 2nd edn, 2010.
9. A. Gendron, Amphibian Ecotoxicology, in *Encyclopedia of Aquatic Ecotoxicology*, ed. J. F. Férard and C. Blaise, Springer, Netherlands, 2013, pp. 21–38.
10. Z. Mikó, J. Ujszegi, Z. Gál and A. Hettyey, Standardize or diversify experimental conditions in ecotoxicology? A case study on herbicide toxicity to larvae of two anuran amphibians, *Arch. Environ. Contam. Toxicol.*, 2017, **73**, 562–569.
11. C. M. Howe, M. Berrill, B. D. Pauli, C. C. Helbing, K. Werry and N. Veldhoen, Toxicity of glyphosate-based pesticides to four North American frog species, *Environ. Toxicol. Chem.*, 2004, **23**, 1928–1938.
12. R. A. Relyea, The lethal impact of Roundup® on aquatic and terrestrial amphibians, *Ecologl. Appl.*, 2005, **15**, 1118–1124.

13. R. A. Relyea and D. K. Jones, The toxicity of Roundup Original Max® to 13 species of larval amphibians, *Environ. Toxicol. Chem.*, 2009, **28**, 2004–2008.
14. G. Linger, C. M. Lehman and J. R. Bidwell, Ecotoxicology of Amphibians and Reptiles in a Nutshell, in *Ecotoxicology of Amphibians and Reptiles*, ed. D. W. Sparling, G. Linder, C. A. Bishop and S. Krest, CRC Press, New York, 2010, pp. 95–130.
15. L. J. Moore, L. Fuentes, J. H. Rodgers, W. W. Bowerman, G. K. Yarrow, W. Y. Chao and W. C. Bridges, Relative toxicity of the components of the original formulation of Roundup® to five North American anurans, *Ecotoxicol. Environ. Saf.*, 2012, **78**, 128–133.
16. H. Kaiser, Origins and introductions of the Caribbean frog, *Eleutherodactylus johnstonei* (Leptodactylidae): management and conservation concerns, *Biodivers. Conserv.*, 1997, **6**, 1391–1407.
17. F. L. Meza-Joya, M. P. Ramírez-Pinilla and J. L. Fuentes, The Direct-Developing Frog *Eleutherodactylus johnstonei* (Eleutherodactylidae) as a Biological Model for the Study of Toxic, Cytotoxic, and Genotoxic Effects of Agrochemicals, in *Ecotoxicology and Genotoxicology: Non-traditional Terrestrial Models*, ed. M. L. Larramendy, Royal Society of Chemistry, Cambridge, 2017, pp. 211–227.
18. https://www.iucnredlist.org/species/56684/3047195. Accessed 18 August 2022.
19. W. E. Duellman and L. Trueb, in *Biology of Amphibians*, Johns Hopkins University Press, Baltimore MD, 2nd edn, 1994.
20. F. L. Meza-Joya, M. P. Ramírez-Pinilla and J. L. Fuentes, Toxic, cytotoxic, and genotoxic effects of a glyphosate formulation (Roundup®SL-Cosmoflux®411F) in the direct-developing frog *Eleutherodactylus johnstonei*, *Environ. Mol. Mutagen.*, 2013, **54**, 362–373.
21. A. Egea-Serrano, R. A. Relyea, M. Tejedo and M. Torralva, Understanding of the impact of chemicals on amphibians: a meta-analytic review, *Ecol. Evol.*, 2012, **2**, 1382–1397.
22. N. Wagner, S. Lötters, M. Veith and B. Viertel, Effects of an environmentally relevant temporal application scheme of low herbicide concentrations on larvae of two anuran species, *Chemosphere*, 2015, **135**, 175–181.
23. W. F. Carvalho, C. Ruiz de Arcaute, J. M. Pérez-Iglesias, M. R. Laborde, S. Soloneski and L. M. Larramendy, DNA damage exerted by mixtures of commercial formulations of glyphosate and imazethapyr herbicides in *Rhinella arenarum* (Anura, Bufonidae) tadpoles, *Ecotoxicology*, 2019, **28**, 367–377.
24. D. K. Jones, J. I. Hammond and R. A. Relyea, Roundup® and amphibians: The importance of concentration, application time, and stratification, *Environ. Toxicol. Chem.*, 2010, **29**, 2016–2025.

25. R. A. Relyea, Amphibians are not ready for Roundup®, in *Wildlife Ecotoxicology*, ed. J. E. Elliott, C. A. Bishop and C. A. Morrissey, Springer, New York, 2011, pp. 267–300.
26. S. M. Weir, J. G. Suski and C. J. Salice, Ecological risk of anthropogenic pollutants to reptiles: evaluating assumptions of sensitivity and exposure, *Environ. Pollut.*, 2010, **158**, 3596–3606.
27. J. D. Lynch and S. B. Arroyo, Risks to Colombian amphibian fauna from cultivation of coca (*Erythroxylum coca*): a geographical analysis, *J. Toxicol. Environ. Health A*, 2009, **72**, 974–985.
28. J. L. Rodríguez-Gil, R. S. Prosser, S. O. Duke and K. Solomon, Ecotoxicology of Glyphosate, its Formulants, and Environmental Degradation Products, in *Rev. Environ. Contam. Toxicol*, ed. J. B. Knaak, Springer Cham, Gewerbestrasse, 2021, vol. **255**, pp. 129–205.
29. K. R. Solomon, A. Anadón, G. Carrasquilla, A. L. Cerdeira, J. Marshall and L. H. Sanin, Coca and poppy eradication in Colombia: environmental and human health assessment of aerially applied glyphosate, *Rev. Environ. Contam. Toxicol.*, 2007, **190**, 43–125.
30. M. H. Bernal, K. R. Solomon and G. Carrasquilla, Toxicity of formulated glyphosate (Glyphos) and Cosmo-Flux to larval and juvenile Colombian frogs. II. Field and laboratory microcosm acute toxicity, *J. Toxicol. Environ. Health A*, 2009, **72**, 966–973.
31. M. H. Bernal, K. R. Solomon and G. Carrasquilla, Toxicity of formulated glyphosate (glyphos) and Cosmo-Flux to larval Colombian frogs. I. Laboratory acute toxicity, *J. Toxicol. Environ. Health A*, 2009, **72**, 961–965.
32. K. L. Gosner, A simplified table for stating anuran embryos and larvae with notes on identification, *Herpetologica,* 1960, **16**, 183–190.
33. https://amphibiansoftheworld.amnh.org. Accessed 12 September 2022.
34. J. Thorley and C. Schwarz, ssdtools: an R package to fit species sensitivity distributions, *J. Open Source Softw.*, **3**, 1082.
35. W. F. Carvalho, F. C. Franco, F. Godoy, D. Folador, J. B. Avelar, F. Nomura, A. D. Cruz, S. M. T. Sabóia-Morais, R. Bastos and D. Silva, Evaluation of genotoxic and mutagenic effects of glyphosate Roundup Original ® in *Dendropsophus minutus* Peters, 1872 tadpoles, *South Am. J. Herpetol*, 2018, **13**, 220–229.
36. R. E. Borges, L. R. de Souza Santos, M. Benvindo-Souza, R. S. Modesto, R. A. Assis and C. de Oliveira, Genotoxic evaluation in tadpoles associated with agriculture in the Central Cerrado, Brazil, *Arch. Environ. Contam. Toxicol.*, 2019, **77**, 22–28.
37. M. W. Gonçalves, C. B. M. de Campos, F. R. Godoy, P. G. Gambale, H. F. Nunes, F. Nomura, R. P. Bastos, A. D. da Cruz and D. M. Silva, Assessing genotoxicity and mutagenicity of three common Amphibian species

inhabiting agroecosystem environment, *Arch. Environ. Contam. Toxicol.*, 2019, **77**, 409–420.

38. A. Lopes, M. Benvindo-Souza, W. F. Carvalho, H. F. Nunes, P. N. de Lima, M. S. Costa, E. J. Benetti, V. Guerra, S. M. T. Saboia-Morais, C. E. Santos, K. Simões, R. P. Bastos and D. M. Silva, Evaluation of the genotoxic, mutagenic, and histopathological hepatic effects of polyoxyethylene amine (POEA) and glyphosate on *Dendropsophus minutus* tadpoles, *Environ. Pollut.*, 2021, **289**, 117911.
39. N. Wagner, W. Reichenbecher, H. Teichmann, B. Tappeser and S. Lötters, Questions concerning the potential impact of glyphosate-based herbicides on amphibians, *Environ. Toxicol. Chem.*, 2013, **32**, 1688–1700.
40. A. Sih, A. Bell and J. C. Johnson, Behavioral syndromes: an ecological and evolutionary overview, *Trends. Ecol. Evol.*, 2004, **19**, 372–378.
41. B. K. Williams and R. D. Semlitsch, Larval responses of three Midwestern anurans to chronic, low-dose exposures of four herbicides, *Archiv. Environ. Contam. Toxicol.*, 2009, **58**, 819–827.
42. L. Weltje, P. Simpson, M. Gross, M. Crane and J. R. Wheeler, Comparative acute and chronic sensitivity of fish and amphibians: a critical review of data, *Environ. Toxicol. Chem.*, 2013, **32**, 984–994.
43. A. P. C. Boccioni, G. Lener, J. Peluso, P. M. Peltzer, A. M. Attademo, C. Aronzon, M. F. Simoniello, L. D. Demonte, M. R. Repetti and R. C. Lajmanovich, Comparative assessment of individual and mixture chronic toxicity of glyphosate and glufosinate ammonium on amphibian tadpoles: a multibiomarker approach, *Chemosphere*, 2022, **309**, 136554.
44. A. Dhawan, M. Bajpayee and D. Parmar, Comet assay: a reliable tool for the assessment of DNA damage in different models, *Cell Biol. Toxicol.*, 2009, **25**, 5–32.
45. L. C. Valencia, A. García, M. P. Ramírez-Pinilla and J. L. Fuentes, Estimates of DNA damage by the comet assay in the direct-developing frog *Eleutherodactylus johnstonei* (Anura, Eleutherodactylidae), *Gene. Mol. Biol.*, 2011, **34**, 681–688.
46. S. Ralph, M. Petras, R. Pandrangi and M. Vrzoc, Alkaline single-cell gel (comet) assay and genotoxicity monitoring using two species of tadpoles, *Environ. Mol. Mutagen.*, 1996, **28**, 112–120.
47. F. Mouchet, L. Gauthier, C. Mailhes, V. Ferrier and A. Devaux, Comparative evaluation of genotoxicity of captan in amphibian larvae (*Xenopus laevis* and *Pleurodeles waltl*) using the comet assay and the micronucleus test, *Environ. Toxicol.*, 2006, **21**(3), 264–277.
48. S. H. Banner, L. N. Ruben and R. O. Johnson, Bleomycin-induced DNA damage and repair in *Xenopus laevis* and *Xenopus tropicalis*, *J. Exp. Zool. A Ecol. Genet. Physiol.*, 2007, **307**, 84–90.

49. C. Clements, S. Ralph and M. Petras, Genotoxicity of select herbicides in *Rana catesbeiana* tadpoles using the alkaline single-cell gel DNA electrophoresis (comet) assay, *Environ. Mol. Mutagen.*, 1997, **29**, 277–288.

50. S. Soloneski, C. Ruiz-de-Arcaute and M. L. Larramendy, Genotoxic effect of a binary mixture of dicamba- and glyphosate-based commercial herbicide formulations on *Rhinella arenarum* (Hensel, 1867) (Anura, Bufonidae) late-stage larvae, *Environ. Sci. Pollut. Res.*, 2016, **23**, 17811–17821.

51. R. C. Lajmanovich, P. M. Peltzer, A. M. Attademo, C. S. Martinuzzi, M. F. Simoniello, C. L. Colussi, A. P. Boccioni and M. Sigrist, First evaluation of novel potential synergistic effects of glyphosate and arsenic mixture on *Rhinella arenarum* (Anura: Bufonidae) tadpoles, *Heliyon*, 2019, **5**, e02601.

52. M. O'Donovan, A critique of methods to measure cytotoxicity in mammalian cell genotoxicity assays, *Mutagenesis*, 2012, **27**, 615–621.

53. M. F. Montalvão and G. Malafaia, Effects of abamectin on bullfrog tadpoles: insights on cytotoxicity, *Environ. Sci. Pollut. Res.*, 2017, **24**, 23411–23416.

54. C. R. Schavinski, M. B. do Santos, J. E. Londero, M. C. da Rocha, A. M. do Amaral, N. Q. Ruiz, G. L. da Silva, V. L. Loro and A. P. Schuch, Effects of isolated and combined exposures of *Boana curupi* (Anura: Hylidae) tadpoles to environmental doses of trichlorfon and ultraviolet radiation, *Mutat. Res. Genet. Toxicol. Environ. Mutagen.*, 2022, **17**, 503549.

55. A. R. Blaustein and J. M. Kiesecker, Complexity in conservation: lessons from the global decline of amphibian populations, *Ecol. Lett.*, 2002, **5**, 597–608.

56. C. L. Ballaré, M. M. Caldwell, S. D. Flint, S. A. Robinson and J. F. Bornman, Effects of solar ultraviolet radiation on terrestrial ecosystems. Patterns, mechanisms, and interactions with climate change, *Photochem. Photobiol. Sci.*, 2011, **10**, 226–241.

57. D.-P. Häder, E. W. Helbling, C. E. Williamson and R. C. Worrest, Effects of UV radiation on aquatic ecosystems and interactions with climate change, *Photochem. Photobiol. Sci.*, 2011, **10**, 242–260.

58. A. P. Schuch, V. M. Lipinski, M. B. Santos, C. P. Santos, S. S. Jardim, S. Z. Cechin and E. L. S. Loreto, Molecular and sensory mechanisms to mitigate sunlight-induced DNA damage in treefrog tadpoles, *J. Exp. Biol.*, 2015, **218**, 3059–3067.

59. A. P. Schuch, N. C. Moreno, N. J. Schuch, C. F. M. Menck and C. C. M. Garcia, Sunlight damage to cellular DNA: Focus on oxidatively generated lesions, *Free Rad. Biol. Med.*, 2017, **107**, 110–124.

60. J. E. L. Londero, M. B. dos Santos and A. P. Schuch, Impact of solar UV radiation on amphibians: focus on genotoxic stress, *Mutat. Res. Genet. Toxicol. Environ. Mutagen.*, 2019, **842**, 14–21.
61. A. P. Schuch and C. F. M. Menck, The genotoxic effects of DNA lesions induced by artificial UV-radiation and sunlight, *J. Photochem. Photobiol. B*, 2010, **99**, 111–106.
62. L. K. Belden and A. R. Blaustein, Exposure of red-legged frog embryos to ambient UV-B radiation in the field negatively affects larval growth and development, *Oecologia*, 2002, **130**, 551–554.
63. L. A. Alton and C. E. Franklin, Drivers of amphibian declines: effects of ultraviolet radiation and interactions with other environmental factors, *Clim. Chang. Responses*, 2017, **4**, 6.
64. M. C. Croteau, M. A. Davidson, D. R. S. Lean and V. L. Trudeau, Global increases in ultraviolet B radiation: potential impacts on amphibian development and metamorphosis, *Physiol. Biochem. Zool.*, 2008, **81**, 743–761.
65. D. Muruzabal, A. Collins and A. Azqueta, The enzyme-modified comet assay: past, present and future, *Food. Chem. Toxicol.*, 2021, **147**, 111865.
66. W. Strober, Trypan Blue Exclusion Test of Cell Viability, *Curr. Protoc. Immunol.*, 2015, **111**, A3.B.1–A3.B.3.
67. N. Ara, M. H. Nur, M. S. Amran, M. I. I. Wahid and M. Ahmed, In vitro antimicrobial and cytotoxic activities of leaves and flowers extracts from *Lippia alba*, *Pak. J. Biol. Sci.*, 2009, **12**, 87–90.
68. A. R. Collins, D. L. Mitchell, A. Zunino, J. Wit and D. Busch, UV-sensitive rodent mutant cell lines of complementation groups 6 and 8 differ phenotypically from their human counterparts, *Environ. Mol. Mutagen.*, 1997, **29**, 152–160.
69. A. Collins, M. Dusinská, M. Franklin, M. Somorovská, H. Petrovská, S. Duthie and N. Vaughan, Comet assay in human biomonitoring studies: reliability, validation, and applications, *Environ. Mol. Mutagen.*, 1997, **30**, 139–146.
70. M. Pitarque, A. Creus, R. Marcos, J. A. Hughes and D. Anderson, Examination of various biomarkers measuring genotoxic endpoints from Barcelona airport personnel, *Mutat. Res.*, 1999, **440**, 195–204.
71. D. J. Bruggeman, J. A. Bantle and C. Goad, Linking teratogenesis, growth, and DNA photodamage to artificial ultraviolet B radiation in *Xenopus laevis* larvae, *Environ. Toxicol. Chem.*, 1998, **17**, 2114–2121.
72. S. Yu, S. Tang, G. D. Mayer, G. P. Cobb and J. D. Maul, Interactive effects of ultraviolet-B radiation and pesticide exposure on DNA photo-adduct accumulation and expression of DNA damage and repair genes in *Xenopus laevis* embryos, *Aquat. Toxicol.*, 2015, **159**, 256–266.
73. S. A. Morison, R. L. Cramp, L. A. Alton and C. E. Franklin, Cooler temperatures slow the repair of DNA damage in tadpoles exposed to

ultraviolet radiation: Implications for amphibian declines at high altitude, *Glob. Change Biol.*, 2020, **26**, 1225–1234.

74. M. P. Lesser, S. L. Turtle, J. H. Farrell and C. W. Walker, Exposure to ultraviolet radiation (290-400 nm) causes oxidative stress, DNA damage, and expression of p53/p73 in laboratory experiments on embryos of the spotted salamander, *Ambystoma maculatum*, *Physiol. Biochem. Zool.*, 2001, **74**, 733–741.
75. O. Merk, K. Reiser and G. Speit, Analysis of chromate-induced DNA-protein crosslinks with the comet assay, *Mutat. Res.*, 2002, **471**, 71–80.
76. N. P. Singh, Microgels for estimation of DNA strand breaks, DNA protein crosslinks and apoptosis, *Mutat. Res.*, 2000, **455**, 111–127.
77. C. O. Everard, D. G. Carrington, H. Korver, R. Burke, J. D. Everard and C. Gravekamp, Leptospires in the whistling frog (*Eleutherodactylus johnstonei*) on Barbados, *J. Trop. Med. Hyg.*, 1990, **93**, 140–145.
78. G. R. Bourne, Reproductive behavior of terrestrial breeding frogs *Eleutherodactylus johnstonei* in Guyana, *J. Herpetol.*, 1997, **31**, 221–229.
79. D. Rödder, Human footprint, facilitated jump dispersal, and the potential distribution of the invasive *Eleutherodactylus johnstonei* Barbour 1914 (Anura Eleutherodactylidae), *Trop. Zool.*, 2009, **22**, 205–217.
80. F. L. Meza-joya, E. P. Ramos-pallares and M. P. Ramírez-pinilla, Ontogeny of the vertebral column of *Eleutherodactylus johnstonei* (Anura: Eleutherodactylidae) reveals heterochronies relative to metamorphic frogs, *Anat. Rec.*, 2013, **296**, 1019–1030.

CHAPTER 11

Morphological Responses as a New Tool to Evaluate the Effects of Agrochemicals and Other Emerging Contaminants in Neotropical Frogs

LILIAN FRANCO-BELUSSI*[a,b], BRUNO S. L. VALVERDE[a], NADIA C. BACH[c], JUAN M. PÉREZ-IGLESIAS[d,e], CLASSIUS DE OLIVEIRA[a], GABRIEL C. CANEVARI[f] AND CARLOS E. FERNANDES[b]

[a]Department of Biological Sciences, São Paulo State University (UNESP), Cristóvão Colombo 2265, 15054-000, São José do Rio Preto, Brazil; [b]Laboratório de Patologia Experimental, Universidade Federal de Mato Grosso do Sul (UFMS), Avenida Costa e Silva s/n, 79070-900, Campo Grande, Brazil; [c]Área de Biología, Departamento de Biología, Facultad de Química, Bioquímica y Farmacia, Universidad Nacional de San Luis (UNSL), Ejército de los Andes 950, San Luis (5700), Argentina; [d]Instituto de Química de San Luis (INQUISAL), Consejo Nacional de Investigaciones Científicas y Técnicas (CONICET), Facultad de Química, Bioquímica y Farmacia (FQByF), Universidad Nacional de San Luis (UNSL), Chacabuco 917, D5700BWS, Juan Martín de Pueyrredón, San Luis, Argentina; [e]Department of Environmental and Production Sciences, National University of the Comechingones (UNLC), Heroes de Malvinas 1587 (CP5881), Villa de Merlo, San Luis, Argentina; [f]Programa de Pós-Graduação em Biodiversidade, Departamento

Issues in Toxicology No. 46
Amphibian Species in Environmental Risk Assessment Strategies
Edited by Marcelo L. Larramendy and Guillermo Eli Liwszyc

Published by the Royal Society of Chemistry, www.rsc.org

de Ciências Biológicas, São Paulo State University (UNESP), 15054-000, São José do Rio Preto, Brazil
*E-mail: lilian.belussi@gmail.com

11.1 The Importance of Morphological Biomarkers

Biological markers are commonly used to assess the effects of xenobiotics because they quantify individual responses of an exposed organism. However, the correct choice of biological markers and adequate bioindicator organisms are essential for determining the effects of contaminants on the biotic matrix. Morphological and histological characteristics are good intermediate response markers, as they integrate biochemical and physiological responses. Morphological biomarkers can detect changes ranging from subcellular and cellular levels to individual abnormalities. Morphological changes in organs and tissues may be defensive or compensatory to maintain tissue and organ functions against degenerative alterations.[1] These biomarkers are adequate tools in laboratory experiments and environmental monitoring programs because they are in an intermediate position at organismal organization levels. As such, they allow for the understanding of how contaminants disrupt physiological functions, predicting their ecological consequences. Thus, the choice of organ/tissue, techniques and a model organism are key factors when evaluating the effects of xenobiotics.

The techniques used to analyse morphological responses determine the levels at which changes in the target tissue will be assessed. Thus, defining standardized morphological tools and comparing different morphological techniques for measuring responses are essential for testing tissue changes. There is still little comparative work on morphological techniques that test the effects of external agents on target organs. Therefore, investigating how the choice of organ and technique influence study outcomes is needed to guide not only ecotoxicological studies but also environmental monitoring programs. Here, we review the application of morphological techniques to evaluate changes in organs and tissues with distinct structural and functional properties using Neotropical anurans selected as an experimental model, *e.g.*, the whistling frog *Leptodactylus fuscus* (Leptodactylidae), the veined tree frog *Trachycephalus typhonius* (Hylidae) (see Figure 11.1), the Chaco tree frog *Boana raniceps* (Hylidae), the Cuyaba dwarf frog *Physalaemus nattereri* (Leptodactylidae) (see Figure 11.2), the pointed belly frog *Leptodactylus podicipinus* (Leptodactylidae) and the dwarf tree frog *Dendropsophus nanus* (Hylidae) (see Figure 11.3).

Figure 11.1 Dorsal view of adults of the whistling frog *Leptodactylus fuscus* (Leptodactylidae) and the veined tree frog *Trachycephalus typhonius* (Hylidae) used as experimental models. Scale = 1 cm.

Figure 11.2 Dorsal view of adults of the Chaco tree frog *Boana raniceps* (Hylidae) and the Cuyaba dwarf frog *Physalaemus nattereri* (Leptodactylidae) used as experimental models. Scale = 1 cm.

11.2 Morphological Techniques to Detect Responses of Target Organs to Xenobiotics

Histological techniques allow accurate quantitative assessments at a low cost. Quantitative techniques are more objective, making the results more reproducible. They also allow comparisons between groups and are useful for testing hypotheses.[2] In addition, different morphological techniques can

Figure 11.3 Dorsal view of adults of the pointed belly frog *Leptodactylus podicipinus* (Leptodactylidae) and the dwarf tree frog *Dendropsophus nanus* (Hylidae) used as experimental models. Scale = 1 cm.

be used to evaluate the effects of environmental stressors. Below, we review four techniques for evaluating target organs and tissues: *bone morphometry*, *histomorphometry*, *structural volumetric density* and *colour-difference analysis*.

11.2.1 Bone Morphometry

Bones are key biomarkers to evaluate the effect of xenobiotics. Xenobiotics can cause endocrine changes, especially in the thyroid gland, which impact larval development and the skeletal system which later affect adults.[3–5] Nonetheless, bones are rarely used as a marker for analysing the effects of environmental stressors, and techniques still need to be standardized to demonstrate their feasibility and elucidate patterns of response.

Bone morphometry has many applications, from understanding macroevolutionary patterns of locomotion,[6] morphofunctional aspects (*e.g.*, locomotor mode)[7] to analysing the effects of environmental contaminants.[8] The methods can also vary, ranging from X-rays for bone visualization,[6] histological sections[8] to clearing and staining the whole organism.[7] Measurements can be performed either manually (*e.g.*, digital calipers or ocular micrometers)[7,8] or using specific software.[6]

11.2.1.1 Bone Morphometry Applied to the Skeleton Analyses

The skeleton has already been used in frogs as an effect biomarker for insecticides, metals and fluoride.[3,5,8–10] However, it is still underused as an analytical tool to evaluate the response to xenobiotics. Advanced techniques that detect sublethal effects are also needed.

The main functions of the skeletal system are to provide shape and support to the body, protect organs and tissues and provide locomotion, together with muscles and joints.[11] The skeleton can be divided into two units: the cranial and postcranial skeletons. The former is solely composed of the skull, while the latter consists of the axial skeleton (spine and ribcage) and appendicular skeleton (limbs, scapular and pelvic girdles).[11] The shapes of skeletal elements of frogs can vary according to their functions and the species' habitat. The appendicular skeleton is adapted to jumping, in which the scapular girdle and forelimbs absorb impacts while the pelvic girdle and hind limbs, which are quite developed and elongated, have a propulsive role.[12–14]

A cleared and stained specimen is used here for bone morphological analysis. This is accomplished first by euthanizing, eviscerating and removing the skin of animals, followed by fixation in non-buffered 10% formalin for 48 h and dehydration in an increasing alcoholic series. Preparation of specimens is followed by clearing and staining according to Taylor and van Dyke's protocol,[15] replacing trypsin with pancreatin. Then, bones are stained with Alizarin red and cartilages with Alcian blue (see Figure 11.4). Afterward, the specimen is photographed in ventral view with a stereomicroscope and analysed with the 'create trace feature' tool of Image-ProPlus 6.0 software. The lengths of the largest bones of the fore (humerus and radius-ulnar) and hind limbs (femur, tibiofibula, calcaneus and astragalus) are measured from the proximal to the distal extremity of the bone that inserts into the epiphyseal cartilage or from the proximal to the distal epiphysis, depending on the developmental stage of the specimen. As bones may be curved or bent, the median sagittal line is used for measurements (see Figure 11.4). The overall size of hind limbs was performed by the sum of the length of the largest bones of the limb.

In the case of metamorphosing and newly metamorphosed specimens (up to developmental Gosner stage 46), the long bones are still ossifying.[16] Ossification occurs up to the extremities of the bone that are inserted into the epiphyseal cartilages because epiphyseal cartilage mineralization is a late process that occurs after metamorphosis and is not essential for bone elongation.[17–19] Therefore, the measurements are made from the proximal to the distal extremity of the bone that inserts into the epiphyseal cartilage in young individuals. For adults whose epiphyses are already mineralized, measurements are made from the proximal to the distal epiphysis.

To demonstrate the effects of phenylpyrazole insecticide fipronil on bone development in frogs, we exposed specimens of the frogs *Leptodactylus fuscus*, *Trachycephalus typhonius* (see Figure 11.1) and the snouted tree frog *Scinax fuscovarius* (Hylidae) to 0.4 mg fipronil L^{-1} during the larval development up to the newly metamorphosed phase (*i.e.*, 45–46 Gosner stage) (G.C. Canevari, unpublished data). Fipronil is an insecticide that alters bone formation.[20] We found that the length of the femur, calcaneus and astragalus of *L. fuscus* and *S. fuscovarius* decreased (see Figure 11.5),

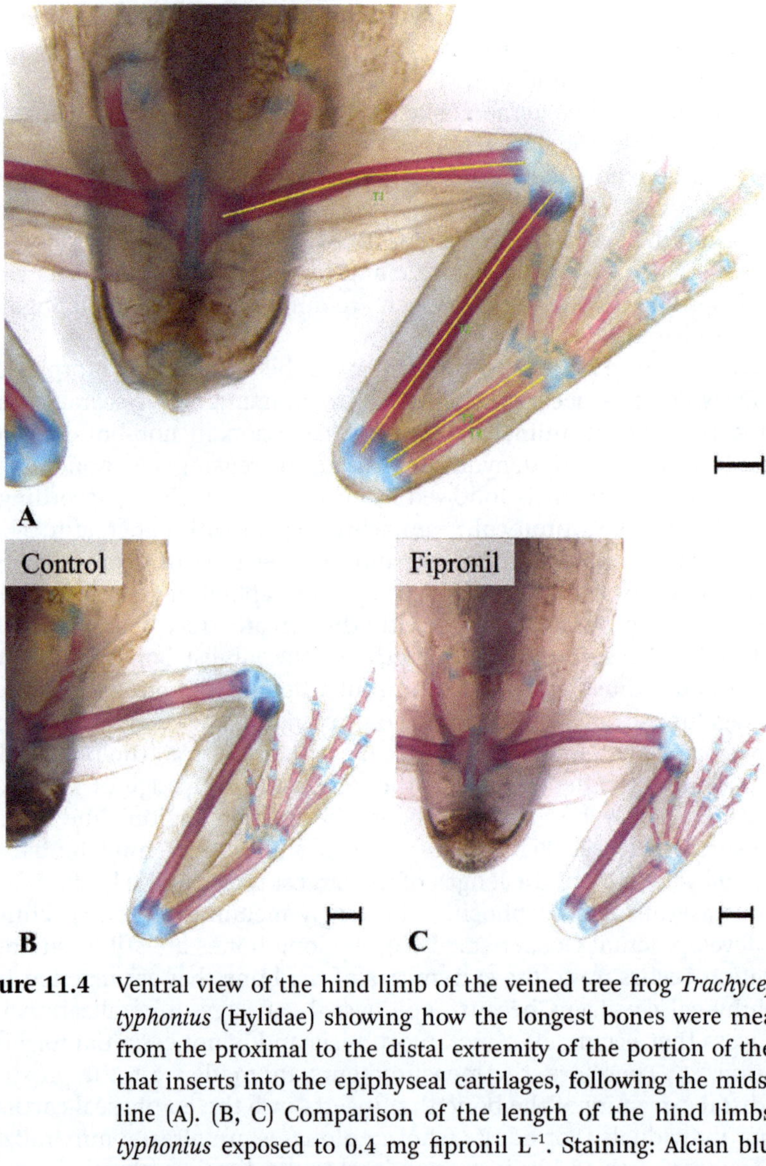

Figure 11.4 Ventral view of the hind limb of the veined tree frog *Trachycephalus typhonius* (Hylidae) showing how the longest bones were measured from the proximal to the distal extremity of the portion of the bone that inserts into the epiphyseal cartilages, following the midsagittal line (A). (B, C) Comparison of the length of the hind limbs of *T. typhonius* exposed to 0.4 mg fipronil L^{-1}. Staining: Alcian blue and Alizarin red. Scale = 1 mm.

but the tibiofibula decreased only in *S. fuscovarius*. However, the bones of *T. typhonius* did not change significantly when analysed separately, but the overall limb size decreased, as observed for *L. fuscus* and *S. fuscovarius* (see Figures 11.4B and C and 11.6). In summary, evaluating the length of long bones is a useful tool for evaluating the sublethal effect of contaminants on frogs. Nevertheless, it becomes important to consider limb size in this process rather than only assessing the size of separate bones individually.

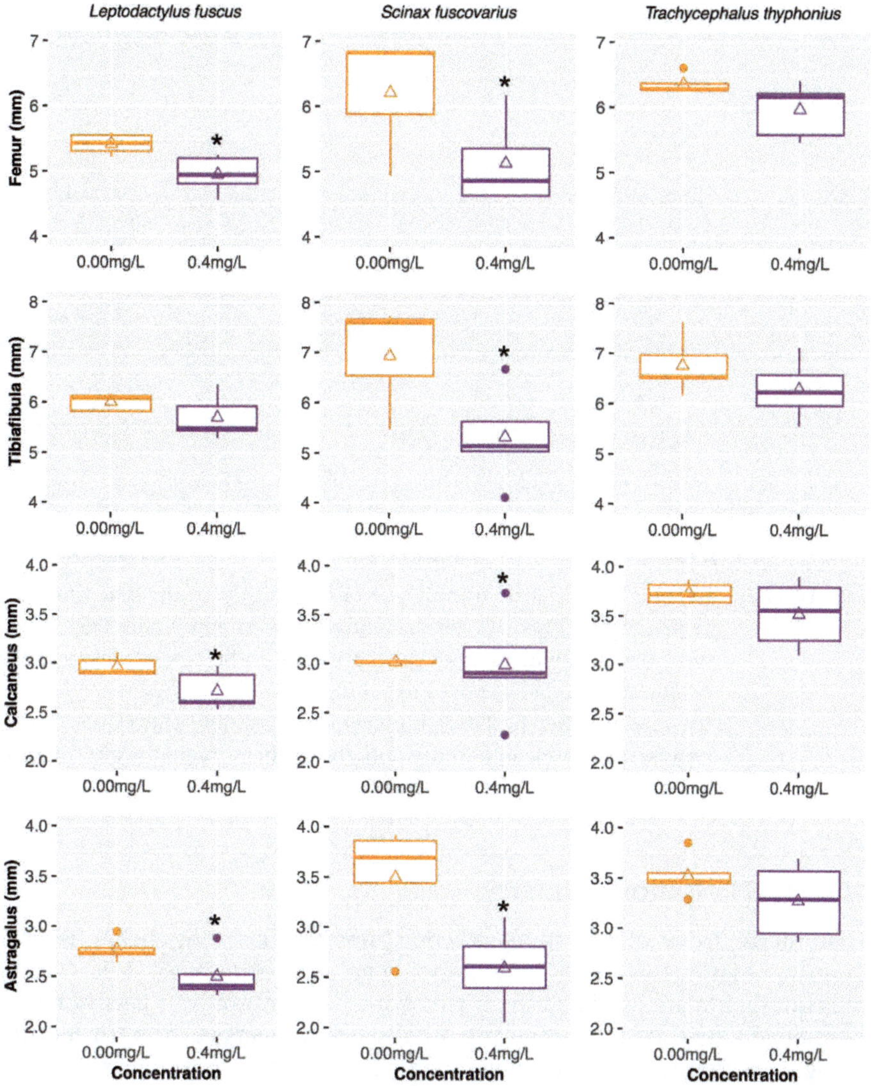

Figure 11.5 Boxplot showing the size of each bone in the hind limbs of the whistling frog *Leptodactylus fuscus* (Leptodactylidae), the snouted tree frog *Scinax fuscovarius* (Hylidae) and the common milk frog *Trachycephalus typhonius* (Hylidae) exposed to 0.4 mg fipronil L^{-1}. The lines represent the median and the boxes represent the first and third quartiles. Δ, represents the sample mean and the dots the outliers. *, significant differences with respect to control values.

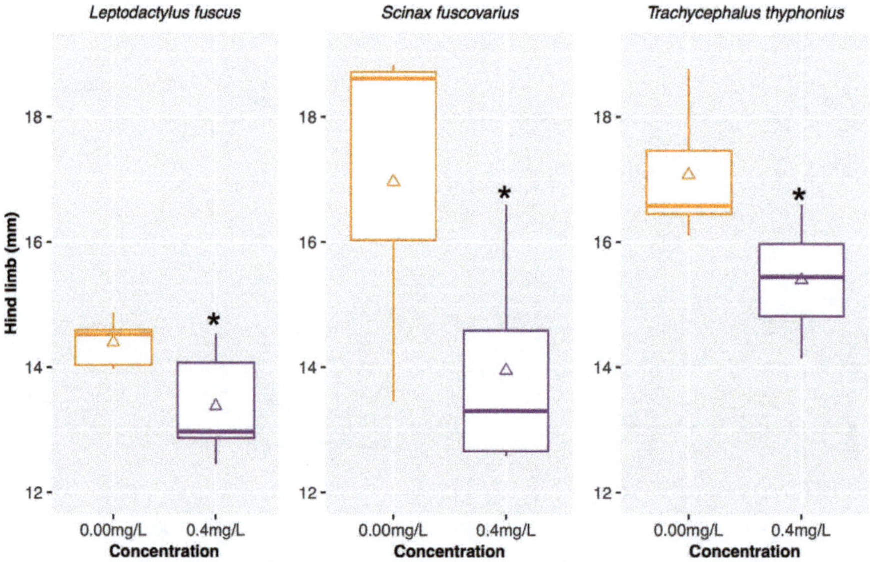

Figure 11.6 Boxplot showing the overall size of hind limbs of the whistling frog *Leptodactylus fuscus* (Leptodactylidae), the snouted tree frog *Scinax fuscovarius* (Hylidae) and the common milk frog *Trachycephalus typhonius* (Hylidae) exposed to 0.4 mg fipronil L^{-1}. The lines represent the median and the boxes represent the first and third quartiles. Δ, represents the mean and the dots the outliers. *, significant differences with respect to control values.

11.2.2 Histomorphometry

Histometry from digitized microscopic images has long been used to quantify histological and histopathological characteristics. Microscopic measurements are as old as the microscope itself. Anton van Leeuwenhoek developed a method for measuring microscopic objects in the 17th century, using sand grains and hair strands of different sizes as references. In 1674, he discovered that the human erythrocyte was 25 000 times smaller than a sand grain, corresponding to 8.5 µm.[21]

Modern histometric studies use dedicated software to obtain and process information from pixels. Pixels are then converted into actual units (*e.g.*, µm) after calibration.[22] This technique combines automated and semi-automated microscopic information with quantitative image analysis and allows the acquisition of unbiased multiparametric data at the single-cell level. This approach has been used to address diverse biological questions and identify and compare quantitative histological features in environmental bioindicators.[23]

11.2.2.1 Histomorphometry Applied to the Skin

The skin is the primary protective barrier against external factors in frogs. The skin may possess mucus secretions containing antimicrobial peptides and a variety of proteins of the innate immune system.[24,25] Frogs have thin, semi-permeable skin with several layers. Here we show histological sections of the skin of two Neotropical species: *Boana raniceps* (see Figure 11.2) and *Leptodactylus podicipinus* (see Figure 11.3). The outermost one is the epidermis (EP), which is multi-stratified and composed of a thin surface layer called *stratum corneum* that is traversed by ducts of dermal glands (see Figure 11.7). The dermis lies right beneath the epidermis and is composed of a superficial layer called *stratum spongiosum* (SS), which is made up of loose connective tissue with glands, blood vessels and pigment cells that give different colours to frogs. The deepest layer of the dermis is the *stratum compactum* (SC), composed of dense connective tissue. The Eberth-Kastschenko (EK) lies between the two layers of the dermis. This is a hydrophilic amorphous matrix, a calcified layer composed of glycosaminoglycans, proteoglycans and minerals, such as Ca. Environmental stressors can alter a frog's skin and decrease its ability to prevent infections.[25]

Due to the skin's structural characteristics, it has an important function in gas exchange.[26,27] Respiration through the skin is possible due to the presence of glands that keep the skin moist and a vascular system under the epidermis.[12,26]

A few studies have evaluated the effects of pollutants and abiotic factors, such as UV radiation, on the frog's skin.[28–30] Rissoli *et al.* experimentally demonstrated that the skin thickness of American bullfrog *Aquarana castebeiana* tadpoles (Ranidae) increases after exposition to glyphosate-based herbicides, which correlates with decreased metabolic rate and oxygen consumption.[29] Fenoglio *et al.* tested the effect of metals on the activity of antioxidant enzymes of keratinocytes in the skin of the edible frog *Rana esculenta* (Ranidae).[28] They found that the epidermis exhibited morphofunctional plasticity in response to contamination. However, little is known about the effects of aquatic xenobiotics or the synergism between these contaminants and abiotic factors on the skin characteristics of Neotropical frogs.

Histomorphometry of the skin layers uses 3 μm histological sections stained with hematoxylin and eosin (HE). For histomorphometry, 10 images per animal are taken at 400× magnification. For demonstration purposes, we will use Image-Pro Plus 6.0 software, but any software that allows linear measurements can be used. The 'measurements' tool can be used for measuring the thickness of skin layers and the total thickness of the skin (see Figure 11.7C). To systematize sampling, five random measurements are taken of each stratum per image, totaling 25 measurements per image.

A previous study (B. S. L. Valverde, unpublished data) exposed the frog *P. nattereri* (see Figure 11.2) to 2 mg benzo[alpha]pyrene kg^{-1} *via* the skin

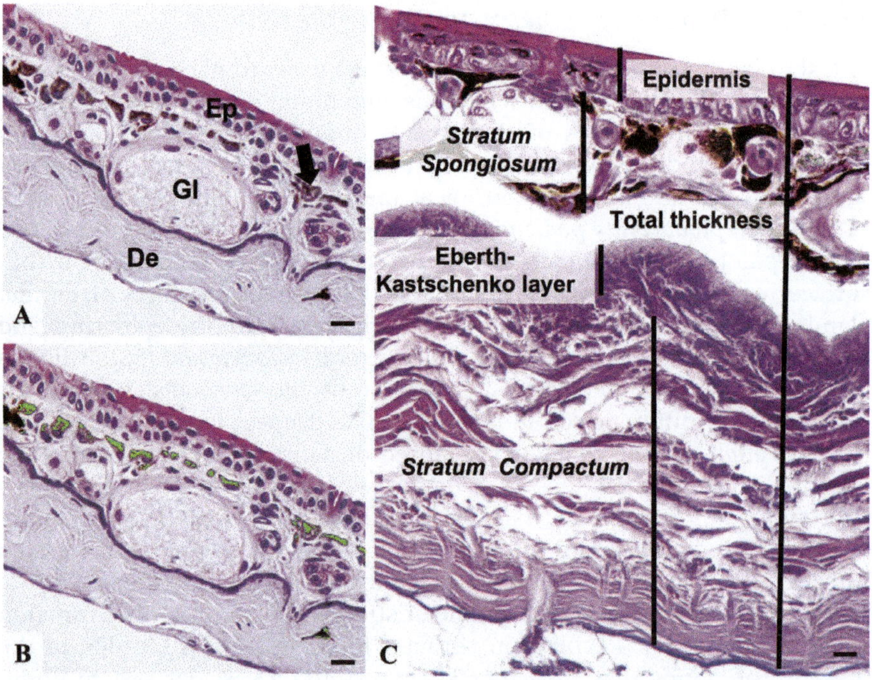

Figure 11.7 Histological section of the skin of the Chaco tree frog *Boana raniceps* (Hylidae) (A). Notice the thin epidermis (Ep), glands (Gl), melanocytes (arrow) that confer the colouration to animals and the dermis (De) below the epidermis. Histological section of the skin of *B. Raniceps* demonstrating the technique of marking the melanin pigment (B) and the skin of the pointed belly frog *Leptodactylus podicipinus* (Leptodactylidae) showing histomorphometry technique (C). Staining: HE. Scale = 10 μm.

for 1 and 7 days. The thickness of the SC and SS and the total thickness of the skin increased in animals exposed for 1 day (see Figure 11.8). At 7 days, the thickness of the EP, SC and EK layers decreased as well as the total skin thickness (see Figure 11.8). Another study found that the thickness of the EP, SS and EK layers and the total skin thickness were thinner in the frog *L. podicipinus* (see Figure 11.3) collected in urban environments (L. Franco-Belussi, unpublished data) (see Figure 11.9). However, the SC was thicker in animals from urban environments (see Figure 11.9). Taken together, these results suggest that the thickness of the EP and EK layers are the most important end-points to be measured in both experimental and environmental monitoring studies.

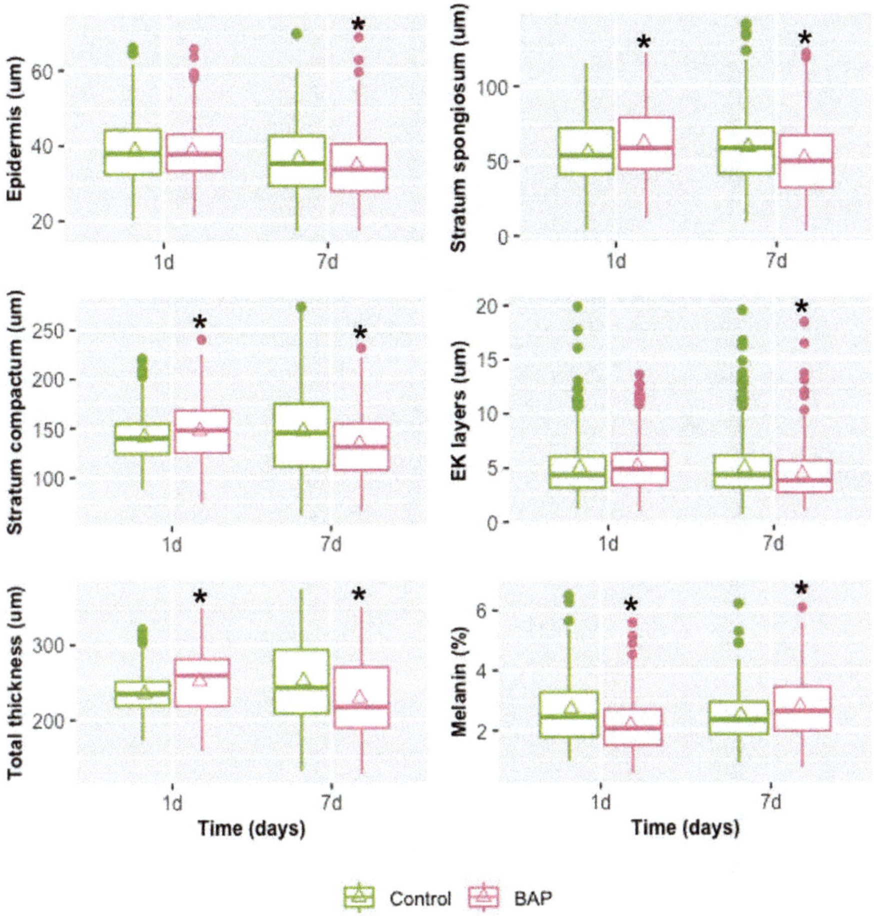

Figure 11.8 Boxplot showing the variation in skin layers of the Cuyaba dwarf frog *Physalaemus nattereri* (Leptodactylidae) cutaneously exposed to 3 mg benzo[alpha]pyrene (BAP) kg^{-1} over a 1- and 7-day exposure period. The lines represent the median and the boxes represent the first and third quartiles. Δ, represents the sample mean and the dots the outliers. *, significant differences with respect to control values from the same experimental time.

11.2.2.2 *Histomorphometry Applied to the Intestine*

The digestive system plays an important role in maintaining homeostasis[12] and providing energy and nutrient absorption.[31] In amphibians, this system has two components: the digestive tract (oral cavity, esophagus, stomach, small intestine, large intestine and cloaca), which is directly involved in the digestive processes and the accessory glands (liver and pancreas), which aid in the digestive process.[12,32] The small intestine consists of an elongated

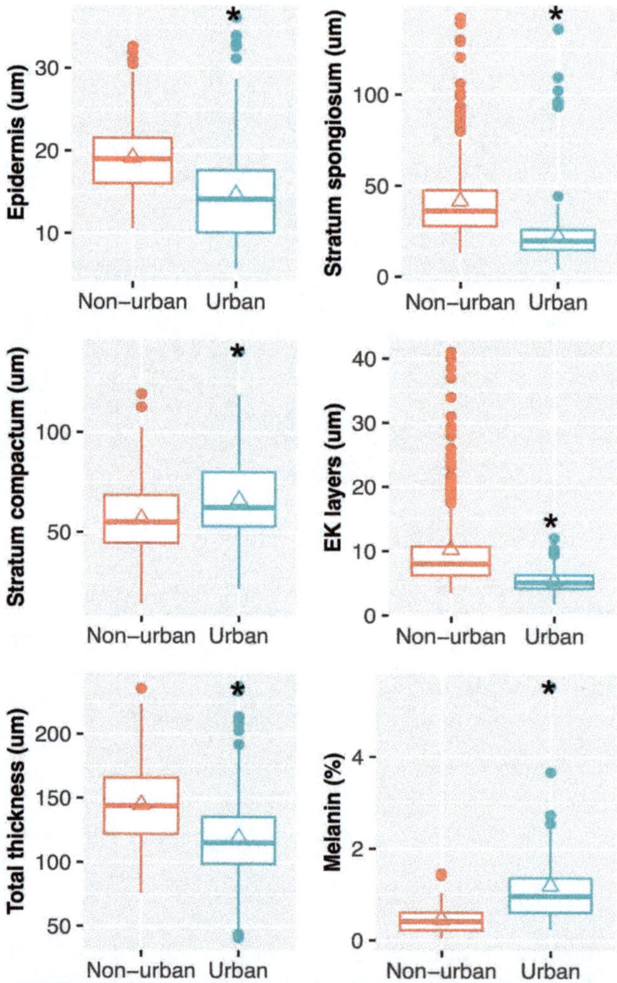

Figure 11.9 Boxplot showing the variation in skin layers of the pointed belly frog *Leptodactylus podicipinus* (Leptodactylidae) from urban and non-urban sites. The lines represent the median and the boxes represent the first and third quartiles. Δ, represents the sample mean and the dots the outliers. *, means significant differences between collection sites.

organ that is uniformly wide and lies between the stomach and the large intestine. It is composed of four layers (or tunics): serosa, muscular, submucosa and mucosa (see Figure 11.10). The mucosa corresponds to the epithelial portion, featuring simple columnar cells with a brush border in the apical portion.[12,33]

Several studies have assessed the effects of chemical compounds on the gut of frogs, such as fertilizers (*e.g.*, nitrate), fungicides (*e.g.*, triadi-

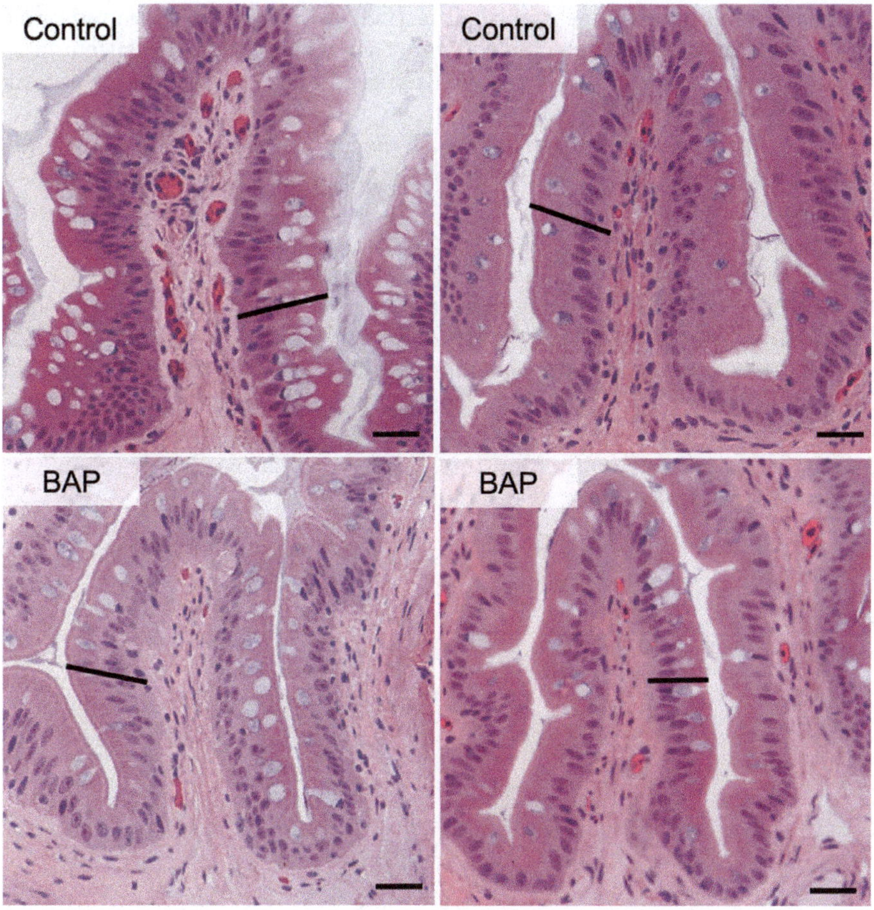

Figure 11.10 Histological sections of the intestine of the Cuyaba dwarf frog *Physalaemus nattereri* (Leptodactylidae) exposed to oral and injectable benzo[alpha]pyrene (BAP) at a concentration of 3 mg kg^{-1} for 3 days. The black bars illustrate the histomorphometric technique. Staining: HE. Scale = 25 μm.

mefon), herbicides (*e.g.*, 2,4-D, atrazine, glyphosate), hydrocarbons (*e.g.*, benzo[alpha]pyrene) and insecticides (*e.g.*, delmatrin, carbaryl).[34–39] The main analytical approach includes recording the occurrence of increased secretion of goblet cells, congestion in the tunica muscularis, degeneration of the villi, detachment, disorganization, vacuolization of the epithelium, oedema, hemorrhage, infiltration of inflammatory cells and necrosis.[34–39] Additionally, changes in the epithelium thickness caused by xenobiotic exposure have also been found.[38]

Histomorphometric analyses of the intestinal epithelium are performed on 2 μm histological sections stained with HE and photographed at 100×

magnification. The thickness of the intestinal epithelium is measured in 25 images per animal with selected software that allows linear measurements. The thickness (or height) of the intestinal epithelium should be measured from the base of the epithelium, which is in contact with the connective tissue, to its apex, including the brush border. A good sampling of the tissue is made by taking 30 random measurements of the epithelium per image (see Figure 11.10).

A previous study (B.S.L. Valverde, unpublished data) found that the intestinal epithelium thickness of *P. nattereri* decreased (see Figures 11.2, 11.10, and 11.11) after a 3-day exposure to 3 mg benzo[alpha]pyrene kg^{-1} orally administered and injected. Studies with another anuran, the white-spotted tree frog *Boana albopunctata* (Hylidae), also reported decreased intestinal epithelium caused by benzo[alpha]pyrene.[38] Then, intestinal epithelium thickness can be used to evaluate a rapid (*e.g.*, 3 days) effect of contaminants, and epithelium intestinal is a good parameter to analyse in the intestine from exposed specimens.

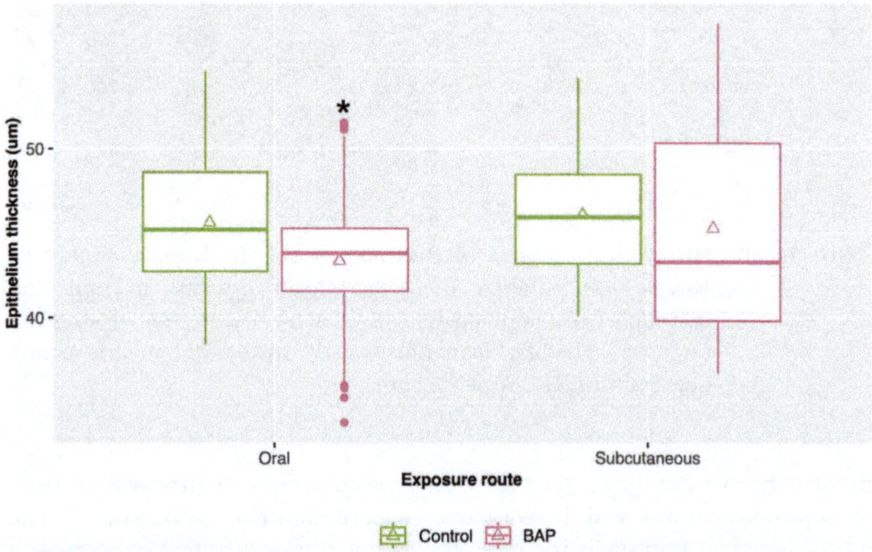

Figure 11.11 Boxplot showing the thickness of the intestinal epithelium of the Cuyaba dwarf frog *Physalaemus nattereri* (Leptodactylidae) exposed to benzo[alpha]pyrene (BAP) orally and injected at a concentration of 3 mg Kg^{-1} for 3 days. The lines represent the median and the box represents the first and third quartile. Δ, represents the sample mean and the dots the outliers. *, means significant differences between control and treatment..

11.2.2.3 Histomorphometry Applied to the Liver

The liver is the largest gland in vertebrates.[40] It stores nutrients absorbed from the digestive tract and releases them into the tissues. Liver metabolism is primarily related not only to the anabolism of proteins, lipids and carbohydrates but also to the catabolism of glycogen and nitrogen. Liver functions also include storage of substances, secretion of bile and detoxification of pollutants and toxic agents.[40] Therefore, the liver plays an essential role in maintaining body homeostasis.[40,41] The histological characteristics of the liver (organization, hepatocyte size and amount of melanomacrophage) vary among species. For example, we show histological sections of the liver of three species: *B. raniceps*, *D. nanus* and *L. podicipinus* (see Figure 11.12).

The liver is heavily affected by environmental stressors due to its detoxification functions, such as aquatic contaminants or abiotic factors that alter the homeostasis of organisms.[42–44] Several studies have evaluated

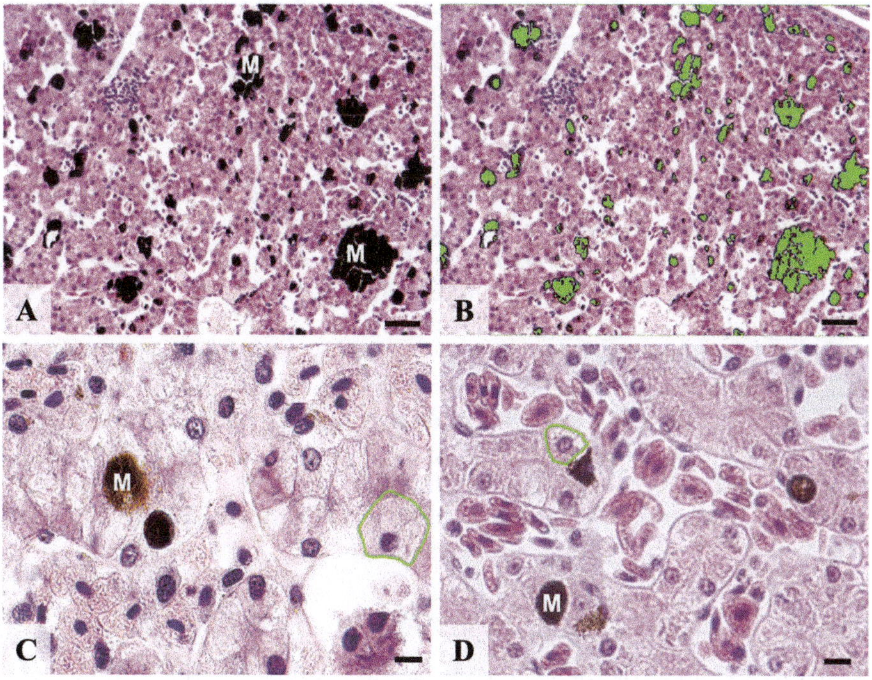

Figure 11.12 Histological liver sections of the Chaco tree frog *Boana raniceps* (Hylidae) demonstrating the technique (A, B) of marking the pigment contained in melanomacrophage (M). Histological sections demonstrating the histomorphometry technique of the hepatocyte area and its nucleus, demonstrating size variation between the dwarf tree frog *Dendropsophus nanus* (Hylidae) (C) and the pointed belly frog *Leptodactylus podicipinus* (Leptodactylidae) (D). Staining: HE. Scale = A–B = 25 μm; C–D = 5 μm.

the effects of metals, insecticides and herbicides on liver enzymes and on biomarkers of frogs, such as melanomacrophage.[43–47] However, studies evaluating the effects of xenobiotics in Neotropical anuran species are scarce, with most of the studies using larvae and reporting biochemical and genotoxic effects.[43,48,49] However, a recent study has quantified structure counts, linear measurements (height, width and length), cell and nucleus area and perimeter, second-order measures (volume, shape factor and nuclear circularity) and the volumetric density of several liver structures.[50] The authors found that liver structures differ between species and the sexes of the same species. These differences are attributed to the bioenergetic and physiological aspects of the species. Additionally, differences in liver architecture may indicate a distinct rate of metabolization of xenobiotics, since the greater the contact area of the sinusoids with hepatocytes, the greater the area of substance uptake.[51] Thus, morphological and comparative studies using different frog species remain key to improving our knowledge of how xenobiotics alter the liver.

The histomorphometry of hepatocytes is often evaluated with 3 μm histological sections stained with HE. Then, 10 images per animal are taken at 1000× magnification. Finally, the area, diameter and perimeter of six hepatocytes and their nuclei are measured per picture using the line artool of software (see Figure 11.12C and D). Subsequently, the volume of the nucleus and hepatocyte is calculated indirectly, according to Leão *et al.*[50]

A previous study found that the area and volume of hepatocytes of the spotted-flanks four-eyed frog *Pleurodema tucumanum* (Leptodactylidae) exposed to glyphosate herbicide for 96 h decreased (J. M. Pérez-Iglesias, unpublished data) as did the volume of the nucleus at both 100 ppm and 1 000 ppm (see Figure 11.13). The nucleus area was smaller only at 100 ppm. Similarly, both hepatocyte area and nucleus volume decreased in the pointed belly frog *L. podicipinus* in urban compared to non-urban environments (L. Franco-Belussi, unpublished data; see Figure 11.14). These results suggest a homogeneous response of the liver to environmental changes and demonstrate the efficiency of both the technique and the biomarkers to evaluate the deleterious effects of contaminants.

11.2.3 Structural Volumetric Density

Structural volumetric density consists in estimating the volume of a structure based on a tissue fragment. It consists in transforming planimetric data, which is 2D, into a tri-dimensional volume of each tissue component. Briefly, a grid is overlaid onto an image with *n* intersections, and the observer counts the number of vertices on the top of the structure of interest. The density of each component is calculated according to Freere and Weibel,[52] Reid,[53] Rocha *et al.*,[54] and Leão *et al.*[50] to obtain the proportion that each structure occupies within the tissue. The output is quantitative data on the internal structures of organs, tissues and cells.[52] For

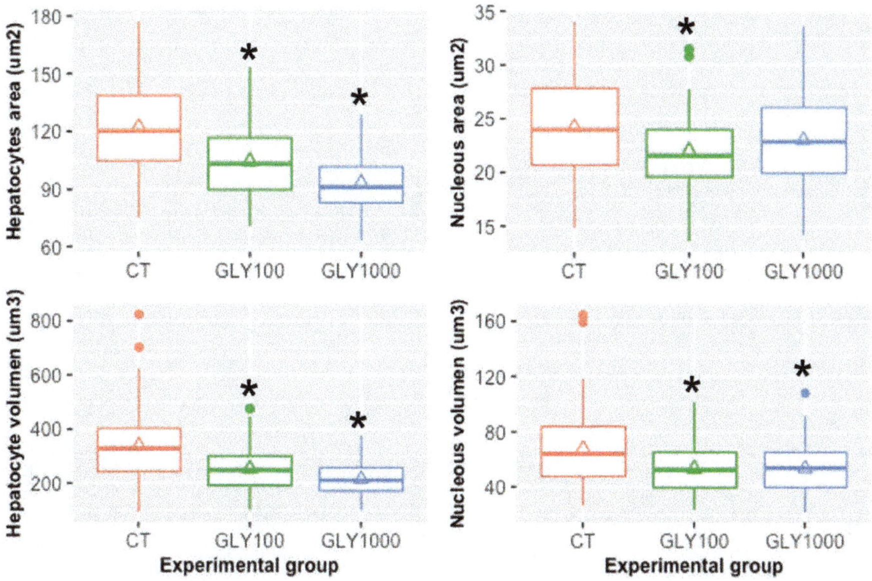

Figure 11.13 Boxplot showing the variation in area and volume of hepatocytes as well as the area and volume of nuclei of the spotted-flanks four-eyed frog *Pleurodema tucumanum* (Leptodactylidae) exposed to the herbicide glyphosate at concentrations of 100 ppm (GLY100) and 1000 ppm (GLY1000). CT is the control group, where the animals were not exposed to the herbicide. The lines represent the median and the boxes represent the first and third quartiles. Δ, represents the sample mean and the dots the outliers. *, significant differences with respect to control values.

this technique to be applied satisfactorily, the acquisition of histological sections must be done in a systematic manner to ensure adequate spacing between histological sections so that there is no overlapping of structures. Additionally, samples of the entire tissue must be evaluated, including the entire extent of the organ, to check for possible historegionalization effects.

11.2.3.1 Structural Volumetric Density Applied to the Kidney

Another organ involved in the metabolism of toxic substances is the kidney. This organ is responsible for water balance and the elimination of harmful metabolic wastes.[55] The kidney and the liver play a key role in the biotransformation and elimination of toxic substances, including environmental contaminants. Structural and functional characteristics of the kidney, such as a large blood flow, an ion transport system, biotransformation enzymes and its ability to concentrate solutes during urine production, make the kidney a target organ for xenobiotics.[56,57]

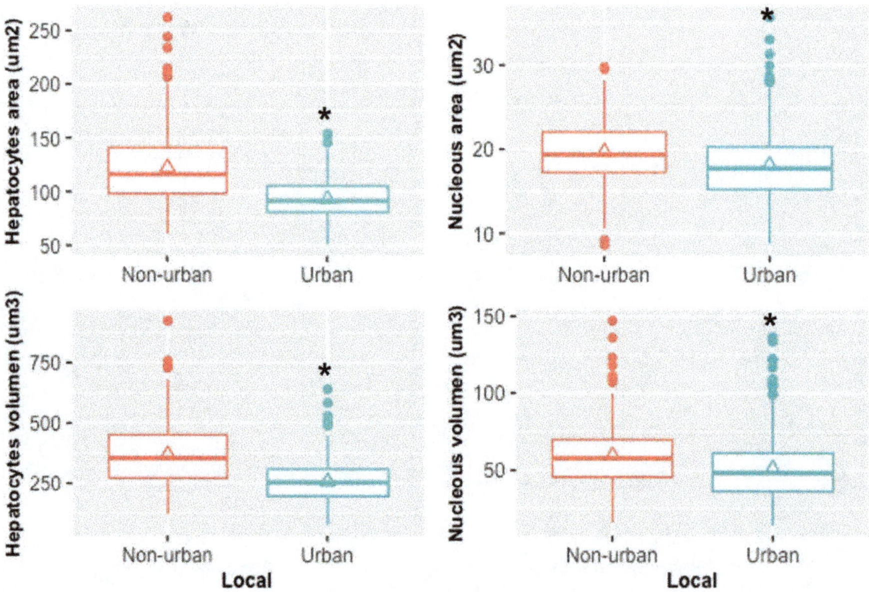

Figure 11.14 Boxplot showing the variation in area and volume of hepatocytes as well as the area and volume of nuclei of the pointed belly frog *Leptodactylus podicipinus* (Leptodactylidae) collected in urban and non-urban environments. The lines represent the median and the box represents the first and third quartile. Δ, represents the sample mean and the dots the outliers. *, means significant differences between collection sites..

The renal tissue of frogs is composed of the renal corpuscles, located in the cortical region of the parenchyma and the distal and proximal contorted tubules (see Figure 11.15). The renal corpuscle is composed of the Bowman's capsule and the glomerulus, which is a complex set of tangled blood capillaries. Between these two structures is the Bowman's space, which is coated by a simple squamous epithelium with large, flattened nuclei. The glomerulus is attached to the inner surface of the capsule by a peduncle (vascular pole), through which the afferent and efferent arterioles of the glomerulus pass. The contorted tubules have a simple cuboidal epithelium, with the proximal one having a narrower lumen and thicker epithelium, while the distal one has a wider lumen and a thinner epithelium. For example, we show histological sections of the kidney of *B. raniceps* in Figure 11.2.

Most studies on the kidneys of frogs are experiments testing the effects of metal accumulation and enzymatic regulation.[47,58] At least two insecticides have already been demonstrated to cause histopathological changes in the kidneys of temperate frog species, *i.e.*, the toad *Bufotes variabilis* (Bufonidae) and the marsh frog *Pelophylax ridibundus*.[45,59] However, no

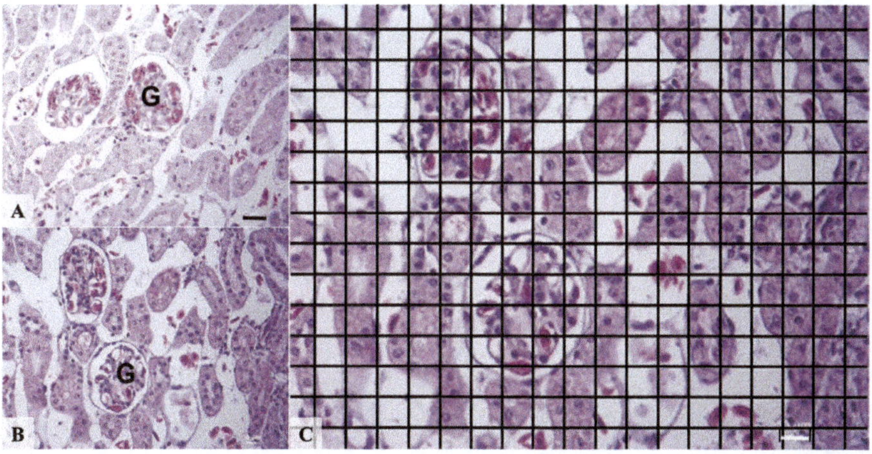

Figure 11.15 Histological sections of the kidney of the Chaco tree frog *Boana raniceps* (Hylidae) (A) and the pointed belly frog *Leptodactylus podicipinus* (Leptodactylidae) (B) demonstrating the arrangement of renal tubules and glomeruli (G). (C) Illustration of the volumetric density technique. Staining A–C: HE. Bars: A–B: 10 μm, C: 25 μm.

study has investigated the synergistic effects of multiple environmental contaminants on Neotropical species.

For the structural volumetric density analysis, 3 μm histological sections stained with HE are needed. Then, 10 images per animal are taken at 400× magnification and analysed using the 'cell counter' tool of ImageJ1.48 software. This analysis can be performed in any software that allows the superimposition of a grid on photomicrography. To calculate the volumetric density of each component of the renal tissue, we used a grid with 252 intersections (see Figure 11.15C). With it, the percentage of each component is calculated using the formula: SDV = [(number of intersections of each component*100)/252], following Freere and Weibel,[52] Reid,[53] Rocha *et al.*,[54] and Leão *et al.*[50] This technique can be applied to several tissues. The user has only to change the number of grid intersections accordingly.

A previous study found that the percentage of renal tubules increased while vessels, glomeruli and interstitium decreased (L. Franco-Belussi, unpublished data) in *L. podicipinus* collected in urban sites when compared to non-urban environments (see Figure 11.16), suggesting a compensatory mechanism due to a low reabsorption rate of metabolites and/or increased water reabsorption due to water imbalance. This technique is important for assessing the effects of contaminants on the tissue as a whole, since it assesses the amount of each structure.

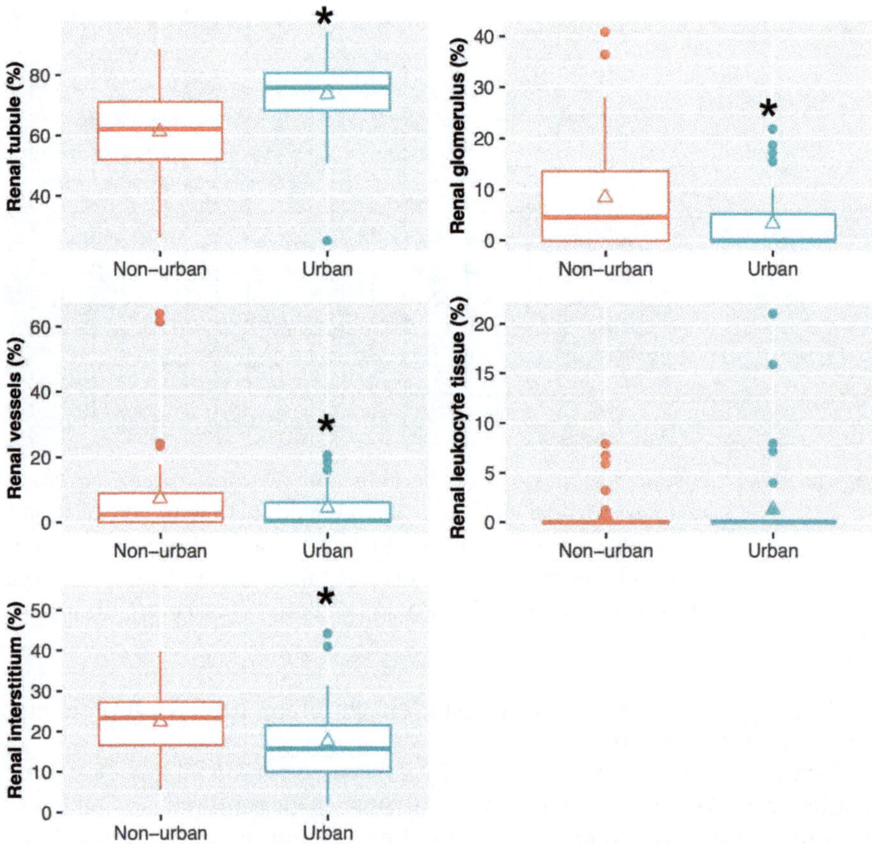

Figure 11.16 Boxplot showing the variation in renal tubules, vessels, glomerulus, interstitium and renal leukocyte tissue of the pointed belly frog *Leptodactylus podicipinus* (Leptodactylidae) from urban and non-urban sites. The lines represent the median and the boxes represent the first and third quartiles. Δ, represents the mean and dots the outliers.*, means significant differences between collection sites.

11.2.4 Colour-difference Analysis

The technique of colour-difference analysis allows for estimating the area that a certain colour occupies in the tissue (*e.g.*, any histological marker that contrasts with the tissue). The technique consists of using a digital tool that quantifies the area (μm^2) of a given structure in relation to the total area of a previously calibrated image in pixels using image analysis software. This technique can quantify both external staining of animals (*e.g.*, nuptial staining of Sticklebacks)[60] and substances in tissues, allowing the standardization of the analysis. This technique is complementary to that of Svensson *et al.*,[61] with which it is possible to quantify three colour properties provided by the software: lightness (L*), redness (a*) and yellowness (b*) using images converted to International Commission on Illumination (CIE, L*a*b*). An important limitation is that it only allows evaluating three-colour spectra, while colour-difference analysis allows evaluating any colour spectrum. This is because colour-difference analysis is based on the use of colour marking performed manually, making it possible to adjust sensitivity to increase accuracy.

11.2.4.1 Colour-difference Analysis Applied to the Skin and Liver

For colour-difference analysis of melanin in the skin and liver, 25 images per animal are taken at 200× stained with HE. One additional staining is needed for the quantification of catabolic substances, such as hemosiderin, which is done using Perl's method for ferric iron.[62] Any software that allows area estimation by colour differences can be used for quantifying the substances (*e.g.*, melanin and hemosiderin). The studies above used the 'count/size' tool of the Image-Pro Plus 6.0 software, with the colour

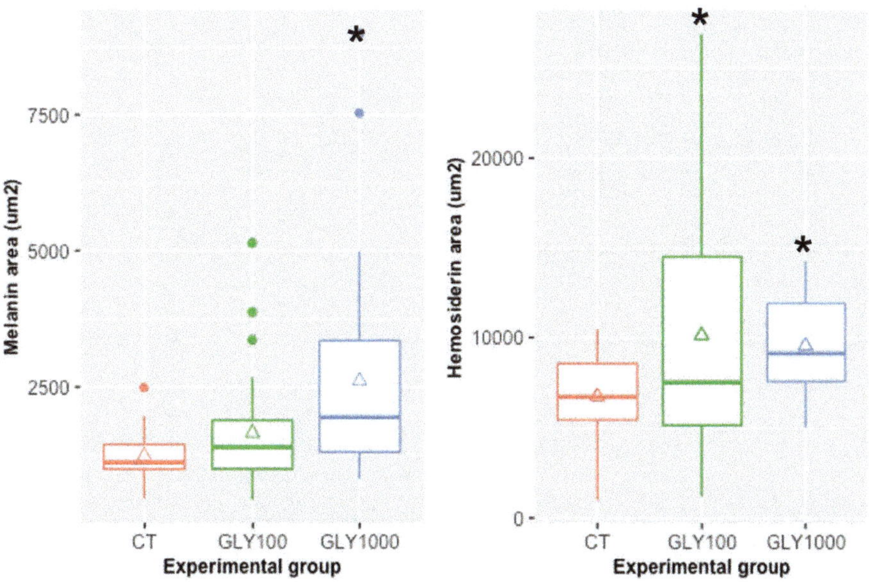

Figure 11.17 Boxplot showing the variation of the area occupied by melanin and hemosiderin in the hepatic melanomacrophage of the spotted-flanks four-eyed frog *Pleurodema tucumanum* (Leptodactylidae) exposed to the herbicide glyphosate at concentrations of 100 ppm (GLY100) and 1000 ppm (GLY1000). CT is the control group, where the animals were not exposed to the herbicide. The lines represent the median and the boxes represent the first and third quartiles. Δ, represents the sample mean and the dots the outliers. *, means significant differences between treatments.

maker (*e.g.*, black/brown colour) determined manually (see Figures 11.7B and 11.12B). If the histological section does not occupy the entire field and the total tissue area varies (see Figure 11.7B), it is necessary to measure the tissue area with the linear measurement so the output is standardized as the ratio pigment colour area : tissue area.

In the skin, cutaneous melanin can also be a morphological marker because it changes in response to environmental alterations. For example, the pigmentation area of *P. nattereri* increased after 7 days of exposure to benzo[alpha]pyrene (B. S. L. Valverde, unpublished data). The same occurred with *L. podicipinus* in urban environments (L. Franco-Belussi, unpublished data; see Figures 11.8 and 11.9).

Melanin and hemosiderin also increased in the liver of *P. tucumanum* exposed to 1000 ppm of glyphosate (see Figure 11.17), while hemosiderin increased at 100 ppm (J. M. Pérez-Iglesias, unpublished data). Similar results were observed in the oven frog *L. latinasus* after 48 h of exposure to 100, 1000 and 10 000 μg glyphosate g^{-1}, in which melanin increased, but hemosiderin decreased.[43] Increased hepatic melanin was also observed in *L. podicipinus* from urban environment-exposed specimens (L. Franco-Belussi, unpublished data; see Figure 11.18). These results

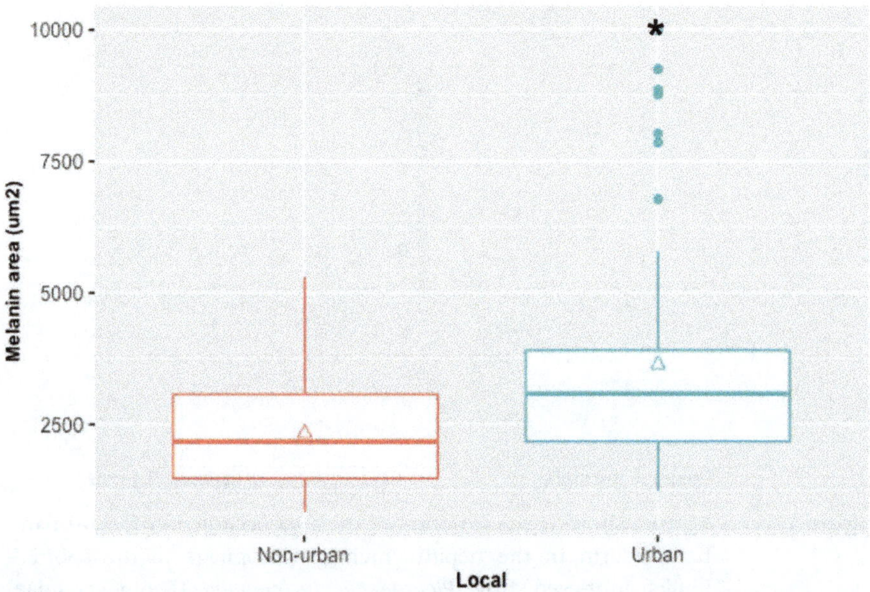

Figure 11.18 Boxplot showing the variation of the area occupied by melanin in liver melanomacrophages of the pointed belly frog *Leptodactylus podicipinus* (Leptodactylidae) collected in urban environments. Lines represent median and the box the first and third quartile. Δ, represents the sample mean and the dots the outliers. *, means significant differences between collection sites..

bring a new perspective on the influence of land use on cellular physiology, demonstrating how environmental changes might affect the internal morphological aspects of animals.[63] Therefore, melanin is a useful biomarker for the response of contaminants in the skin and liver, confirming the proprieties of the molecule and the utility of colour-difference analyses.

11.3 Final Remarks

Different techniques have been applied to evaluate the responses of organs and tissues to environmental stressors in frogs. Although most of the studies analysed specific organs because of their characteristics or functions, they are complementary. Techniques are a toolbox that allows evaluating the systemic effects of compounds in the body and are even more useful if used together with biochemical and genotoxic techniques to generate a comprehensive scenario of the responses of organisms to stressors. Additionally, ideally, researchers must evaluate different organs by applying the techniques reviewed here to understand whether they are coupled or not. Morphological techniques are excellent tools for testing the effects of contaminants in both laboratory experiments and field studies for distinct species and life stages. The application of histomorphometric techniques and colour-difference analysis, useful tools for evaluating different tissues, has proven to be applicable to the liver and skin of different species. We observed similarities in tissue responses of animals living in urban environments and experimental exposure for 96 h or more, demonstrating that the effects of environmental stressors are permanent and tissue responses homogeneous.

Acknowledgements

This study was funded in part by Fundação de Amparo à Pesquisa do Estado de São Paulo to C.O. (FAPESP, 2013/02067-5; 2015/12006-9; 2018/01078-7). L.F.B. was supported by a post-doctoral fellowship from São Paulo State University (PROPe UNESP 13/2022). C.O. and C.E.S.F. have continuously been supported by Conselho Nacional de Desenvolvimento Científico e Tecnológico (Proc#304552/2019-4 and #310058/2020-1, respectively). C.E.S.F. received a grant from Fundação de Apoio ao Desenvolvimento do Ensino, Ciência e Tecnologia do Estado de Mato Grosso do Sul (#71/700.136). The authors wish to thank Taynara Leão for her help in field work. Diogo B. Provete kindly translated and reviewed the manuscript.

References

1. C. A. O. Ribeiro and M. F. Narciso, in *Pollution and fish health in tropical ecosystems*, ed. E. A. Almeida and C. A. O. Ribeiro, CRC Press, Boca Raton, 2013, pp. 206–242.
2. C. A. Mandarim-de-Lacerda, C. Fernandes-Santos and T. D. Aguila, in *Histologyprotocols: methods in molecular biology*, ed. T. Hewitson and I. Darby, Humana Press, Totowa, 2009, pp. 211–225.
3. H. Zhao, L. Chai and H. Wang, Effects of fluoride on metamorphosis, thyroid and skeletal development in *Bufo gargarizans* tadpoles, *Ecotoxicology*, 2013, **22**, 1123–1132.
4. S. V. Smirnov and A. B. Vassilieva, Thyroid hormones in the skeletogenesis and accessory sources of endogenous hormones in *Xenopus laevis* (Amphibia; Anura) ontogeny: Experimental evidence, *Dokl. Biol. Sci.*, 2014, **455**, 136–138.
5. W. Wang, Y. Yang, A. Chen, X. Song and L. Chai, Inhibition of metamorphosis, thyroid gland, and skeletal ossification induced by hexavalent chromium in *Bufo gargarizans* Larvae, *Enviro. Toxic. Chem.*, 2021, **40**, 2474–2483.
6. M. E. Jorgensen and S. M. Reilly, Phylogenetic patterns of skeletal morphometrics and pelvic traits in relation to locomotor mode in frogs,*J. Evol. Biol.*, 2013, **26**, 929–943.
7. T. G. Petrović, T. D. Vukov and N. T. Kolarov, Morphometric ratio analyses: locomotor mode in anurans, *C. R. Biol.*, 2017, **340**, 250–257.
8. R. Alvarez, M. P. Honrubia and M. P. Herráez, Skeletal malformations induced by the insecticides ZZ-Aphox® and Folidol® during larval development of *Rana perezi*, *Arch. Environ. Contam. Toxicol.*, 1995, **28**, 349–356.
9. H. Yang, R. Liu, Z. Liang, R. Zheng, Y. Yang, L. Chai and H. Wang, Chronic effects of lead on metamorphosis, development of thyroid gland, and skeletal ossification in *Bufo gargarizans*, *Chemosphere*, 2019, **236**, 124251.
10. N. Sum, H. Wang, Z. Ju and H. Zhao, Effects of chronic cadmium exposure on metamorphosis, skeletal development, and thyroid endocrine disruption in Chinese toad *Bufo gargarizans* tadpoles, *Environ. Toxicol. Chem.*, 2018, **37**, 213–223.
11. K. V. Kardong, in *Vertebrados: anatomia comparada, função e evolução*, Roca, São Paulo, 5th edn, 2014.
12. W. E. Duellman and L. Trueb, in *Biology of amphibians*, The Johns Hopkins University Press, Baltimore and London, 2nd edn, 1994.
13. J. Klaczko, A. C. M. Ferreira, A. L. Falcão, G. Dillenburg, I. F. Oliveira, P. S. Wanderley, P. P. S. Queiroz, W. Coelho, V. Slobodianand and A. Sebben, in *Atlas fotográfico de anatomia comparada de vertebrados:*

Volume IV – Sistemas esquelético e muscular, Universidade de Brasília, Brasília, 2019.

14. B. S. Emerson, in *Functional Vertebrate Morphology*, ed. M. Hildebrand, D. M. Branble, K. F. Liem and D. B. Wake, Harvard University Press, Cambridge, MA and London, 1985, pp. 58–72.
15. W. R. Taylor and G. C. Van Dyke, Revised procedures for staining and clearing small fishes and other vertebrates for bone and cartilage study, *Cybium*, 1985, **9**, 107–119.
16. K. L. Gosner, A simplified table for staging anuran embryos and larvae with notes on identification, *Herpetologica*, 1960, **16**, 183–190.
17. S. L. Felisbino and H. F. Carvalho, Growth cartilage calcification and formation of bone trabeculae are late and dissociated events in the endochondral ossification of *Rana catesbeiana*, *Cell Tissue Res.*, 2001, **306**, 319–323.
18. S. L. Felisbino and H. F. Carvalho, The epiphyseal cartilage and growth of long bones in *Rana catesbeiana*, *Tissue Cell*, 1999, **31**, 301–307.
19. S. Miura, K. Hanaoka and S. Togashi, Skeletogenesis in *Xenopus tropicalis*: Characteristic bone development in an anuran amphibian, *Bone*, 2008, **43**, 901–909.
20. J. dos Santos Mendonça, L. Q. L. Hirano, A. L. Q. Santos, D. D. e Silva, C. Castro Pereira, M. S. Costa and L. G. Vieira, The exposure *in ovo* of embryos belonging to Amazonian turtle species *Podocnemis expansa* (Testudines) to commercial glyphosate and fipronil formulations impairs their growth and changes their skeletal development, *Sci. Total. Environ.*, 2022, **842**, 156709.
21. G. A. Meijer, J. A. Beliën, P. J. van Diest and J. P. Baak, Origins of image analysis in clinical pathology, *J. Clin. Pathol.*, 1997, **50**, 365–370.
22. R. Marcos, R. A. F. Monteiro and E. Rocha, The use of design-based stereology to evaluate volumes and numbers in the liver: a review with practical guidelines, *J. Anat.*, 2012, **220**, 303–317.
23. E. T. Arena, C. T. Rueden, M. C. Hiner, S. Wang, M. Yuan and K. W. Eliceiri, Quantitating the cell: turning images into numbers with ImageJ, *Wiley Interdiscip. Rev. Dev. Biol.*, 2017, **6**, e260.
24. L. A. Rollins-Smith, J. P. Ramsey, J. D. Pask, L. K. Reinert and D. C. Woodhams, Amphibian immune defenses against chytridiomycosis: impacts of changing environments, *Integr. Comp. Biol.*, 2011, **51**, 552–562.
25. A. Catenazzi, State of the world's amphibians, *Annu. Rev. Environ. Res.*, 2015, **40**, 91–119.
26. J. Czopek, Quantitative studies on the morphology of respiratory surfaces in amphibians, *Acta. Anat.*, 1965, **62**, 296–323.
27. G. J. Tattersall and G. R. Ultsch, Physiological ecology of aquatic overwintering in ranid frogs, *Biol. Rev.*, 2008, **83**, 119–140.

28. C. Fenoglio, A. Grosso, E. Boncompagni, G. Milanesi, C. Gandini and S. Barni, Morphofunctional evidence of changes in principal and mitochondria-rich cells in the epidermis of the frog *Rana esculenta* living in a polluted habitat, *Arch. Environ. Contam. Toxicol.*, 2006, **51**, 690–702.
29. R. Z. Rissoli, F. C. Abdalla, M. J. Costa, F. T. Rantin, D. J. McKenzie and A. L. Kalinin, Effects of glyphosate and the glyphosate based herbicides Roundup Original® and Roundup Transorb® on respiratory morphophysiology of bullfrog tadpoles, *Chemosphere*, 2016, **156**, 37–44.
30. L. Franco-Belussi, H. Nilsson Sköld and C. de Oliveira, Internal pigment cells respond to external UV radiation in frogs, *J. Exp. Biol.*, 2016, **219**, 1378–1383.
31. A. Sebben, A. C. M. Ferreira, A. K. Amorim, G. Dillenburg, H. R. Silva, I. F. Oliveira, L. A. Campos, L. B. Nascimento, P. S. Wanderley, V. Slobodian, W. Coelho and J. Klaczko, in *Atlas fotográfico de anatomia comparada de vertebrados: Volume II – sistema digestório*, Universidade de Brasília, Brasília, 2019.
32. L. J. Vittand and J. P. Caldwell, in *Herpetology: An introductory biology of amphibians and reptiles*, Academic Press, 4th edn, 2014.
33. B. S. L. Valverde, L. Z. Fanali, L. Franco-Belussi and C. de Oliveira, Comparative morphology of the digestive tract of two Neotropical tree frogs (Genus *Boana*), *Zool. Anz.*, 2019, **281**, 44–52.
34. S. V. Krishnamurthy, D. Meenakumari, H. P. Gurushankara and V. Vasudev, Nitrate-induced morphological anomalies in the tadpoles of *Nyctibatrachus major* and *Fejervarya limnocharis* (Anura: Ranidae), *Turk. J. Zool.*, 2008, **32**, 239-244.
35. J. R. Lenkowski, G. Sanchez-Bravo and K. A. McLaughlin, Low concentrations of atrazine, glyphosate, 2,4-dichlorophenoxyacetic acid, and triadimefon exposures have diverse effects on *Xenopus laevis* organ morphogenesis, *J. Environ. Sci.*, 2010, **22**, 1305–1308.
36. Ö. Çakici, Carbaryl-induced histopathologic alterations in the digestive tract of the Levantine frog, *Pelophylax bedriagae* (Anura: Ranidae), *Toxicol. Pathol.*, 2014, **42**, 1032–1040.
37. Ö. Çakıcı, Histopathological study of toxic effects of carbaryl on digestive tract of *Bufotes variabilis* (Anura: Bufonidae), *Environ. Sci. Pollut. Res.*, 2016, **23**, 13432–13437.
38. L. Z. Fanali, B. S. L. Valverde, L. Franco-Belussi, D. B. Provete and C. de Oliveira, Response of digestive organs of *Hypsiboas albopunctatus* (Anura: Hylidae) to benzo[α]pyrene, *Amphib. Reptil.*, 2017, **38**, 175–185.
39. F. Alnoaimi, H. Dane and T. Şişman, Histopathologic and genotoxic effects of deltamethrin on marsh frog, *Pelophylax ridibundus* (Anura: Ranidae), *Environ. Sci. Pollut. Res.*, 2021, **28**, 3331–3343.
40. M. Hildebrandand and G. Goslow, in *Analise da Estrutura dos Vertebrados*, Atheneu, São Paulo, 2nd edn, 2006.

41. A. M. Rappaport, in *Disease of the Liver*, ed. L. Schiff and J. B. Lippincott, Philadelphia, 1963, pp. 1–46.
42. L. R. S. Santos, L. Franco-Belussi, R. Zieri, R. E. Borges and C. de Oliveira, Effects of Thermal Stress on Hepatic Melanomacrophages of *Eupemphix nattereri* (Anura), *Anat. Rec.*, 2014, **297**, 864–875.
43. J. M. Pérez-Iglesias, G. S. Natale, L. Franco-Belussi, C. de Oliveira, L. Moreno and S. Tripoli, Effects of glyphosate on hepatic tissue evaluating melanomacrophages and erythrocytes responses in Neotropical anuran *Leptodactylus latinasus*, *Environ. Sci. Pollut. Res. Int.*, 2016, **23**, 9852–9861.
44. N. C. Bach, D. J. Marino, G. S. Natale and G. M. Somoza, Effects of glyphosate and its commercial formulation, Roundup® Ultramax, on liver histology of tadpoles of the neotropical frog, *Leptodactylus latrans* (Amphibia: Anura), *Chemosphere*, 2018, **202**, 289–297.
45. A. Păunescu, C. M. Ponepal, V. T. Grigorean and M. Popescu, Histopathological changes in the liver and kidney tissues of marsh frog (*Pelophylax ridibundus*) induced by the action of Talstar 10EC insecticide, *Analele Univ. din Oradea, Fasc. Biol.*, 2012, **19**, 5–10.
46. C. de Oliveira, L. Franco-Belussi, L. Z. Fanali and L. R. S. Santos, in *Ecotoxicology and Genotoxicology: Non-traditional Terrestrial Models*, ed. M. L. Larramendy, The Royal Society of Chemistry, London, 2017, pp. 125–142.
47. C. S. Carvalho, H. S. M. Utsunomiya, T. Pasquoto, R. Lima, M. J. Costa and M. N. Fernandes, Blood cell responses and metallothionein in the liver, kidney and muscles of bullfrog tadpoles, *Lithobates catesbeianus*, following exposure to different metals, *Environ. Pollut.*, 2017, **221**, 445–452.
48. V. Sotomayor, T. S. Chiriotto, A. M. Pechen and A. Venturino, Biochemical biomarkers of sublethal effects in *Rhinella arenarum* late gastrula exposed to the organophosphate chlorpyrifos, *Pestic. Biochem. Phys.*, 2015, **119**, 48–53.
49. T. G. Santos, R. Melo, D. G. Costa-Silva, M. E. M. Nunes, N. R. Rodrigues and J. L. Franco, Assessment of water pollution in the Brazilian Pampa biome by means of stress biomarkers in tadpoles of the leaf frog *Phyllomedusa iheringii* (Anura: Hylidae), *PeerJ.*, 2015, **3**, e1016.
50. T. Leão, M. Siqueira, S. Marcondes, L. Franco-Belussi, C. de Oliveira and C. E. Fernandes, Comparative liver morphology associated with the hepatosomatic index in five Neotropical anuran species, *Anat. Rec.*, 2021, **304**, 860–871.
51. D. E. Hinton, H. Segner and T. Braunbeck, in *Target organ toxicity in marine and freshwater teleosts*, ed. D. Schlenk and W. H. Benson, CRC Press, London, 2001, pp. 224–268.

52. R. H. Freere and E. R. Weibel, Practical stereological methods for morphometric cytology, *J. Cell. Biol.*, 1966, **30**, 23–38.
53. I. M. Reid, Morphometric methods in veterinary pathology: a review, *Vet. Pathol.*, 1980, **17**, 522–543.
54. E. Rocha, R. A. Monteiro and C. A. Pereira, Liver of the brown trout, *Salmo trutta* (Teleostei, Salmonidae): a stereological study at light and electron microscopic levels, *Anat. Rec.*, 1997, **247**, 317–328.
55. B. K. Larsen and E. J. Perkins Jr., in *Target organ toxicity in marine and freshwater teleosts*, ed. D. Schlenk and W. H. Benson, CRC Press, London, 2001, pp. 90–150.
56. W. Pfaller and G. Gstraunthaler, Nephrotoxicity testing *in vitro* - what we know and what we need to know, *Environ. Health Perspect.*, 1998, **106**, 559–569.
57. T. R. Van Vleet and R. G. Schnellmann, Toxic nephropathy: environmental chemicals, *Semin. Nephrol.*, 2003, **23**, 500–508.
58. N. S. Loumbourdis, I. Kostaropoulos, B. Theodoropoulou and D. Kalmanti, Heavy metal accumulation and metallothionein concentration in the frog *Rana ridibunda* after exposure to chromium or a mixture of chromium and cadmium, *Environ. Pollut.*, 2007, **145**, 787–792.
59. Ö. Çakıcı, Histopathologic changes in liver and kidney tissues induced by carbaryl in *Bufotes variabilis* (Anura: Bufonidae), *Exp. Toxicol. Pathol.*, 2015, **67**, 237–243.
60. L. Franco-Belussi, L. Z. Fanali and C. de Oliveira, UV-B affects the immune system and promotes nuclear abnormalities in pigmented and non-pigmented bullfrog tadpoles, *J. Photochem. Photobiol. B, Biol.*, 2018, **180**, 109–117.
61. P. A. Svensson, E. Forsgren, T. Amundsen and H. N. Sköld, Chromatic interaction between egg pigmentation and skin chromatophores in the nuptial coloration of female two-spotted gobies, *J. Exp. Biol.*, 2005, **208**, 4391–4397.
62. L. Franco-Belussi, A. M. L. Castrucci and C. de Oliveira, Responses of melanocytes and melanomacrophages of *Eupemphix nattereri* (Anura: Leiuperidae) to Nle^4,$D\text{-}Phe^7$-α-melanocyte stimulating hormone and lipopolysaccharides, *Zoology*, 2013, **116**, 316–324.
63. L. Franco-Belussi, D. B. Provete, R. E. Borges, C. de Oliveira and L. R. S. Santos, Idiosyncratic liver pigment alterations of five frog species in response to contrasting land use patterns in the Brazilian Cerrado, *Peer J.*, 2020, **8**, e9751.

CHAPTER 12

Genotoxic, Biochemical and Physiological Biomarkers Triggered by Agrochemicals in Neotropical Anuran Species

JUAN MANUEL PÉREZ-IGLESIAS*[a,b], NADIA C. BACH[c], FABRICIO D. CID[c], LILIAN FRANCO-BELUSSI[d,e], CELESTE RUIZ DE ARCAUTE[f] AND RAQUEL SALLA[g]

[a]Laboratorio de Química Analítica Ambiental, Consejo Nacional de Investigaciones Científicas y Técnicas (CONICET), Facultad de Ciencias de la Salud, Universidad Nacional de San Luis (UNSL), Ejército de los Andes 950 (D5700BWS), San Luis, Argentina; [b]Departamento de Ciencias Ambientales y Producción, Universidad Nacional de Los Comechingones (UNLC), Héroes de Malvinas 1587 (CP5881), Villa de Merlo, San Luis, Argentina; [c]Área de Biología, Facultad de Química, Bioquímica y Farmacia, Universidad Nacional de San Luis (UNSL), Ejército de los Andes 950 (D5700BWS), San Luis, Argentina; [d]Department of Biological Sciences, São Paulo State University (UNESP), Cristóvão Colombo 2265, 15054-000, São José do Rio Prêto, Brazil; [e]Laboratório de Patologia Experimental, Universidade Federal de Mato Grosso do Sul (UFMS), Avenida Costa e Silva s/n, 79070-900, Campo Grande, Brazil; [f]Cátedra de Citología, Facultad de Ciencias Naturales y Museo, Universidad Nacional de La Plata (UNLP), Calle 64 N° 3, B1904AMA La Plata, Buenos Aires, Argentina; [g]Federal University of São Carlos (Campus Sorocaba), João Leme dos Santos Highway, km 110 -SP-264, Itinga, Sorocaba, São Paulo, Brazil
*E-mail: jpereziglesias@unlc.edu.ar

Issues in Toxicology No. 46
Amphibian Species in Environmental Risk Assessment Strategies
Edited by Marcelo L. Larramendy and Guillermo Eli Liwszyc

Published by the Royal Society of Chemistry, www.rsc.org

12.1 The Use of Agrochemicals in the Neotropical Region

Since the "Green Revolution" started 50 years ago, modern agriculture has been associated with a large use of agrochemicals, such as herbicides, insecticides and fungicides, among other agents.[1,2] In the group of the most employed agrochemicals applied in the Neotropical region, we can mention four classes of herbicides: (1) non-selective post-emergence [*e.g.*, glyphosate (GLY) and paraquat], (2) selective pre- and post-emergence [*e.g.*, atrazine (ATZ) and clomazone (CMZ)], (3) selective post-emergence [*e.g.*, 2,4-D, dicamba (DIC), glufosinate-ammonium (GLA) and picloram (PCM)] and (4) synthetic compounds termed as imidazolinone herbicides, including selective pre-emergence groups [*e.g.*, flurochloridone (FLC)], to name a few. Besides, other agrochemicals such as organochlorine (OCs), organophosphate (OPEs), pyrethroid (PYRs) and neonicotinoid (NEOs) insecticides should also be mentioned in the list.[3,4]

In Latin America, these agricultural practices have neither been subject to a critical evaluation nor strict official regulation procedures. Also lacking is adequate information regarding their impact and mitigation measures to be implemented in countries where agrochemicals are intensively used.[5] These deficiencies, coupled with deforestation and fragmentation of natural habitats produced by the current agricultural model, lead us to think about the possible risk to the health of the human population, environment and other non-target living species, such as anurans.[5]

It is well known that amphibians are the most vulnerable group of vertebrates, with approximately 41% of the species threatened worldwide.[6,7] In this context, the contamination of ecosystems (both terrestrial and aquatic) by agrochemicals stands out as a major stress factor influencing the global decline in amphibian populations.[7–9] Several studies have reported the adverse effects caused by these agents in anuran populations both globally and in Neotropical species in particular. The range goes from disruption in trophic relationships and alterations in survival rates to metamorphosis in those populations with genetic disorders.[8,9] In this regard, amphibians are excellent bioindicators of environmental health status[7,8] and are considered "new-age biological models" for ecotoxicological assessments of the health of a given ecosystem.[10–12]

The use of biomarkers in native non-conventional anuran models allows for obtaining a toxicological profile of the biota's response to environmental contaminants. Bearing in mind that different aquatic species are not equally susceptible to the same pollutant and even the same species is not equally susceptible throughout its life cycle, underpins the importance of the use of different biological models.[13,14] This, in turn, highlights the interactions between biotic matrices and environmental stressors that will enable us to reveal realistic scenarios in risk assessment programmes.[3,4] When evaluating the adverse effects of agrochemicals in anurans, it is useful to apply different biomarkers such as molecular, biochemical, histological and physiological ones, which can provide more accurate information during the complete life cycle of this taxon. This approach will help us better understand the effects of these environmental stressors on the decline of amphibian populations.[8,13,14] If we consider the concept of biomarkers on a scale of ecological levels (see Figure 12.1), we can find three large groups of biomarkers that include physiological, cytogenetic and biochemical levels (see Sections 12.2.1, 12.2.2 and 12.2.3). In turn, these groups of biomarkers can be integrated for their application in the evaluation of adverse effects in reproductive scales (see Section 12.2.4).[15]

Neotropics are the most biodiverse regions on our planet and have the largest number of anurans worldwide, with approximately more than 3030 species reported so far.[16] As mentioned previously, this is coupled with the fact that the Neotropical region is the largest zoogeographical area impacted by modern agriculture associated with the use of genetically modified organisms and agrochemicals at a large scale.[1] In this context and focusing on the importance of evaluating biomarkers in native anurans, studies on Neotropical species have only covered 1.2% of the total 36 species evaluated so far. We have noticed a growing trend in the use of biomarkers in Neotropical anurans in the last 10 years (see Figure 12.2). These studies consider different life stages of anurans (embryos, tadpoles and adults), demonstrating the effectiveness and the importance of the use of different biomarkers for evaluating pollutant(s) exposure, detecting adverse effects at relevant environmental concentrations and for environmental risk assessment programmes.

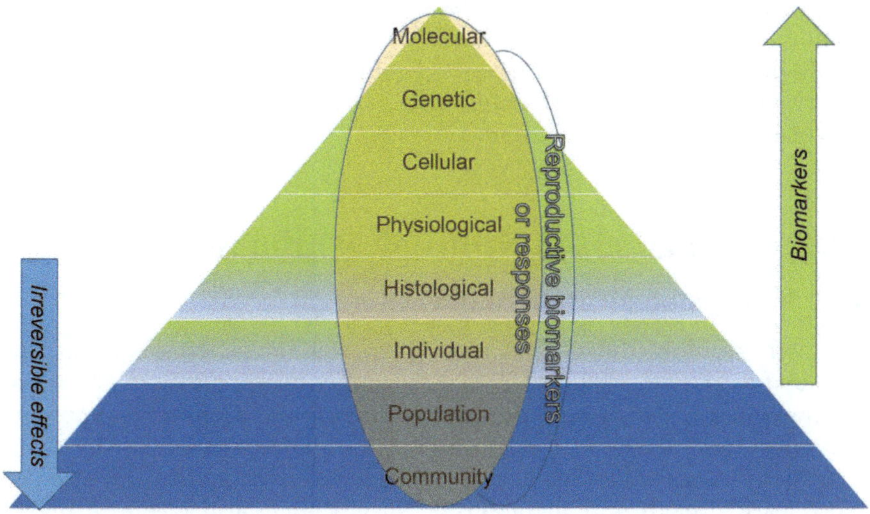

Figure 12.1 Different levels of biological organisation for biomarkers evaluation: a multibiomarker approach concept. The reproduction biomarkers or responses cover all levels.

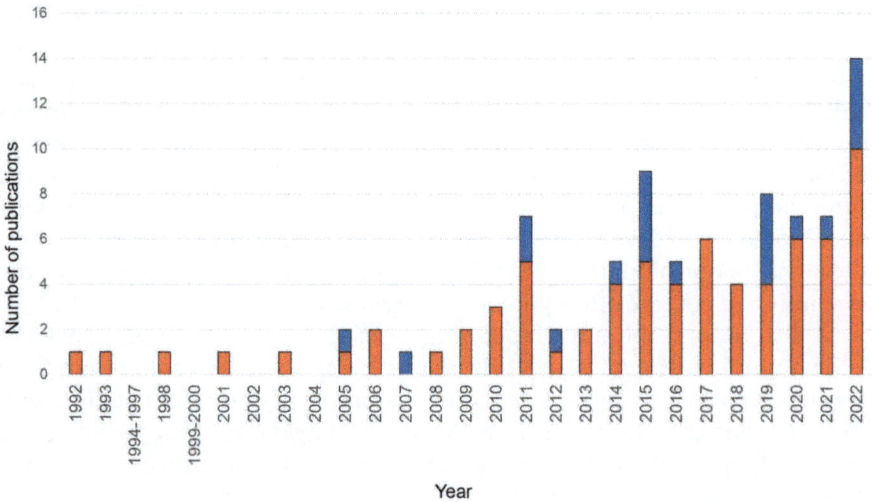

Figure 12.2 Trends in the use of biomarkers during the period 1992–2022 in Neotropical anurans according to scientific databases. Cobalt blue section bars represent biomonitoring studies, while red section bars represent laboratory bioassays.

Biomarkers of exposure, effect and susceptibility are needed to relate the presence of pollutants in the environment with their action in organisms.[3,4,13] In this context, they can assist in assessing the health status of amphibian populations by acting as sublethal endpoints of intoxication.[13] The different biomarkers used in Neotropical anurans in the next sections of this chapter will be listed according to the type of agrochemicals involved. In addition, some important concepts about the advantages and disadvantages of using biomarkers at different ecotoxicological scales will also be mentioned and discussed.

12.2 Biomarkers in Neotropical Anurans

12.2.1 Physiological Biomarkers

Although recently the use of amphibians as bioindicators or sentinel species has grown in the field of environmental monitoring,[17,18] studies of physiological biomarkers aimed at evaluating the ecotoxicity of contaminants are still scarce.[19–28] Physiological biomarkers can involve cardiac alterations related to crucial mechanisms such as homeostasis, metamorphosis, growth and metabolism. Specifically, cardiac effects involve relaxation of the heart, heart rate alterations, reduced atrium size and organ malformation, as observed in two well-known anuran experimental models such as the American bullfrog *Aquarana catesbeiana* tadpoles and the African clawed frog *Xenopus laevis*.[21–23,27,28,31,32] Another group of physiological biomarkers are endocrine-disrupting chemicals (EDCs), which are actively present in several pesticides and affect reproduction and development.[29] It is worth mentioning that amphibians' development is highly susceptible to environmental contaminants since thyroid hormones and their receptors are frequently altered by many toxicants.[29] The studies with amphibian model species showed that distinct agrochemicals have decreased metamorphic rates.[29,30] Table 12.1 summarises the main physiological biomarkers of amphibians applied in toxicological studies of agrochemicals.

12.2.1.1 Case Study: Herbicides

Studies on herbicides showed that cardiac physiological biomarkers reflect the effects of agrochemical use. Specifically, some pesticides have been reported as cardiotoxic in model amphibians.[22,23] Focusing on Neotropical species, EDC biomarkers have been shown to be effective after exposure to commercial formulations of butachlor (Machete EC) in the Neotropical cane toad *Rhinella marina*. This report evidenced alterations in the development of the thyroid gland and consequently decreased the rates of metamorphosis at butachlor concentrations ranging between 0.002 and 0.2 mg L^{-1}.[26] Another study employing the snouted tree frog *Scinax nasicus* and the two-coloured oval frog *Elaschistocleis bicolor* reported that the herbicide DIC in its formulation Cowboy Elite Surcos® (20% active ingredient) increased the production of thyroid hormones in concentrations ranging between 0.01875 and 20 mg L^{-1}.[25] These studies demonstrate the high potential of endocrine biomarkers to be used as sentinel indicators in response to agrochemicals.

12.2.1.2 Case Study: Insecticides

Cardiac effects have been detected in Neotropical species, but to the best of our knowledge, only one study has been carried out.[24] This study shows a decrease in cardiac activity (bradycardia) in the common lesser "escuercito", also called the American ground frog, *Odontophrynus asper* (= *O. americanus*) tadpoles exposed at sublethal concentrations (between 0.1 and 10 mg L^{-1}) to the commercial insecticide pyriproxyfen Dragon® (2% active ingredient of the pyridine-based pesticide).[24]

The above-described review demonstrates that although physiological biomarkers are, in fact, very sensitive, early and effective indicators of environmental changes, their use in ecotoxicological studies is still quite limited. It is worth noting that some authors suggest evaluating, for agrochemical exposure, novel biomarkers in adults, related to permeable and highly vascularised skin, which acts both as an osmoregulatory and as a respiratory organ.[32] However, the wide variation among the different effects of agrochemicals on the anuran models demonstrates their sensitivity but lack of specificity as biomarkers to stressors. This limitation is even

Table 12.1 Physiological biomarkers commonly used for the evaluation of agrochemicals in anurans. Letters indicate: commercial formulation (F), pure active ingredient (AI), herbicide (H), insecticide (I), embryos (E), tadpoles (T) and adults (A).

Agrochemical							
Type	**Presentation**	**Class**	**Exposure stage**	**Exposure period (days)**	**Species**	**Biomarker**	**Ref.**
Glyphosate	F	H	T	2	*Aquarana catesbeiana*	Cardiac: heart rate, ventricular inotropism and chronotropism	22
Chlorpyrifos	AI	I	T	5	*Xenopus laevis*	Cardiac: heart rate	31
Methyl parathion	F	I	T	2	*Aquarana catesbeiana*	Cardiac: heart rate, ventricular inotropism, ventricular mass	23
Azocyclotin	AI	I	T	21	*Xenopus laevis*	Endocrine: metamorphic rate	30
Pyriproxyfen	F	I	T	2	*Odontophrynus laevis*	Cardiac: heart rate	24
Dicamba	AI	H	T	2	*Scinax nasicus*, *Elachistoclei bicolor*	Endocrine: thyroid hormone levels	25
Imazapyr	F	H	E	4	*Xenopus laevis*	Cardiac: oedema	28
Butachlor	AI	H	T	12	*Rhinella marina*	Endocrine: metamorphic rate, thyroid development	26
Chlorpyrifos	AI	I	A	45	*Xenopus laevis*	Osmoregulatory: decreased function	32
Glyphosate	F	H	E	3.5	*Xenopus laevis*	Cardiac: heart malformations	27

more significant when it comes to Neotropical species. In addition, the present lack of knowledge limits our understanding of the toxicodynamical mechanisms involved in agrochemicals.[25,26]

12.2.2 Cytogenetic Biomarkers

The term *genotoxicity* refers to any physical or chemical agent capable of inducing damage in the chromosomes or DNA, altering its normal structure. Moreover, the term *cytotoxicity* encompasses the toxic effects that cause damage and cell death in a target tissue or organ.[3,33] Different bioassays can be performed to visualise and quantify these alterations and have an important role to play in the prediction potential of certain cytotoxic xenobiotics and genotoxicity, which consequently may trigger a carcinogenesis process. Cytogenetic bioassays have the advantage of being reproducible and, depending on the type of tissue, pointing out the possibility of reparation of certain induced damages.[34] It is often necessary to employ a set of bioassays to cover different biological systems, as there is no single assay that detects all genotoxic agents or all types of cyto- or genotoxic damage.[35] In anurans, the first cytogenetic studies were performed 30 years ago using *micronucleus* (MN) and *single-cell gel electrophoresis* (SCGE, comet assay) assays.[36,37] Specifically, for Neotropical anurans, the first MN studies date from 1993[38] but only from 2005 onwards were they massively employed. Most used techniques in Neotropical anurans include MN assay, SCGE assay, mitotic index and cellular viability. New studies have been encouraged to propose novel biomarkers applied to cyto- and genotoxicity studies in Neotropical anurans, such as the viability technique[39] and heterophil/lymphocyte (H/L) ratio.[40]

12.2.2.1 Micronucleus Assay

This methodology allows for the evaluation of the damage induced by a xenobiotic at the chromosomal level and can be employed in different types of proliferating cells by inducing clastogenic or aneugenic damage. It is considered an indirect biomarker of chromosomal damage since at least one cell division is required to visualise MNs. In addition to MNs, several nuclear abnormalities have been described, such as binucleated cells, nuclear buds, lobed and notched nuclei.[41,42] Specifically in amphibians, the analysis of MNs in the circulating erythrocytes of individuals exposed to xenobiotics has become a widely employed technique.[41,42] The advantages of this bioassay include simplicity, low cost and the possibility of evaluating chromosomal instability or mitotic status.[43] Among its limitations, we can include its restriction to cells undergoing the first mitotic division after an injury, performed in nucleated somatic cells, providing an indirect estimate of damage.[35]

12.2.2.2 Single-cell Gel Electrophoresis (SCGE) Assay

The technique is a sensible, rapid, simple and visual methodology employed to provide a direct estimate of damage on single- and double-strand breaks in DNA, alkali-sensitive sites, DNA–DNA and DNA–protein crosslinks, as well as single-strand breaks associated with DNA repair mechanisms.[44–47] SCGE variants have been described, including *in-situ* hybridisation techniques with different fluorochromes (FISH), variable field electrophoresis and the addition of restriction endonucleases within the methodology.[48] The latter is one of the variants adopted worldwide and proposes the inclusion of restriction enzymes, such as Endonuclease III and formamidopyrimidine-DNA glycosylase (FPG), that detect specific DNA lesions.[46] The SCGE assay has become a valuable biomarker in amphibian genetic status evaluation.[37,44–47]

12.2.2.3 Mitotic Index

The cytotoxicity induced by exposure to agrochemicals has also been analysed by employing the mitotic index as a biological endpoint. A gradually decreasing mitotic index value is indicative of toxicity upon division of the cells evaluated. We are aware of only one report that proposes this novel cytogenetic biomarker.[49]

12.2.2.4 Cell Viability

Cell viability is a measure of the proportion of live, healthy cells within a population. Typically, cell viability assays provide a readout of cell health through the measurement of metabolic activity, ATP content or cell proliferation. Its application to Neotropical anurans has been poorly estimated. To the best of our knowledge, Gonçalves *et al.* evaluated sublethal concentrations of the ATZ-based herbicide formulation Atanor 50SC® on cell death of the lesser tree frog *Dendropsophus minutus* tadpoles at different stages of development.[50]

12.2.2.5 Case Study: Herbicides

To the best of our knowledge, there are 23 studies employing cytogenetic biomarkers to detect effects induced by herbicides in Neotropical anurans.[15,38,41,50–53] Moreover, when analysing the data, 47% of these studies have GLY as the evaluated xenobiotic, either as an active ingredient in commercial formulations or in the form of pesticide mixtures. The other 45% is focused on other widely employed herbicides, such as 2,4-D, FLC, imazethapyr (IMZT) and glufosinate ammonium (GLA). In this sense, it is worth mentioning that only two studies were performed evaluating the cytogenetic effects induced by ATZ, the first and second most employed herbicide in Brazil and the Neotropical region, respectively.[15,50]

Considering the type of biomarker used, 65% of the studies employ the MN assay, demonstrating its efficacy for the evaluation of stress induced by agrochemicals on anuran Neotropical species.[40,49,52,54–66] On the other hand, an increase in studies employing the SCGE assay on Neotropical anurans adults and tadpoles has been observed, reaching 56% of the studies performed since 2014.[51] Curiously, this biomarker has not been employed in the region for biomonitoring studies. However, Gonçalves *et al.* demonstrated the importance and efficacy of SCGE in three Neotropical species inhabiting agroecosystem environments.[39] Micronucleous and SCGE assays were performed to evaluate several herbicides as ATZ formulations at sublethal concentrations, detecting cytogenetic damage from 1.5 to 19 mg L^{-1} in *D. minutus* and the Cope's toad *R. diptycha* tadpoles after acute exposure.[15,50] Glufosinate-ammonium and FLC commercial formulations were tested on *R. arenarum* tadpoles, detecting increase in the frequency of MNs to 7.5 mg GLA L^{-1} and 0.71 mg FLC L^{-1} and DNA damage since 0.71 mg FLC L^{-1}.[51,55] Also, when the binary mixture of herbicides was assayed in *R. arenarum* tadpoles, the results showed DNA damage for combinations of GLY-IMZT and GLY-DIC at 5% of $LC_{50\ 96hrs}$ concentrations of each herbicide.[63,64] Added to this, MNs showed an increase in the blacksmith tree frog *Boana faber* and the South American common frog *Leptodactylus latrans* exposed to mixtures of GLY-2,4-D at concentrations of 0.065 mg GLY L^{-1} and 0.004 mg 2,4-D L^{-1}.[66] In addition, the cyto- and genotoxic effects of pure GLY were detected at 0.00125 mg GLY L^{-1} or higher in *D. minutus*.[53] In other Neotropical species, such as the barker frog *Physalaemus cuvieri* and the graceful dwarf frog *P. gracilis*, GLY formulation (Original Roundup Glyphosate®) was shown to produce an increase in MNs starting at 1 mg in chronic exposure.[65] In addition, our working group was the first to use these techniques in combination to evaluate cyto- and genotoxicity in both larvae and adults of the Montevideo tree frog *Boana pulchella*[41,58,59,62] and the oven frog *L. latinasus* exposed to IMZT-based herbicide formulation Pivot® H at environmentally relevant concentrations between intervals of 0.07 and 10 mg IMZT L^{-1}.[60,61] This pioneering work in the region demonstrated the usefulness of using biomarkers in species with different life habits and probable dissimilar contaminants pathways. We have shown that the response of cytogenetic biomarkers in tadpoles and adults is clearly different because they are exposed in a different fashion to agrochemicals. In tadpoles of *L. latinasus,* the tested sublethal concentrations between 0.07 and 0.22 mg IMZT L^{-1} product of a runoff were sufficient to produce an increase in MNs and DNA damage in acute exposure, whereas in adults with a direct application exposure, 10 mg IMZT L^{-1} was necessary to produce DNA damage after 96 h. Finally, we demonstrate the need to apply different cytogenetic biomarkers because they are not only sensitive at different concentrations but also provide different information. Added to this, we applied for the first time the modified SCGE technique and found that *B. pulchella* tadpoles exhibit oxidative damage at 0.39 mg IMZT

L^{-1}.[58] This type of research allowed us to reveal important aspects about the specific mode of action of the herbicide (toxicodynamics), being DNA its target site and, thus, proving that it produces oxidative damage in the DNA of Neotropical tadpoles. This technique has also been successful in assessing oxidative damage in DNA from *R. arenarum* adults exposed to effective concentrations (20 mg L^{-1}) of GLY and 2,4-D. To conclude this section, although works with modified SCGE are incipient, it has been demonstrated that they are important biomarkers for understanding the unknown effects of agrochemicals.

12.2.2.6 Case Study: Insecticides

To the best of our knowledge, only 15 studies employing this biomarker in Neotropical anurans have been reported so far.[34,38,40,42,67–76] Of these studies, approximately 35%, 20%, and 13% apply cytogenetic biomarkers in OP insecticides, in both NEO and PYR (including fourth-generation PYR) and in carbamates, respectively. Approximately 6% corroborated its response when OC were assayed. It should be mentioned that the most widely used cytogenetic biomarkers in Neotropical anurans for the evaluation of insecticides are MNs, representing almost 67% of the studies. Regarding SCGE, in the Neotropical region, only two studies have used this indicator of direct DNA damage in a modern insecticide such as imidacloprid (IMI) in commercial formulation and active compound.[42,76] New studies in Neotropical anurans should address the use of these globally validated biomarkers for studies on insecticides, as they provide greater sensitivity than the most widely used MN assay. Among insecticides, IMI was evaluated in *B. pulchella,* which produces genotoxicity in acute exposures.[42,76] Chronic exposures in the South American spotted grass frog *L. luctator* and the Barker frog *P. cuvieri* tadpoles[74] produce MNs that increase at 0.1 mg IMI L^{-1}. Added to this, *B. pulchella* tadpoles showed an increase in MNs following exposure to 0.005 mg endosulfan L^{-1} and 58.52 mg pirimicarb L^{-1} after 96 h of exposure.[67,72] For pirimicarb, it is important to note that acute exposure between 80 and 160 mg L^{-1} produces significant increases in MNs in *R. arenarum* tadpoles.[70] Similarly, cypermethrin showed an increase in MNs in *O. asper* (= *O. americanus*) tadpoles after acute exposure to 0.005 mg L^{-1},[68] and the organophosphorate chlorpyrifos (CPY) increased MNs in Carvalho's escuerzo *O. carvalhoi* tadpoles (at a relevant concentration of 0.1 mg L^{-1})[73] and induce oxidative pyrimidine damage in *R. arenarum* tadpoles (acute exposure of 0.01 mg L^{-1}).[40] These works represent evidence that cytogenetic biomarkers as well as mortality can be used to assess the response of Neotropical anurans to insecticides.[68,73,77] It should be noted that our studies in *B. pulchella* not only were the first to combine both cytogenetic methodologies but also revealed that the SCGE assay is more sensitive than MNs in detecting DNA damage at early exposure times and at low concentrations of the agrochemical of interest, in this case between 30 and 37 mg IMI L^{-1}.

12.2.2.7 Case Study: Fungicides

Studies employing cytogenetic biomarkers in fungicides are scarce. Only Asis *et al.* found cytotoxic effects in *L. luctator* tadpoles induced by the fungicide formulation Elatus® containing a mixture of azoxistrobin and benzovindiflupyr at relevant environmental concentrations ranging from 0.01 to 0.05 mg L^{-1} during 96 h.[78] It is important to mention that only MN analysis was carried out in conjunction with the evaluation of other nuclear abnormalities, demonstrating the usefulness of the bioassay and the wide variety of exposure scenarios for this biomarker. However, it would be important to expand studies on fungicides to evaluate the response of native anurans, as they are the third most used pesticide in the Neotropical region.

It is noteworthy that Neotropical anurans are valid models for evaluating the agrochemical impact and that the MN and SCGE assays have been probed as useful tools. Although the aforementioned techniques have shown that they can be employed as valid endpoints for detecting pesticide-induced deleterious effects in Neotropical native anurans, it would still be important to focus on studies that try to find out how species not employed as biological matrices so far respond to these agents.

12.2.2.8 Case Study: In Situ Biomonitoring Studies

In addition, field studies at agrochemical sites have used MN and SCGE assays as biomarkers in tadpoles of *B. albopunctatus,*[39] *D. minutus,*[39,79,80] *R. arenarum,*[81] the San Luis snouted tree frog *Scinax fuscovarius*[39,80] and adults of *L. luctator.*[82] As previously mentioned, MNs are the most widely used and validated cytogenetic biomarkers in field evaluations.[81–87] Once again, it should be noted that recent studies have proposed potential novel cyto- and genotoxicity biomarkers in Neotropical anurans, such as the TUNEL assay, to evaluate the induction of apoptotic cells,[69] the frequency of erythroblasts,[72] the H/L ratio[40] or even the expression of c-Fos y Mek genes on Neotropical anurans embryos.[71]

12.2.3 Biochemical Biomarkers

Understanding biochemical mechanisms allows us to predict the effects of several unknown environmental stressors based upon their similarity in biochemical mode of action to well-known pollutants.[3] Specifically for anuran, enzymatic variations related to oxidative stress and cholinergic pathways are the most employed biochemical biomarkers. Agrochemicals are generally linked to oxidative stress mediated directly by the generation of reactive oxygen species (ROS).[3,4] Changes in these enzymes might be observed in individuals at contaminated sites or those exposed to stressors under laboratory conditions.[3] Besides, the measurement of antioxidant enzymes could be used as a biomarker of oxidative stress.[3,13] Another point to consider is the enhancement in enzyme concentration and/or activity related to the developmental stage of the specimens.[8] Among the enzymes commonly involved, we can name catalase (CAT), superoxide dismutase (SOD), glutathione peroxidase (GPx), glutathione *S*-transferase (GST) and 7-ethoxyresorufin-*O*-deethylase (EROD) as enzymatic antioxidant biomarkers. On the other hand, reduced glutathione (GSH) and lipid peroxidation (TBARS or LPO) can be grouped as non-enzymatic antioxidant biomarkers. Finally, acetylcholinesterase (AChE), butyrylcholinesterase (BChE) and carboxylesterases (CabE) are grouped as cholinergic-stress enzyme biomarkers.

12.2.3.1 *Case Study: Herbicides*

To the best of our knowledge, 15 works employing tadpoles or adults are reported for the evaluation of the effects of herbicides.[25,49,52,54,60–62,88–96] So far, biochemical biomarker responses have been most tested in the herbicide GLY, in approximately 55% of the cases. The trend indicates that this class of biomarkers has been extensively evaluated in IMZT and GLA in around 20% of the studies. Focusing on enzymes, the most employed biomarkers were ChE and GST in 87% of the studies. Interestingly, enzymes directly related to oxidative stress (*e.g.*, LPO, CAT and SOD) have been poorly studied; only 45% of the cases have attempted to envisage the ROS potential of herbicides. Unfortunately, no studies are evaluating the response of EROD to Neotropical anuran post-agrochemical exposures. In particular, decreases in the activities of GST, ChE and CbE have been reported in different commercial formulations of GLY (ranging from 2 to 120 mg GLY L^{-1}),[89] metsulfuron-methyl, bispyribac-sodium and PCM (ranging from 0.0097 to 160 mg herbicide L^{-1}) after acute exposure in *R. arenarum* tadpoles.[54] In other studies, the commercial herbicide-based CMZ formulation (Gamit® 360CS) showed alterations in biochemical antioxidant biomarkers by an increase in GST, CAT, SOD and G6PDH activity when the Cuyaba dwarf frog *Eupemphix nattereri* and *R. diptycha* tadpoles were exposed in acute bioassays to concentrations ranging between 0.01 and 0.1 mg L^{-1}.[93] Acetylcholinesterase (AChE) and butyrylcholinesterase (BChE) were inhibited when GA was evaluated in *B. pulchella* tadpoles after acute exposure of 3.5 to 15 mg L^{-1} of the commercial formulation Liberty®,[92] coupled with an increase of GST activity reported at 20 mg of 2,4-D L^{-1}.[40] IMZT was evaluated in two species with different life modes (*B. pulchella* and *L. latinasus*) and in their two life stages (tadpoles and adults).[60–62] If we focus on these results obtained by our group, we can corroborate the importance of using a battery of biochemical biomarkers according to the exposure situation, the species in question, and the life stage of the Neotropical anurans. In this case, we demonstrate that antioxidant enzyme responses vary according to anuran life stages when exposed to the same agrochemical. To note, in *L. latinasus* tadpoles, it was necessary to use a sublethal concentration of 0.15 mg L^{-1} to induce an increase in GST activity;

for *L. latinasus* adults, it was necessary to use real acute exposure scenarios of 1000 mg IMZT L^{-1} to trigger increases in AChE and 10 mg IMZT L^{-1} to trigger the antioxidant response with concomitant CAT increase and inhibition of GST. In contrast, *B. pulchella* adults showed CAT inhibition and increased GST antioxidant response in acute exposure scenarios at 10 and 100 mg IMZT L^{-1}, although it is noteworthy that AChE showed the same response as *L. latinasus* for the same scenario assayed.

12.2.3.2 Case Study: Insecticides

To the best of our knowledge, 27 works employing Neotropical anurans reported evaluations on the effects of insecticides in pure formulations or mixtures of active ingredients on biochemical biomarkers.[40,71,75,95,97–116] In Neotropical species, the most widely assayed insecticides are OP, with approximately 85% of the studies and only 8% on carbamates and 1% on OC. Among insecticides in *R. arenarum* tadpoles, pure malathion[100–102] induces increases in AChE and CAT and decreases in TBARS and GR at concentrations between 4 and 20 mg L^{-1} in both chronic and acute bioassays; pure azinphos-methyl[99,103,108,109] and carbaryl[99,103] induce inhibition of AChE, CabE and antioxidant responses of GSH, CAT and SOD, while an increase was observed in GST and GR at concentrations between 0.2 and 20 mg L^{-1} in chronic *in situ* or acute *ex situ* exposure. Pure and commercial formulations of OPE insecticide CPY, at environmentally relevant concentrations, in acute and chronic exposures, induce inhibition of GST, ChE and CabE enzymes and an increase in anti-ROS production enzymes such as GSH and CAT in tadpoles.[71,110,111] Evaluations in adults induce inhibition of BChE and CabE and an increase in CAT.[105] Also, the commercial formulation PYR-Trisada®, containing a mixture of the synthetic PYR insecticides deltamethrin and tetramethrin, was evaluated in acute exposures.[106] CabE and AChE inhibition were detected at a concentration range of 0.0003125–0.00125% in *R. arenarum* tadpoles. Other studies were performed employing novel species, such as *X. laevis* for the evaluation of OPE insecticide fenitrothion-based formulation in a recovery assay to sublethal concentrations (0.5 to 1.5 mg L^{-1}) that caused inhibition of BChE and AChE,[104] *S. fuscovarius* for the evaluation of pure diazinon in acute exposure to sublethal concentrations (1 to 3 mg L^{-1}) that caused inhibition of CbE and AChE.[107] Chlorpyrifos induced reduction in AChE and CabE activity and GST increase in tadpoles of *B. pulchella* to environmentally relevant concentrations (0.05 to 5 mg CPY L^{-1}).[113] *Physalaemus gracilis* was also evaluated with CPY, which induces AChE inhibition and anti-ROS production enzyme (SOD and GST) increase at relevant concentrations of 0.9 mg L^{-1} of the commercial formulation Klorpan 480EC.[97] Cypermethrin induces AChE, BChE and CAT inhibition, and SOD and GST increase at concentrations of 0.006 mg L^{-1} of the commercial formulation Cyptrin 250CE.[98] Fipronil induces AChE, BChE and SOD inhibition, and anti-ROS enzymes (CAT and GST) increase at relevant concentrations of 0.026 mg L^{-1}

at the commercial formulation Terra Forte®,[86] and the same insecticide in its pure ingredient promotes G6PDH, CAT increase and MDA inhibition in *E. nattereri* at doses starting at 0.035 mg kg^{-1}.[112] Finally, some authors have begun to test these biomarkers in new insecticides of natural origin, such as *Bacillus thurigensis*,[91] derivatives of natural origin, such as pyriproxyfen,[56] and spinosad[117] or even in molluscicides.[118]

12.2.3.3 Case Study: Fungicides

To the best of our knowledge, only one study has been performed at the time of publication of this book to evaluate the acute and chronic effects of pure broad-spectrum fungicide chlorothalonil on biochemical biomarkers using the red-eyed tree frog *Agalychnis callidryas*, the meadow tree frog *Isthmohyla pseudopuma* and the common Mexican tree frog *Smilisca baudinii* tadpoles from Costa Rica at concentrations ranging from 0.0025 to 0.1 mg L^{-1}.[119] In this case, at higher concentrations, an increase in muscle ChE activity was detected in *I. pseudopuma* and the liver GST activity increased in *S. baudinii*.[119]

12.2.3.4 Case Study: Biomonitoring In Situ Studies

For Neotropical anurans that live exposed to agrochemicals, studies are focused on biomonitoring or mesocosm. The biochemical biomarker mostly used in all reports published is AChE, followed by GST in 90% and CAT in 65% of cases. Specifically, several biomonitoring studies were performed on adults of *R. diptycha*,[120] *L. chaquensis*,[121] *B. pulchella*,[122,123] *R. arenarum*,[122] the Uruguayan harlequin frog *Lysapsus limellium*,[124] *L. luctator* and *L. latinasus*[125] that report alterations in the most commonly used enzymes, such as ChEs, CbE, GR, GST and CAT. *Rhinella arenarum* tadpoles have been employed in bioassays to evaluate biochemical biomarkers in sediments containing agrochemicals using different enzymes, such as esterases and antioxidants, as effective biomarkers.[81] Recently, a mesocosm study with tadpoles of *E. nattereri* and *S. fuscovarius* exposed to agrochemicals in the field reported adverse responses using the same set of biomarkers.[126] Finally, AChE and GST showed an increase in the Rufous frog *L. mystacinus* tadpoles, but BChE, AChE and GST decreased their activity in the striped snouted tree frog *S. squalirostris* tadpoles from agricultural areas.[34]

For this class of biomarkers, the wide variety of enzymes commonly used to evaluate agrochemicals in Neotropical anurans is worth noting. This situation has allowed us to learn more about which agrochemicals induce ROS and the enzymatic mechanisms required to counteract them.

12.2.4 Reproductive Biomarkers at the Physiological, Biochemical and Genetic Levels

Reproductive biomarkers are measurable changes that directly or indirectly affect reproductive success and, in turn, involve a series of responses stemming from the molecular level up to the highest biological organisation scale (see Figure 12.1). This results in an integrative tool that links the consequences of reproductive success on populations in anthropogenically disturbed ecosystems.[3,4,13] As most anuran species have their reproductive peak in spring, in many cases coinciding with the period of application of many agrochemicals, they would be especially affected by agents during the reproduction phase, either by direct application or by runoff after intense rains.[127]

Reproductive biomarkers can be evaluated at physiological and biochemical levels in response to agrochemicals that mimic endogenous molecules that produce hormone alterations and endocrine disruption, mainly in the hypothalamic-pituitary-gonadal axis. Furthermore, it has also been shown that other metabolic alterations could indirectly affect amphibian reproduction.[29,127] EDCs' main effects linked to reproduction can be oestrogenic, anti-oestrogenic, androgenic, anti-androgenic and progestogenic.[13,29] These effects can be evidenced, for example, through circulating levels of sex steroids,[128,129] induction of hepatic biosynthesis of vitellogenin,[128,129] alterations on gametogenesis through gonadal histology,[130] or simply by gonad size (*e.g.*, gonadosomatic index).[121,123] There are also differences in the expression of genes related to sexual cycles in adults or sexual differentiation in anuran tadpoles.[131,132] In addition, perturbations of reproductive behaviour (*e.g.*, calling in males) can be considered a reproductive EDC biomarker,[133,134] including alterations in secondary male characteristics such as nuptial pads that become greater and blackish in the breeding season.[133–135] Another commonly evaluated endpoint in anurans is fertility. The indirect measurement of fertility can be done by counting viable gametes in the gonads, evaluating both sperm morphology and the number of viable eggs.[136,137] If DNA damages induced by agrochemicals are produced in germinal cells, they become inheritable and are passed on to the next generation.[3,13] To the best of our knowledge, there are no known studies evaluating the genotoxic effects on germ cells of agrochemical-exposed Neotropical anurans.

12.2.4.1 Case Studies: Insecticides

To date, there are no reports evaluating reproductive biomarkers in Neotropical anurans exposed to insecticides. Recently, genotoxic effects have been found in germ cells of adult *R. arenarum* males exposed to the NEO insecticide IMI in realistic exposure scenarios using the sperm SCGE assay (Bach and Cid, unpublished data; see Figure 12.3). The aforementioned study also included the first biochemical evaluations in sperm

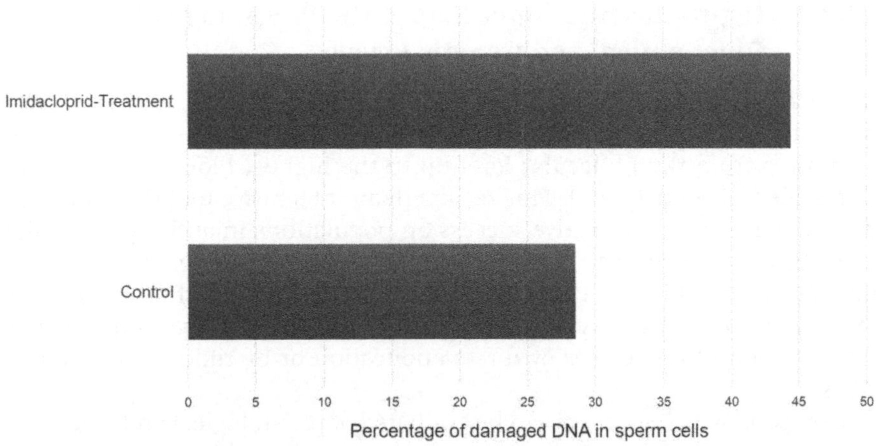

Figure 12.3 DNA damage evaluated by the single-cell gel electrophoresis (SCGE) assay in spermatozoa from the South American common toad *Rhinella arenarum* exposed to the neonicotinoid insecticide imidacloprid.

using CAT and LPO as biomarkers, which demonstrated oxidative stress in anuran germinal cells after IMI exposure.

12.2.4.2 Biomonitoring In Situ Studies

Alterations in spermatogenesis have only been reported in histological analyses of several cell types in the testes of adult individuals of *R. fernandezae*, the Sanborn's tree frog *D. sanborni*, *L. limellum*, the pigmy toad *R. bergi*, *P. cuvieri*, *D. minutus* and *B. albopunctata* that inhabit different ecosystems in which the presence of agrochemicals has not been determined yet.[90,138,139]

12.3 Perspectives

12.3.1 Trends in Neotropical Anurans

We have observed a significant increase in the number of studies using biomarkers for agrochemical evaluations in Neotropical anurans during the last twenty years (see Figures 12.4A and 12.4B). These studies show the usefulness of deploying a battery of various biomarkers in different agrochemical exposure scenarios. In herbicides, most of the studies are focused on the use of biomarkers for GLY; therefore, new studies should focus more on applying biomarkers to evaluate the effects of new herbicides. On the contrary, studies on insecticides have been focused on accompanying the current agricultural model (*e.g.*, NEO). Unfortunately, studies with fungicides using biomarkers are scarce. In this sense, further

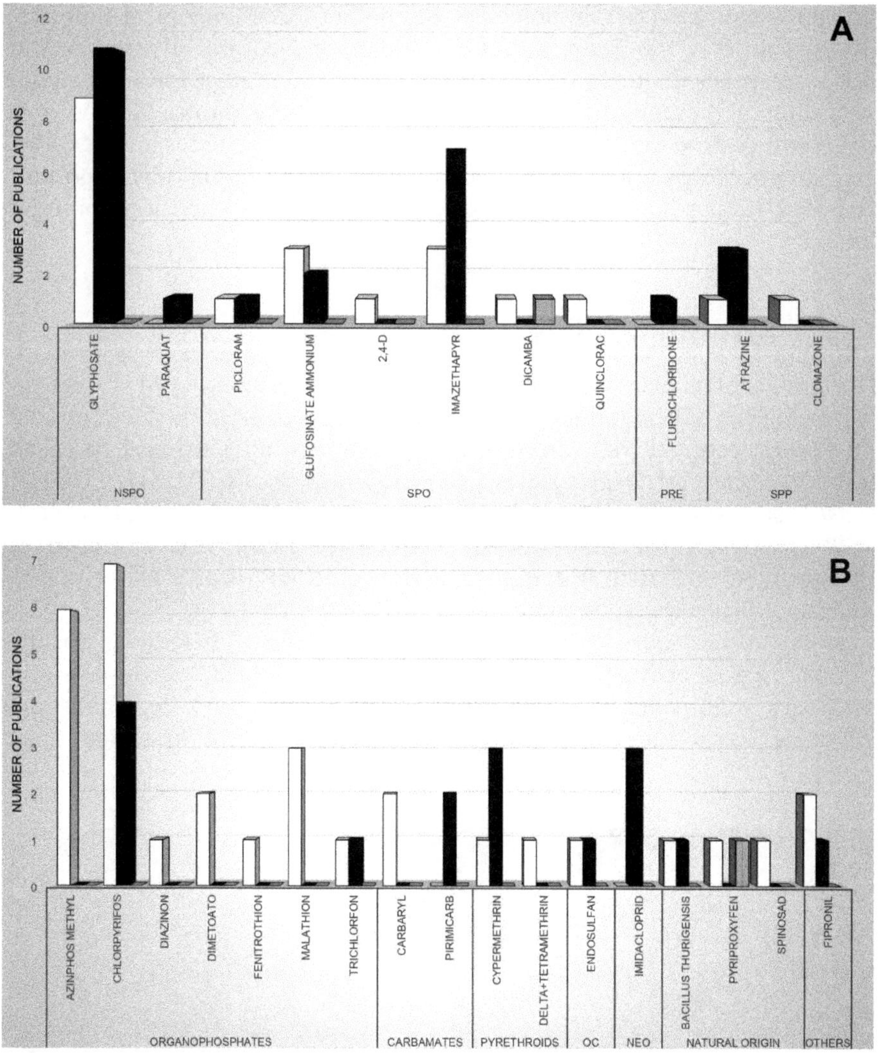

Figure 12.4 Different types of biomarkers used in Neotropical anurans to assess the effects of herbicides (A) and insecticides (B). White bars represent biochemical, black bars cytogenetic and grey bars physiological biomarkers. NSPO, non-selective post-emergent; SPO, systemic post-emergent; SPRE, selective pre-emergent; SPP, selective pre- and post-emergent herbicides; OC, organochlorines; NEO, neonicotinoids; OP, organophosphate insecticides.

studies are necessary to determine whether these biomarkers are suitable for fungicide risk assessment in Neotropical anurans or if new biomarkers should be pursued. To be factored into the drawing of conclusions, it is important to note the present lack of knowledge about the effects of agrochemicals on the physiological biomarkers of Neotropical anurans. Having said this, the future use of biomarkers in Neotropical anurans and risk assessment strategies is promising, both in bioassays with agrochemicals and in biomonitoring studies in agricultural regions. Finally, it is important to note that some terms used, such as cytogenetics and genotoxicity, do not always refer to the same biomarkers, and it would be important to review their generalised use to avoid confusion and speak a common language.

12.4 Anuran Models

The studies previously mentioned show that the most employed species for evaluating the effects induced by agrochemicals is *R. arenarum,* a terrestrial species. However, there is an increasing trend of studies using *B. pulchella* and *S. nasicus* (arboreal species), *L. luctator*, *O. laevis, P. cuvieri* and *P. gracilis* (semi-aquatic species), *L. latinasus, E. nattereri* (cavicolous species) or *R. diptycha* (terrestrial species) (see Figure 12.5). This approach of employing species with different habitats would allow for a better understanding of the effect of agrochemicals on native biota.

Most studies were performed employing species belonging to the Pampas region of Argentina (see Figure 12.6). It is also worthwhile and necessary to consider studies in species that inhabit other regions where the use of agrochemicals is also intensive, *e.g.*, the Great South American

Figure 12.5 Main Neotropical adult anurans proposed as models for risk assessment studies after agrochemical exposure. Images depict *Boana pulchella* (A), *Leptodactylus latinasus* (B), *Odontophrynus laevis* (= *O. americanus*) (C) and *L. luctator* (D).

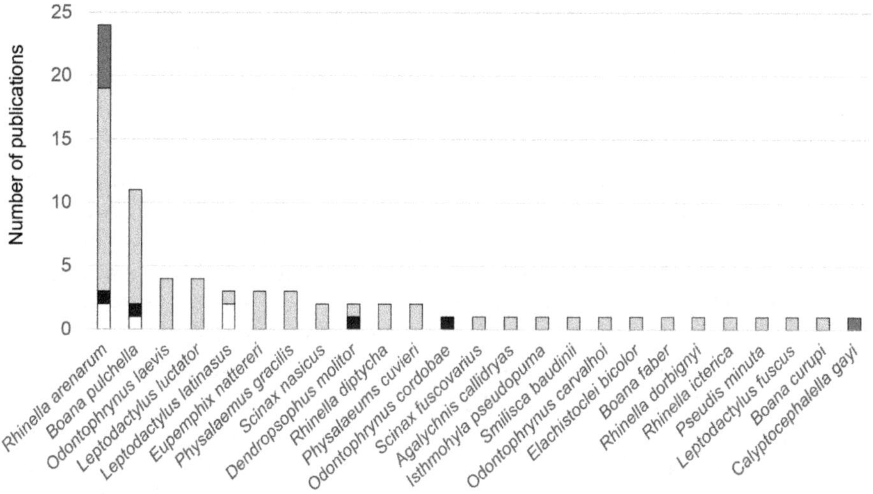

Figure 12.6 Neotropical anuran life stages employed as models for risk assessment studies after agrochemical exposure. Bars represent: adults (white), metamorphs (black), tadpoles (light grey) and embryos (dark grey).

Chaco shared by Argentina, Bolivia, Brazil and Paraguay. It would also be a *sine qua non* requirement to establish unified protocols for the biomarker assessment to be able to compare results from different laboratories, as the techniques employed generally differ between species and life stages of Neotropical anurans. Finally, as we emphasised in our work with IMZT, it is necessary to evaluate agrochemicals in both tadpoles and adults to obtain results that are more representative of what is happening in the region.

12.5 Lack of Linking Biomarkers: A Multibiomarker Approach

Biomarkers are sensitive tools that indicate that environmental stressors have entered an organism and have been distributed among tissues, in turn causing detrimental effects.[3,15] The comprehensive application of a battery of biomarkers could improve the interpretation of the effects induced by agrochemicals and assist in environmental risk assessment, management and the decision-making process before irreversible damage occurs in anuran populations or, even worse, in Neotropical ecosystems.[3,4,8,13] In this context, the "multibiomarker approach" to evaluate and biomonitor the environmental quality of water and soil is recommended by modern ecotoxicology/toxicology. A long pathway has yet to be travelled to arrive at this point, as previously discussed.[3,13,15,60–62] Our works have incorporated the current most worldwide employed approaches and, for the first time, tried to integrate several biomarkers at different ecotoxicological levels (*e.g.*, individual, biochemical and cytogenetical). The use of both life stages of Neotropical anurans (*e.g.*, *B. pulchella* and *L. latinasus*) exposed to a novel herbicide, such as IMZT at relevant environmental concentrations, helps us in our endeavours to seek explanations that biomarkers alone do not provide and/or make for a more realistic risk assessment. This demonstrates that a holistic and integrative view is an important tool required for this task.[60–62] Our work clearly showed that depending on the xenobiotic concentration, a different biomarker should also be called in for a complete agrochemical biomonitoring programme since responses in Neotropical anuran species are different *vis-à-vis* the same agent(s). This situation would allow us to not only predict the presence of such environmental stressors but also enable us to take the necessary steps and actions to avoid irreversible effects. Undoubtedly, future work with Neotropical anurans and agrochemicals should consider this approach to improve the analysis and refine results further. This, in turn, would allow for the drafting of regulations that would ensure the responsible use of these chemicals in the environment.

Acknowledgements

J.M.P.I. was supported by the National Council of Scientific and Technical Research (CONICET), Argentina, and funds provided by Dr. César Almeida and Dr. Patricia Gonzalez from UNSL-(PROICO 02–2418), Argentina. N.B. was supported by Universidad Nacional de San Luis (UNSL), Argentina, and by Dr. Fabricio Cid from UNSL (PROICO 02–0820), Argentina. L.F.B. was supported by a post-doctoral fellowship from São Paulo State University (UNESP, AGRUP PROPe 13/2022), Brazil.

References

1. P.Pingali, Green revolution: impacts, limits, and the path ahead, *Proc. Natl. Acad. Sci. USA.*, 2012, **109**, 12302–12308.
2. V. Shiva, in *The violence of the green revolution: Third world agriculture, ecology, and politics*, University Press of Kentucky, Penang, Malaysia, 1993.
3. M. C. Newman, in *Fundamentals of Ecotoxicology. The Science of Pollution*, CRC Press, Taylor & Francis Group, New York, 5th edn, 2019.
4. C. Walker, in *Ecotoxicology: effects of pollutants on the natural environment*, CRC Press, Taylor & Francis Group, New York, 2014.
5. S. López, D. Aiassa, S. Benítez, R. Lajmanovich, F. Manas, G. Poletta, N. Sánchez, F. Simoniello and A. Carrasco, Pesticides used in South

American GMO-based agriculture: A review of their effects on humans and animal models, *Adv. Mol. Toxicol.*, 2012, **6**, 41–75.

6. J. E. Houlahan, C. S. Findlay, B. R. Schmidt, A. H. Meyer and S. L. Kuzmin, Quantitative evidence for global amphibian population declines, *Nature*, 2000, **404**, 752–755.
7. A. Egea-Serrano, R. A. Relyea, M. Tejedo and M. Torralva, Understanding of the impact of chemicals on amphibians: A meta-analytic review, *Ecol. Evol.*, 2012, **2**, 1382–1397.
8. D. Sparling, G. Linder, C. Bishop and S. Krest, in *Ecotoxicology of Amphibians and Reptiles*, SETAC Books, Boca Raton, FL, 2010.
9. N. T. Halstead, T. A. McMahon, S. A. Johnson, T. R. Raffel, J. M. Romansic, P. W. Crumrine and J. R. Rohr, Community ecology theory predicts the effects of agrochemical mixtures on aquatic biodiversity and ecosystem properties, *Ecol. Lett.*, 2014, **17**, 932–941.
10. C. J. De Garady and R. S. Halbrook, Using anurans as bioindicators of PCB contaminated streams, *J. Herpetol.*, 2006, **40**, 127–130.
11. C. A. Brühl, T. Schmidt, S. Pieper and A. Alscher, Terrestrial pesticide exposure of amphibians: An underestimated cause of global decline?, *Sci. Rep.*, 2013, **3**, 1135.
12. R. Kendall, T. Lacher, G. Cobb and S. Cox, in *Wildlife Toxicology: Emerging Contaminant and Biodiversity Issues*, Taylor & Francis Group, CRC Press, Boca Raton, FL, 2010.
13. R. Gupta, in *Biomarkers in Toxicology*, Elsevier, Academic Press, 2nd edn, 2019.
14. R. Connon, J. Geist and I. Werner, Effect-based tools for monitoring and predicting the ecotoxicological effects of chemicals in the aquatic environment, *Sensors*, 2012, **12**, 12741–12771.
15. J. M. Pérez-Iglesias, L. Franco-Belussi, G.Natale and C. de Oliveira, Biomarkers at different levels of organisation after atrazine formulation (SIPTRAN 500SC®) exposure in *Rhinella schineideri* (Bufonidae) Neotropical tadpoles, *Environ. Poll.*, 2019, **244**, 733–746.
16. D. Frost, Amphibian Species of the World: an Online Reference. Version 6.1 (02/02/2023). Electronic Database, American Museum of Natural History, New York, USA, 2023, Available from: https://-amphibiansoftheworld.amnh.org.
17. W.Hopkins, Amphibians as models for studying environmental change, *ILAR J.*, 2007, **48**, 270–277.
18. V. Langlois, Amphibian Toxicology: A rich but underappreciated model for ecotoxicology research, *Arch. Environ. Contam. Toxicol.*, 2021, **80**, 661–662.
19. M. Ortiz-Santaliestra, J. Maia, A. Egea-Serrano and I. Lopes, Validity of fish, birds and mammals as surrogates for amphibians and reptiles in pesticide toxicity assessment, *Ecotoxicology*, 2018, **27**, 819–833.

20. M. Daam, M. Moutinho, E. Espíndola and L. Schiesari, Lethal toxicity of the herbicides acetochlor, ametryn, glyphosate and metribuzin to tropical frog larvae, *Ecotoxicology*, 2019, **28**, 707–715.
21. H. Lillywhite, K. Zippel and A. Farrell, Resting and maximal heart rates in ectothermic vertebrates, *Comp. Biochem. Physiol.*, 1999, **124**, 369-382.
22. M.Costa, D.Monteiro, A. O. Neto, F.Rantin and A.Kalinin, Oxidative stress biomarkers and heart function in bullfrog tadpoles exposed to Roundup Original®, *Ecotoxicolology*, 2008, **17**, 153–163.
23. M. Costa, L. Ribeiro, R. Salla, F. Gamero, L. Alves and E. Silva-Zacarin, Effects of the organophosphorus pesticide Folisuper 600 (methyl parathion) on the heart function of bullfrog tadpoles, *Lithobates catesbeianus* (Shaw, 1802), *Braz. J. Biol.*, 2015, **75**, 163–168.
24. R. Lajmanovich, P. Peltzer, C. Martinuzzi, A. Attademo, A. Bassó and C. Colussi, Insecticide pyriproxyfen (Dragón®) damage biotransformation, thyroid hormones, heart rate, and swimming performance of *Odontophrynus americanus* tadpoles, *Chemosphere*, 2019, **220**, 714–722.
25. A. Attademo, R. Lajmanovich, P. Peltzer, A. Boccioni, C. Martinuzzi, F. Simonielo and M. Repetti, Effects of the emulsifiable herbicide Dicamba on amphibian tadpoles: an underestimated toxicity risk?, *Environ. Sci. Pollut. Res.*, 2021, **28**, 31962–31974.
26. M. Shuman-Goodier, G. Singleton, A. Forsman, S. Hines, N. Christodoulides, K. Daniels and C. Propper, Developmental assays using invasive cane toads, *Rhinella marina*, reveal safety concerns of a common formulation of the rice herbicide, butachlor, *Environ. Pollut.*, 2021, **272**, 115955.
27. H. Flach, A. Lenz, S. Pfeffer, M. Kühl and S.Kühl, Impact of glyphosate-based herbicide on early embryonic development of the amphibian *Xenopus laevis*, *Aquat. Toxicol.*, 2022, **244**, 106081.
28. O. Babalola, J. Truter and J. Van Wyk, Lethal and teratogenic impacts of imazapyr, diquat dibromide, and glufosinate ammonium herbicide formulations using frog embryo teratogenesis assay-*Xenopus* (FETAX), *Arch. Environ. Contam. Toxicol.*, 2021, **80**, 708–716.
29. V. Trudeau, P. Thomson, W. Zhang, S. Reynaud, L. Martin and V. Langlois, Agrochemicals disrupt multiple endocrine axes in amphibians, *Mol. Cell. Endocrinol.*, 2020, **513**, 110861.
30. M. Li, C. Cao, S. Li, W. Gui and G. Zhu, Thyroid endocrine disruption of azocyclotin to *Xenopus laevis* during metamorphosis, *Environ. Toxicol. Pharmacol.*, 2016, **43**, 61–67.
31. F. Watson, H. Schmidt, Z. Turman, N. Hole, H. Garcia, J. Gregg and E. Fradinger, Organophosphate pesticides induce morphological abnormalities and decrease locomotor activity and heart rate in *Danio rerio* and *X. laevis*, *Environ.Toxicol. Chem.*, 2014, **33**, 1337–1345.

32. F. Álvarez-Vergara, J. Sanchez-Hernandez and P. Sabat, Biochemical and osmoregulatory responses of the African clawed frog experimentally exposed to salt and pesticide, *Comp. Biochem. Physiol.*, 2022, **258**, 109367.
33. M. Mudry and M. Carballo, in *Genética Toxicológica*, Buenos Aires, Argentina, 2006.
34. A. Attademo, P. Peltzer, R. Lajmanovich, M. Cabagna, C. Junges and A. Basso, Biological endpoints, enzyme activities, and blood cell parameters in two anuran tadpole species in rice agroecosystems of mid-eastern Argentina, *Environ. Monit. Assess.*, 2014, **186**, 635–649.
35. C. Ruiz de Arcaute, M. Laborde, S. Soloneski and M. L. Larramendy, in *Principios de Ecotoxicología, Genotoxicidad y carcinogénesis*, ed. P. Carriquiriborde, Libros de Cátedra, Universidad Nacional de La Plata, 2021.
36. Z. Rudek and M. Rozek, Induction of micronuclei in tadpoles of *Rana temporaria* and *Xenopus laevis* by the pyrethroid Fastac 10 EC, *Mutat. Res.*, 1992, **298**, 25–29.
37. C. Clements, S. Ralph and M. Petras, Genotoxicity of select herbicides in *Rana catesbeiana* tadpoles using the alkaline single-cell gel DNA electrophoresis (comet) assay, *Environ. Mol. Mutagen.*, 1997, **29**, 277–288.
38. W. Venegas, I. Hermosilla, L. Quevedo and G. Montoya, Genotoxic and teratogenic effect of pentachlorophenol, pollutant present in continental water bodies in the South of Chile, *Bull. Environ. Contam. Toxicol.*, 1993, **51**, 107–114.
39. M. Gonçalves, C. de Campos, F. Godoy, P. Gambale, H. Nunes, F. Nomura and D. Melo e Silva, Assessing genotoxicity and mutagenicity of three common amphibian species inhabiting agroecosystem environment, *Arch. Environ. Contam. Toxicol.*, 2019, **77**, 409–420.
40. R. Lajmanovich, A. Attademo, M. Simoniello, G. Poletta, C. Junges, P. Peltzer and M. Zenklusen, Harmful effects of the dermal intake of commercial formulations containing chlorpyrifos, 2,4-D, and glyphosate on the common toad *Rhinella arenarum* (Bufonidae), *Wat. Air Soil Pollut.*, 2015, **226**, 1–12.
41. J. M. Pérez-Iglesias, S. Soloneski, N. Nikoloff, G.Natale and M. L. Larramendy, Toxic and genotoxic effects of the imazethapyr-based herbicide formulation Pivot H® on montevideo tree frog *Hypsiboas pulchellus* tadpoles (Hylidae), *Ecotoxicol. Environ. Saf.*, 2015, **119**, 15–24.
42. C. Ruiz de Arcaute, J. M. Pérez-Iglesias, N. Nikoloff, G. S. Natale, S. Soloneski and M. L. Larramendy, Genotoxicity evaluation of the insecticide imidacloprid on circulating blood cells of Montevideo tree frog *Hypsiboas pulchellus* tadpoles (Anura, Hylidae) by comet and micronucleus bioassays, *Ecol. Indic.*, 2014, **45**, 632–639.

43. T. Cavaş and S. Ergene-Gözükara, Micronuclei, nuclear lesions and interphase silver-stained nucleolar organizer regions (AgNORs) as cyto-genotoxicity indicators in *Oreochromis niloticus* exposed to textile mill effluent, *Mutat. Res.*, 2003, **538**, 81–91.

44. T. Çavaş, The use of fish as model aquatic organisms in genotoxicity studies, in *Ecotoxicology and Genotoxicology, Non-traditional Aquatic Models*, ed. M. L. Larramendy, Royal Society of Chemistry, Cambridge, 2017, pp. 243–277.

45. N. Singh, Microgel electrophoresis of DNA from individual cells, in *Technologies for detection of DNA damage and mutations*, Springer, 1996, pp. 3–24.

46. A. Collins, P. Møller, G. Gajski, S. Vodenková, A. Abdulwahed, D. Anderson, E. Bankoglu, S. Bonassi, E. Boutet-Robinet, G. Brunborg, C. Chao, M.Cooke, C. Costa, S. Costa, A. Dhawan, J. de Lapuente, C. D. Bo, J. Dubus, M. Dusinska, S. J. Duthie, N.El Yamani, B. Engelward, I. Gaivão, L. Giovannelli, R. Godschalk, S. Guilherme, K. Gutzkow, K. Habas, A. Hernández, O. Herrero, M. Isidori, A. N. Jha, S. Knasmüller, I.Kooter, G. Koppen, M. Kruszewski, C. Ladeira, B. Laffon, M. L. Larramendy, L.Hégarat, A. Lewies, A. Lewinska, G. E. Liwszyc, A.López de Cerain, M. Manjanatha, R. Marcos, M. Milić, V.Moraes de Andrade, M. Moretti, D. Muruzabal, M. Novak, R. Oliveira, A.-K. Olsen, N. Owiti, M. Pacheco, A. K. Pandey, S. Pfuhler, B. Pourrut, K. Reisinger, E. Rojas, E. Rundén-Pran, J. Sanz-Serrano, S. Shaposhnikov, V. Sipinen, K. Smeets, H. Stopper, J. P. Teixeira, V. Valdiglesias, M. Valverde, F. van Acker, F.-J. van Schooten, M. Vasquez, J. F. Wentzel, M. Wnuk, A. Wouters, B. Žegura, T. Zikmund, S. A. S. Langie and A. Azqueta, Measuring DNA modifications with the comet assay: A compendium of protocols, *Nat. Protoc.*, 2023, **18**, 929–989.

47. R. Tice, E. Agurell, D. Anderson, B. Burlinson, A. Hartmann, H. Kobayashi, Y. Miyamae, E. Rojas, J. Ryu and Y. Sasaki, Single cell gel/comet assay: Guidelines for *in vitro* and *in vivo* genetic toxicology testing, *Environ. Mol. Mutagen.*, 2000, **35**, 206–221.

48. K. Sorensen, J. Stucki, E. Wagner and M. Plewa, Modulation of the genotoxicity of pesticides reacted with redox-modified smectite clay, *Environ. Mutagen.*, 2005, **46**, 174–181.

49. R. Lajmanovich, P. Peltzer, A. Attademo, C. Martinuzzi, M. Simoniello, C. Colussi and M. Sigrist, First evaluation of novel potential synergistic effects of glyphosate and arsenic mixture on *Rhinella arenarum* (Anura: Bufonidae) tadpoles, *Heliyon*, 2019, **5**, e02601.

50. M. Gonçalves, C. Campos, V. Batista, A. da Cruz, P. Marco, R. Bastos and D. M. Silva, Genotoxic and mutagenic effects of Atrazine Atanor 50 SC on *Dendropsophus minutus* Peters, 1872 (Anura: Hylidae) developmental larval stages, *Chemosphere*, 2017, **182**, 730–737.

51. N. Nikoloff, G. S. Natale, D. Marino, S. Soloneski and M. L. Larramendy, Flurochloridone-based herbicides induced genotoxicity effects on *Rhinella arenarum* tadpoles (Anura: Bufonidae), *Ecotoxicol. Environ. Saf.*, 2014, **100**, 275–281.
52. A. Boccioni, G. Lener, J. Peluso, P. Peltzer, A. Attademo, C. Aronzon and R. Lajmanovich, Comparative assessment of individual and mixture chronic toxicity of glyphosate and glufosinate ammonium on amphibian tadpoles: A multibiomarker approach, *Chemosphere*, 2022, **309**, 136554.
53. A. Lopes, M. Souza, W. Carvalho, H. Nunes, P. Lima, M. Costa and D. M. Silva, Evaluation of the genotoxic, mutagenic, and histopathological hepatic effects of polyoxyethylene amine and glyphosate on *Dendropsophus minutus* tadpoles, *Environ. Pollut.*, 2021, **289**, 117911.
54. R. Lajmanovich, C. Junges, A. Attademo, P. Peltzer, M. Cabagna and A. Basso, Individual and mixture toxicity of commercial formulations containing glyphosate, metsulfuron-methyl, bispyribac-sodium, and picloram on *Rhinella arenarum* tadpoles, *Wat. Air Soil Poll.*, 2013, **224**, 1–13.
55. R. C. Lajmanovich, M. C. Cabagna-Zenklusen, A. M. Attademo, C. M. Junges, P. M. Peltzer, A. Bassó and E. Lorenzatti, Induction of micronuclei and nuclear abnormalities in tadpoles of the common toad (*Rhinella arenarum*) treated with the herbicides Liberty® and glufosinate-ammonium, *Mutat. Res.*, 2014, **769**, 7–12.
56. B. Bosch, F. Mañas, N. Gorla and D. Aiassa, Micronucleus test in post metamorphic *Odontophrynus cordobae* and *Rhinella arenarum* (Amphibia: Anura) for environmental monitoring, *J. Toxicol. Environ. Health*, 2011, **3**, 155–163.
57. J. M. Pérez-Iglesias, L. Franco-Belussi, L. Moreno, S. Tripole, C. de Oliveira and G. Natale, Effects of glyphosate on hepatic tissue evaluating melanomacrophages and erythrocytes responses in neotropical anuran *Leptodactylus latinasus*, *Environ. Sci. Pollut. Res.*, 2016, **23**, 9852–9861.
58. J. M. Pérez-Iglesias, C. Ruiz de Arcaute, G. Natale, S. Soloneski and M. L. Larramendy, Evaluation of imazethapyr-induced DNA oxidative damage by alkaline Endo III- and Fpg-modified single-cell gel electrophoresis assay in *Hypsiboas pulchellus* tadpoles (Anura, Hylidae), *Ecotoxicol. Environ. Saf.*, 2017, **142**, 503–508.
59. J. M. Pérez-Iglesias, G. S. Natale, S. Soloneski and M. L. Larramendy, Are the damaging effects induced by the imazethapyr formulation Pivot® H in *Boana pulchella* (Anura) reversible upon ceasing exposure?, *Ecotoxicol. Environ. Saf.*, 2018, **148**, 1–10.
60. J. M. Pérez-Iglesias, J. C. Brodeur and M. L. Larramendy, An imazethapyr-based herbicide formulation induces genotoxic, biochemical, and

individual organizational effects in *Leptodactylus latinasus* tadpoles, *Environ. Sci. Pollut. Res.*, 2020, **27**, 2131–2143.

61. J. M. Pérez-Iglesias, L. Z. Fanali, L. Franco-Belussi, G. S. Natale, C. de Oliveira, J. C. Brodeur and M. L. Larramendy, Multiple level effects of imazethapyr on *Leptodactylus latinasus* (Anura) adult frogs, *Arch. Environ. Contam. Toxicol.*, 2021, **81**, 492–506.
62. J. M. Pérez-Iglesias, G. S. Natale, J. C. Brodeur and M. L. Larramendy, Realistic scenarios of pesticide exposure alters multiple biomarkers in *Boana pulchella* (Anura) adult frogs, *Ecotoxicology*, 2023, **32**, 309–320.
63. S. Soloneski, C. Ruiz de Arcaute and M. L. Larramendy, Genotoxic effect of a binary mixture of dicamba- and glyphosate-based commercial herbicide formulations on *Rhinella arenarum* (Anura, Bufonidae) late-stage larvae, *Environ. Sci. Pollut. Res.*, 2016, **23**, 17811–17821.
64. W. Carvalho, C. Ruiz de Arcaute, J. Pérez-Iglesias, M. Laborde, S. Soloneski and M. L. Larramendy, DNA damage exerted by mixtures of commercial formulations of glyphosate and imazethapyr herbicides in *Rhinella arenarum* (Anura, Bufonidae) tadpoles, *Ecotoxicology*, 2019, **28**, 367–377.
65. J. Herek, L. Vargas, S.Trindade, C. Rutkoski, N. Macagnan, P. Hartmann and M. Hartmann, Genotoxic effects of glyphosate on *Physalaemus* tadpoles, *Environ. Toxicol. Pharmacol.*, 2021, **81**, 103516.
66. F. Pavan, C. Samojeden, C. Rutkoski, A. Folador, S. Da Fré, C. Müller and M. Hartmann, Morphological, behavioral and genotoxic effects of glyphosate and 2,4-D mixture in tadpoles of two native species of South American amphibians, *Environ. Toxicol. Pharmacol.*, 2021, **85**, 103637.
67. R. C. Lajmanovich, M. Cabagna, P. M. Peltzer, G. A. Stringhini and A. M. Attademo, Micronucleus induction in erythrocytes of the *Hyla pulchella* tadpoles (Amphibia: Hylidae) exposed to insecticide endosulfan, *Mutat. Res.*, 2005, **587**, 67–72.
68. M. Cabagna, R. Lajmanovich, P. Peltzer, A. Attademo and E. Ale, Induction of micronuclei in tadpoles of *Odontophrynus americanus* (Amphibia: Leptodactylidae) by the pyrethroid insecticide cypermethrin, *Toxicol. Environ. Chem.*, 2006, **88**, 729–737.
69. V. H. Casco, M. F. Izaguirre, L. Marín, M. N. Vergara, R. C. Lajmanovich, P. Peltzer and A. P. Soler, Apoptotic cell death in the central nervous system of *Bufo arenarum* tadpoles induced by cypermethrin, *Cell. Biol. Toxicol.*, 2006, **22**, 199–211.
70. J. V. Candioti, G. Natale, S. Soloneski, A. Ronco and M. L. Larramendy, Sublethal and lethal effects on *Rhinella Arenarum* (Anura, Bufonidae) tadpoles exerted by the pirimicarb-containing technical formulation insecticide Aficida®, *Chemosphere*, 2010, **78**, 249–255.

71. V. Sotomayor, T. S. Chiriotto, A. M. Pechen and A. Venturino, Biochemical biomarkers of sublethal effects in *Rhinella arenarum* late gastrula exposed to the organophosphate chlorpyrifos, *Pestic. Biochem. Physiol.*, 2015, **119**, 48–53.
72. G. Natale, J. Vera-Candioti, C. Ruiz de Arcaute, S. Soloneski, M. L. Larramendy and A. Ronco, Lethal and sublethal effects of the pirimicarb-based formulation Aficida® on *Boana pulchella* tadpoles (Anura, Hylidae), *Ecotoxicol. Environ. Saf.*, 2018, **147**, 471–479.
73. M. Silva, R. Fraga, P. Nishiyama, N. Costa, S. Silva, T. Queiroz and F. Juncá, Genotoxic effect of the insecticide Chlorpyrifos on the erythrocytes of *Odontophrynus carvalhoi* tadpoles (Odontophrynidae), *Ecotoxical. Environ. Saf.*, 2020, **15**, 9–13.
74. C. Samojeden, F. Pavan, C. Rutkoski, A. Folador, S. da Fré, C. Müller and M. Hartmann, Toxicity and genotoxicity of imidacloprid in the tadpoles of *Leptodactylus luctator* and *Physalaemus cuvieri* (Anura: Leptodactylidae), *Sci. Rep.*, 2022, **12**, 11926.
75. C. Schavinski, M. Santos, J. Londero, M. da Rocha, A. do Amaral, N. Ruiz and A. Schuch, Effects of isolated and combined exposures of *Boana curupi* (Anura: Hylidae) tadpoles to environmental doses of trichlorfon and ultraviolet radiation, *Mutat. Res.*, 2022, **883**, 503549.
76. J. M. Pérez-Iglesias, C. Ruiz de Arcaute, N. Nikoloff, L. Dury, S. Soloneski, G. S. Natale and M. L. Larramendy, The genotoxic effects of the imidacloprid-based insecticide formulation Glacoxan Imida on Montevideo tree frog *Hypsiboas pulchellus* tadpoles (Anura, Hylidae), *Ecotoxicol. Environ. Saf.*, 2014, **104**, 120–126.
77. R. Lajmanovich, E. Lorenzatti, P. De la Sierra, F. Marino, G. Stringhini and P. Peltzer, Reduction in the mortality of tadpoles (*Physalaemus biligonigerus*; Leptodactylidae) exposed to cypermethrin in presence of aquatic ferns, *Fresenius Environ. Bull*, 2003, **12**, 1558–1561.
78. R. A. Assis, M. Benvindo-Souza, C. G. Araújo-Santos, R. E. Borges, I. D. Santos-Filho, L. C. Oliveira and L. R. Santos, Mutagenic effect of a commercial fungicide on *Rana catesbeiana* and *Leptodactylus latrans* tadpoles, *An. Acad. Bras. Ciênc.*, 2022, **94**(suppl 4), e20210161.
79. M. W. Gonçalves, T. B. Vieira, N. M. Maciel, W. F. Carvalho, L. S. Lima, P. G. Gambale and D. de Melo e Silva, Detecting genomic damages in the frog *Dendropsophus minutus*: preserved versus perturbed areas, *Environ. Sci. Pollut. Res.*, 2015, **22**, 3947–3954.
80. R. E. Borges, L. R. Santos, M. Benvindo-Souza, R. S. Modesto, R. A. Assis and C. de Oliveira, Genotoxic evaluation in tadpoles associated with agriculture in the Central Cerrado, Brazil, *Arch. Environ. Contam. Toxicol.*, 2019, **77**, 22–28.
81. J. Peluso, C. Aronzon, A. Chehda, A. Boccioni, P. Peltzer, E. De Geronimo and R. Lajmanovich, Environmental quality and ecotoxicity of

sediments from the lower Salado River basin (Santa Fe, Argentina) on amphibian larvae, *Aquat. Toxicol.*, 2022, **253**, 106342.

82. T. Ascoli-Morrete, N. Bandeira, E. Signor, H. Gazola, I. Homrich, R. Biondo and N. Zanella, Bioaccumulation of pesticides and genotoxicity in anurans from southern Brazil, *Environ. Sci. Pollut. Res.*, 2022, **29**, 45549–45559.
83. M. Babini, C. Bionda, N. Salas and A. Martino, Adverse effect of agroecosystem pond water on biological endpoints of common toad (*Rhinella arenarum*) tadpoles, *Environ. Monit. Assess.*, 2016, **188**, 1–14.
84. F. Pollo, C. Bionda, Z. Salinas, N. Salas and A. Martino, Common toad *R. arenarum* (Hensel, 1867) and its importance in assessing environmental health: test of micronuclei and nuclear abnormalities in erythrocytes, *Environ. Monit. Assess.*, 2015, **187**, 1–9.
85. M. S. Babini, C. L. Bionda, N. E. Salas and A. L. Martino, Health status of tadpoles and metamorphs of *Rhinella arenarum* (Anura, Bufonidae) that inhabit agroecosystems and its implications for land use, *Ecotox. Environ. Saf.*, 2015, **118**, 118–125.
86. F. E. Pollo, P. R. Grenat, M. A. Otero, S. Babini, N. E. Salas and A. L. Martino, Evaluation in situ of genotoxic and cytotoxic response in the diploid/polyploid complex *Odontophrynus* (Anura: Odontophrynidae) inhabiting agroecosystems, *Chemosphere*, 2019, **216**, 306–312.
87. J. Peluso, C. Aronzon and C. Pérez Coll, Assessment of environmental quality of water bodies next to agricultural areas of Buenos Aires province (Argentina) by means of ecotoxicological studies with *Rhinella arenarum*, *J. Environ. Sci. Health*, 2019, **54**, 655–664.
88. L. Machado Reichert, D. Oliveira, J. Papaleo, A. Valgas and G. Oliveira, Biochemical and body condition markers in *Rhinella icterica* tadpoles exposed to atrazine, glyphosate, and quinclorac based herbicides in ecologically relevant concentrations, *Environ. Toxicol. Pharmacol.*, 2022, **93**, 103884.
89. R. Lajmanovich, A. Attademo, P. Peltzer, C. Junges and M. Cabagna, Toxicity of four herbicide formulations with glyphosate on *Rhinella arenarum* (Bufonidae) tadpoles: B-esterases and glutathione S-transferase inhibitors, *Arch. Environ. Contam. Toxicol.*, 2011, **60**, 681–689.
90. W. Rezende, L. Santos, L. Franco-Belussi and C. de Oliveira, Testicular morphometric changes in Neotropical anurans from agroecosystems, *Environ. Pollut.*, 2021, **271**, 116265.
91. R. Lajmanovich, C. Junges, M. Zenklusen, A. Attademo, P. Peltzer, M. Maglianese and A. Beccaria, Toxicity of *Bacillus thuringiensis israelensis* in aqueous suspension on the South American common frog *Leptodactylus latrans* (Anura: Leptodactylidae) tadpoles, *Environ. Res.*, 2015, **136**, 205–212.

92. P. Peltzer, C. Junges, A. Attademo, A. Bassó, P. Grenón and R. Lajmanovich, Cholinesterase activities and behavioral changes in *Hypsiboas pulchellus* (Anura: Hylidae) tadpoles exposed to glufosinate ammonium herbicide, *Ecotoxicology*, 2013, **22**, 1165–1173.
93. J. Freitas, A. Felício, F. Teresa and E. Almeida, Combined effects of temperature and clomazone (Gamit®) on oxidative stress responses and B-esterase activity of *Physalaemus nattereri* (Leiuperidae) and *Rhinella schneideri* (Bufonidae) tadpoles, *Chemosphere*, 2017, **185**, 548–562.
94. J. Freitas, T. da Silva Pinto, M. Yoshii, L. da Silva, L. de Palma Lopes, A. Ogura and E. Espíndola, Realistic exposure to fipronil, 2,4-D, vinasse and their mixtures impair larval amphibian physiology, *Environ. Pollut.*, 2022, **299**, 118894.
95. A. Bassó, S. Devin, P. Peltzer, A. Attademo and R. Lajmanovich, The integrated biomarker response in three anuran species larvae at sublethal concentrations of cypermethrin, chlorpyrifos, glyphosate, and glufosinate-ammonium, *J. Environ. Sci. Health, B*, 2022, **57**, 687–696.
96. N. Wingen, G. Cubas and G. Oliveira, A preliminary approach to the impact of a commercial formulation of glyphosate (Roundup®) in ecologically relevant concentrations on *Pseudis minuta* tadpoles, *Ecotox. Environ. Cont.*, 2022, **17**, 70–81.
97. C. Rutkoski, N. Macagnan, A. Folador, V. Skovronski, A. do Amaral, J. Leitemperger and M. Hartmann, Morphological and biochemical traits and mortality in *Physalaemus gracilis* (Anura: Leptodactylidae) tadpoles exposed to the insecticide chlorpyrifos, *Chemosphere*, 2020, **250**, 126162.
98. C. Rutkoski, N. Macagnan, A. Folador, V. Skovronski, A. do Amaral, J. Leitemperger and M. Hartmann, Cypermethrin- and fipronil-based insecticides cause biochemical changes in *Physalaemus gracilis* tadpoles, *Environ. Sci. Pollut. Res.*, 2021, **28**, 4377–4387.
99. A. Ferrari, C. Lascano, A. Pechen de D'Angelo and A. Venturino, Effects of azinphos methyl and carbaryl on *Rhinella arenarum* larvae esterases and antioxidant enzymes, *Comp. Biochem. Physiol.*, 2011, **153**, 34–39.
100. A. Venturino, L. Gauna, R. Bergoc and A. Pechen de D'Angelo, Effect of exogenously applied polyamines on malathion toxicity in the toad *Bufo arenarum* Hensel, *Arch. Environ. Contam. Toxicol.*, 1992, **22**, 135–139.
101. A. Venturino, O. L. Anguiano, L. Gauna, C. Cocca, R. M. Bergoc and A. M. Pechen de D'Angelo, Thiols and polyamines in the potentiation of malathion toxicity in larval stages of the toad *Bufo arenarum*, *Comp. Biochem. Physiol.*, 2001, **130**, 191–198.
102. A. Ferrari, L. Anguiano, C. Lascano, V. Sotomayor, E. Rosenbaum and A. Venturino, Changes in the antioxidant metabolism in the

embryonic development of the common South American toad *Bufo arenarum*: differential responses to pesticide in early embryos and autonomous-feeding larvae, *J. Biochem. Mol. Toxicol.*, 2008, **22**, 259–267.

103. A. Ferrari, C. Lascano, O. Anguiano, A. Pechan de D'Angelo and A. Venturino, Antioxidant responses to azinphos methyl and carbaryl during the embryonic development of the toad *Rhinella (Bufo) arenarum* Hensel, *Aquat. Toxicol.*, 2009, **93**, 37–44.

104. R. Lajmanovich, A. Attademo, P. Peltzer and C. Junges, Inhibition and recovery of cholinesterases in *Odontophrynus americanus* tadpoles exposed to fenitrothion, *J. Environ. Biol.*, 2009, **30**, 923–926.

105. R. Lajmanovich, P. Peltzer, A. Attademo, C. Colussi and C. Martinuzzi, Blood biomarkers of common toad following chlorpyrifos dermal exposure, *Interdiscip. Toxicol.*, 2018, **11**, 148–154.

106. R. C. Lajmanovich, P. M. Peltzer, C. S. Martinuzzi, A. M. Attademo, A. Bassó, M. I. Maglianese and C. L. Colussi, B-esterases and behavioral biomarkers in tadpoles exposed to pesticide Pyrethroid-TRISADA®, *Toxicol. Environ. Health Sci.*, 2018, **10**, 237–244.

107. P. Leite, T. Margarido, D. de Lima, D. Rossa-Feres and E. de Almeida, Esterase inhibition in tadpoles of *Scinax fuscovarius* (Anura, Hylidae) as a biomarker for exposure to organophosphate pesticides, *Environ. Sci. Pollut. Res.*, 2010, **17**, 1411–1421.

108. C. Lascano, A. Ferrari and A. Venturino, Sublethal concentrations of azinphos-methyl induce biochemical and morphological alterations in *Rhinella arenarum* embryos, *Chem. Ecol.*, 2011, **27**, 557–568.

109. E. Rosenbaum, L. Duboscq, J. Soleño, C. Montagna, A. Ferrari and A. Venturino, Response of biomarkers in amphibian larvae to *in situ* exposures in a fruit-producing region in North Patagonia, Argentina, *Environ. Toxicol. Chem.*, 2012, **31**, 2311–2317.

110. N. Liendro, A. Ferrari, M. Mardirosian, C. I. Lascano and A. Venturino, Toxicity of the insecticide chlorpyrifos to the South American toad *Rhinella arenarum* at larval developmental stage, *Environ. Toxicol. Pharmacol.*, 2015, **39**, 525–535.

111. A. M. Attademo, J. C. Sanchez-Hernandez, R. C. Lajmanovich, P. M. Peltzer and C. Junges, Effect of diet on carboxylesterase activity of tadpoles (*Rhinella arenarum*) exposed to chlorpyrifos, *Ecotox. Environ. Saf.*, 2017, **135**, 10–16.

112. H. S. Gripp, J. S. Freitas, E. A. Almeida, M. C. Bisinoti and A. B. Moreira, Biochemical effects of fipronil and its metabolites on lipid peroxidation and enzymatic antioxidant defense in tadpoles (*Eupemphix nattereri*: Leiuperidae), *Ecotox. Environ. Saf.*, 2017, **136**, 173–179.

113. E. Barreto, C. Salgado Costa, P. Demetrio, C. Lascano, A. Venturino and G. Natale, Sensitivity of *Boana pulchella* (Anura: Hylidae) tadpoles

to environmentally relevant concentrations of chlorpyrifos: effects at the individual and biochemical levels, *Environ. Toxicol. Chem.*, 2020, **39**, 834–841.

114. C. Martinuzzi, A. Attademo, P. Peltzer, T. Mac Loughlin, D. Marino and R. Lajmanovich, Comparative toxicity of two different dimethoate formulations in the common toad (*Rhinella arenarum*) tadpoles, *Bull. Environ. Contam. Toxicol.*, 2020, **104**, 35–40.
115. M. Acquaroni, G. Svartz and C. Pérez Coll, Acute, chronic and neurotoxic effects of dimethoate pesticide on *Rhinella arenarum* throughout the development, *J. Environ. Sci. Health*, 2022, **57**, 142–152.
116. M. Méndez-Rivera, F. Mena, M. Pinnock-Branford, C. Ruepert, M. D. Barquero, R. R. Jiménez and G. Alvarado, Effects of the insecticide β-endosulfan on tadpoles of *Isthmohyla pseudopuma* (Anura: Hylidae), *Aquat. Toxicol.*, 2022, **250**, 106231.
117. M. Bahl, J. Brodeur, C. Salgado Costa, M. D'Andrea, J. Sansiñena, D. Marino and G. Natale, Lethal and sublethal effects of the natural and healthy spinosad-based formulation Tracer™ on tadpoles of two neotropical species, *Environ. Sci. Pollut. Res.*, 2021, **28**, 13524–13535.
118. A. M. Attademo, R. Lajmanovich, P. Peltzer and C. Junges, Acute toxicity of metaldehyde in the invasive rice snail *Pomacea canaliculata* and sublethal effects on tadpoles of a non-target species (*Rhinella arenarum*), *Water Air Soil Pollut.*, 2016, **227**, 1–12.
119. M. Méndez, P. Obando, M. P. Branford, C. Ruepert, L. Castillo, F. Mena and G. Alvarado, Acute, chronic and biochemical effects of chlorothalonil on *Agalychnis callidryas*, *Isthmohyla pseudopuma* and *Smilisca baudinii* tadpoles, *Environ. Sci. Pollut. Res.*, 2016, **23**, 21238–21248.
120. A. Attademo, P. Peltzer, R. Lajmanovich, M. Cabagna and G. Fiorenza, Plasma B-esterase and glutathione S-transferase activity in the toad *Chaunus schneideri* (Amphibia, Anura) inhabiting rice agroecosystems of Argentina, *Ecotoxicology*, 2007, **16**, 533–539.
121. A. Attademo, M. Cabagna, R. Lajmanovich, P. Peltzer, C. Junges and A. Bassó, B-esterase activities and blood cell morphology in the frog *Leptodactylus chaquensis* (Leptodactylidae) on rice agroecosystems from Santa Fe Province (Argentina), *Ecotoxicology*, 2011, **20**, 274–282.
122. J. C. Brodeur, R. P. Suarez, G. S. Natale, A. E. Ronco and M. E. Zaccagnini, Reduced body condition and enzymatic alterations in frogs inhabiting intensive crop production areas, *Ecotoxicol. Environ. Saf.*, 2011, **74**, 1370–1380.
123. J. Brodeur, J. Vera Candioti, S. Soloneski, M. L. Larramendy and A. Ronco, Evidence of reduced feeding and oxidative stress in common tree frogs (*Hypsiboas pulchellus*) from an agroecosystem experiencing severe Ddought, *J. Herpetol.*, 2012, **46**, 72–78.

124. A. Attademo, P. Peltzer, R. Lajmanovich, M. Cabagna-Zenklusen, C. Junges, E. Lorenzatti and P. Grenón, Biochemical changes in certain enzymes of *Lysapsus limellium* (Anura: Hylidae) exposed to chlorpyrifos, *Ecotox. Environ. Saf.*, 2015, **113**, 287–294.

125. J. Brodeur, M. Damonte, D. Rojas, D. Cristos, C. Vargas, M. Poliserpi and A. Andriulo, Concentration of current-use pesticides in frogs from the Pampa region and correlation of a mixture toxicity index with biological effects, *Environ. Res.*, 2022, **204**, 112354.

126. L. Girotto, I. Freitas, M. Yoshii, B. Goulart, C. Montagner, L. Schiesari and J. Freitas, Using mesocosms to evaluate the impacts of pasture intensification and pasture-sugarcane conversion on tadpoles in Brazil, *Environ. Sci. Pollut. Res.*, 2023, **30**, 21010–21024.

127. D. Norris and K. Lopez, in *Hormones and reproduction of vertebrates (Vol. 1)*, Academic Press, Elsevier, 2011.

128. T. McDaniel, P. Martin, J. Struger, J. Sherry, C. Marvin, M. McMaster, S. Clarence and G. Tetreault, Potential endocrine disruption of sexual development in free ranging male northern leopard frogs (*Rana pipiens*) and green frogs (*Rana clamitans*) from areas of intensive row crop agriculture, *Aquat. Toxicol.*, 2008, **88**, 230–242.

129. F. Orton and C. R. Tyler, Do hormone-modulating chemicals impact on reproduction and development of wild amphibians?, *Biol. Rev.*, 2015, **90**, 1100–1117.

130. K. McCoy, C. Amato, L. Guillette Jr. and C. Mary, Giant toads (*Rhinella marina*) living in agricultural areas have altered spermatogenesis, *Sci. Total. Environ.*, 2017, **609**, 1230–1237.

131. P. Duarte-Guterman, V. Langlois, K. Hodgkinson, B. Pauli, G. Cooke, M. Wade and V. Trudeau, The aromatase inhibitor fadrozole and the 5-reductase inhibitor finasteride affect gonadal differentiation and gene expression in *Silurana tropicalis*, *Sex. Dev.*, 2009, **3**, 333–341.

132. F. Orton, M. Säfholm, E. Jansson, Y. Carlsson, A. Eriksson, J. Fick, T. Uren Webster, T. McMillan, M. Leishman, B. Verbruggen, T. Economou, C. R. Tyler and C. Berg, Exposure to an anti-androgenic herbicide negatively impacts reproductive physiology and fertility in *Xenopus tropicalis*, *Sci. Rep.*, 2018, **8**, 9124.

133. T. B. Hayes, V. Khoury, A. Narayan, M. Nazir, A. Park, T. Brown, L. Adame, E. Chan, D. Buchholz, T. Stueve and S. Gallipeau, Atrazine induces complete feminization and chemical castration in male African clawed frogs (*Xenopus laevis*), *Proc. Natl. Acad. Sci. USA.*, 2010, **107**, 4612–4617.

134. A. Schwendiman and C. Propper, A common environmental contaminant affects sexual behavior in the clawed frog, *Xenopus tropicalis*, *Physiol. Behav.*, 2012, **106**, 520–526.

135. M. Kvarnryd, R. Grabic, I. Brandt and C. Berg, Early life progestin exposure causes arrested oocyte development, oviductal agenesis and sterility in adult *Xenopus tropicalis* frogs, *Aquat. Toxicol.*, 2011, **103**, 18–24.
136. M. Babini, C. Bionda, Z. Salinas, N. Salas and A. Martino, Reproductive endpoints of *Rhinella arenarum* (Anura, Bufonidae): Populations that persist in agroecosystems and their use for the environmental health assessment, *Ecotox. Environ. Saf.*, 2018, **154**, 294–301.
137. G. van der Horst, Status of sperm functionality assessment in wildlife species: From fish to primates, *Animals*, 2021, **11**, 1491.
138. L. Sanchez, R. Lajmanovich, P. Peltzer, A. Manzano, C. Junges and A. Attademo, First evidence of the effects of agricultural activities on gonadal form and function in *Rhinella fernandezae* and *Dendropsophus sanborni* (Amphibia: Anura) from Entre Ríos Province, Argentina, *Acta. Herpetol.*, 2014, **9**, 75–88.
139. L. Curi, P. Peltzer, A. Attademo and R. Lajmanovich, Alterations in gonads and liver tissue in two Neotropical anuran species commonly occurring in rice fields crops, *Water Air Soil Pollut.*, 2021, **232**, 203.

CHAPTER 13

Amphibian Species as Experimental Models in Environmental Risk Assessment Studies – Epilogue and Final Remarks

MARCELO L. LARRAMENDY*[a,b] AND
GUILLERMO ELI LIWSZYC[c]

[a]Cátedra de Citología, Facultad de Ciencias Naturales y Museo, Universidad Nacional de La Plata, Calle 64 N° 3, B1904AMA La Plata, Argentina; [b]Consejo Nacional de Investigaciones Científicas y Técnicas (CONICET), Buenos Aires, Argentina; [c]Former Guest Scientist at the Department of Pathology, University of Helsinki, Helsinki, Finland
*E-mail: marcelo.larramendy@gmail.com

In recent times, the growing impact of anthropogenic activities has generated huge quantities of toxic residues that can have direct or indirect detrimental effects on the quality of our environment. The increase in their jeopardizing effects can exert short-, medium- and long-term consequences, affecting not only human but also environmental health in general. Furthermore, these disturbances are usually capable, in most cases, of

Issues in Toxicology No. 46
Amphibian Species in Environmental Risk Assessment Strategies
Edited by Marcelo L. Larramendy and Guillermo Eli Liwszyc

Published by the Royal Society of Chemistry, www.rsc.org

reducing the biodiversity of native fauna and flora, which will, in turn, encourage the resistance and emergence of new pests and diseases.

This book, *Amphibian Species in Environmental Risk Assessment Strategies*, intends to provide an outline, relevant examples and stimulate practical discussions on the use of this group of vertebrates, the Class Amphibia, within the scientific challenges faced by the ecotoxicology risk assessment (ERA) academic world. This book endows relevant tools that may be of use in the analysis prior to the implementation of decisions leading to actions, which will hopefully reduce environmental health risks against environmental factors that may adversely impact human and other living species' health or cause ecological imbalances between them and our environment.

Our aim was to compile information from a diversity of sources into a single volume. The rationale is to give some real-life examples to widen the concept that the use of non-conventional animal models, far from being a scientific conundrum, may instead provide useful avenues to be explored and further developed as a way to provide help and real answers to some of the actual problems the whole planet is dealing with. Concomitantly, these concrete examples extend concepts of hazardous factors to living species worldwide, which may stimulate new research ideas and trends in their respective fields.

This volume comprises three major sections: (1) *Amphibian Species as Experimental Models in Environmental Risk Assessment Studies. General Aspects*; (2) *Amphibian Species as Experimental Models in Environmental Risk Assessment Studies* and (3) *Amphibian Species as Experimental Models in Environmental Risk Assessment Studies. Epilogue, and Final Remarks*.

In the first and general frame section, *Amphibian Species as Experimental Models in Environmental Risk Assessment Studies. General Aspects,* readers will find Chapter 1 to be a broad-spectrum chapter. It provides background and introductory information about the concept and implications of this hot topic in our current world, environmental risk assessment, while highlighting the nature of some environmental stressors and examples of the most versatile and validated methods of analysis. In general terms, the narrative focuses on the worldwide accepted concept that ERA is a scientific practise that identifies and evaluates the likelihood of an activity or substance to threaten the environment, in general and/or living organisms, natural habitats and ecosystems, in particular. In other words, the term employed describes the overall process(es) and/or method(s) allowing the identification of hazards and risk factors that have the potentiality to cause harm (hazard identification) as well as to analyse and evaluate the risk associated with that particular hazard (risk analysis and risk evaluation). It also suggests several, if not all, appropriate ways to eliminate or minimize the hazard or control the risk when this cannot be eliminated (risk control).

The second frame section, *i.e., Amphibian Species as Experimental Models in Environmental Risk Assessment Studies*, includes 11 chapters (Chapters 2–12) in which these vertebrates are used as biotic matrices.

Actually, all extant amphibian species are some of the most successful vertebrates on Earth. The Superclass Tetrapoda is divided into four classes of vertebrate animals with four limbs. Reptiles, birds and mammals are amniotes, the eggs of which are either laid or carried by the female and surrounded by several membranes, some of which are impervious. In contrast, lacking these membranes, amphibians require water bodies for reproduction, although some species have developed various strategies to protect or bypass the vulnerable aquatic larval stage.[1–3] They are not found inhabiting seas with the exception of one or two frogs that live in brackish water in mangrove swamps, *e.g.*, the Anderson's salamander (*Ambystoma andersoni*), a neotenic salamander from the Zacapu Lagoon region in the State of Michoacán (Mexico). On land, amphibians are restricted to moist habitats because of the need to keep their skin damp.

Modern amphibians have a simplified anatomy compared to their ancestors due to paedomorphosis, caused by two evolutionary trends: miniaturization and an unusually large genome, which result in slower growth and developmental rates compared to other vertebrates.[4,5] Another reason for their size is related to their rapid metamorphosis, which apparently has only evolved in the ancestors of Lissamphibia. Since remodelling of the feeding apparatus suggests that they do not feed during the metamorphosis period, this event has to proceed at a faster pace the smaller the individual is and takes place at an early larval stage. In this regard, it is well known that the largest species of salamanders do not go through metamorphosis.[6] Amphibians that lay eggs on land often go through the whole metamorphosis inside the egg. An anamniotic terrestrial egg measures less than 1 cm in diameter in order to enable proper diffusion, a limiting factor of post-hatching growth size.[7] The smallest amphibian (and vertebrate) species in the world is a microhylid frog from eastern Papua New Guinea, *Paedophryne amauensis*. It has an average length of 7.7 mm and is part of a genus that contains four of the world's 10 smallest frog species.[8] On the other hand, the largest living amphibian is the 1.8 m Chinese giant salamander *Andrias davidianus*.[9] Clearly a great deal smaller than the largest amphibian that has ever existed, the extinct 9 m *Prionosuchus plummeri*, a crocodile-like Temnospondyl dating to 270 million years ago from the middle Permian of Brazil.[10] On the other hand, the largest living frog is the Goliath frog *Conraua goliath* from Africa, which can reach up to 32 cm and weighs 3 kg.[9]

Amphibians are ectothermic vertebrates. Their metabolic rate is low, and as a result, their food and energy requirements are limited. In the adult state, they have tear ducts and movable eyelids, and most species have ears that can detect airborne or ground vibrations. They have muscular tongues, which in many species can protrude. Modern amphibians have fully ossified vertebrae with articulate processes. Their ribs are usually short and may be fused to the vertebrae. The skulls are mostly broad and short and are often incompletely ossified. Their skin contains little keratin and lacks scales, apart from a few fish-like scales in certain caecilians. The skin contains many mucous glands and, in some species, poisonous glands. The hearts of amphibians have three chambers, two atria and one ventricle. They have a urinary bladder and nitrogenous waste products are excreted primarily as urea. Amphibians breathe by means of a pump action in which air is first drawn into the buccopharyngeal region through the nostrils. These are then closed and the air is forced into the lungs by contraction of the throat. The respiratory function is also supplemented with gas exchange through the skin.

Amphibians are notable among vertebrates for their diversity of chromosomes and genomes. The chromosomal complement has been determined for at least 1193 (14.5%) of the approximately 8200 known diploid species, including 963 anurans, 209 salamanders and 21 caecilians. In general terms, the karyotypes of diploid amphibians are characterized by 20–26 bi-armed chromosomes. Amphibians also have very large genomes compared to other taxa of vertebrates and corresponding variation in genome size, ranging from 0.95 to 11.5 pg in frogs, from 13.89 to 120.56 pg in salamanders, and from 2.94 to 11.78 pg in caecilians.[11] These large genome sizes have prevented whole-genome sequencing of amphibians, although a number of genomes have been published recently. The 1.7-GB draft genome of *Xenopus tropicalis* was the first to be reported in 2010.[11] Compared to some salamanders, this frog's genome is tiny. In comparison, the genome of the Mexican axolotl is 32 Gb, which is more than 10 times larger than the human genome, *i.e.*, 3 Gb.[12]

In typical amphibian development, eggs are laid in water and larvae are adapted to an aquatic lifestyle. Frogs, toads and salamanders all hatch from the egg as larvae with external gills. Most amphibians go through metamorphosis. This process is induced and stimulated by the thyroxine blood concentration and counter-regulated by prolactin. Specific events are dependent on the threshold values of these hormones in different tissues. Besides, most embryonic development is carried out outside the parental body and, thus, is subject to many scenarios influenced by specific environmental circumstances. For this reason, tadpoles can have horny ridges instead of teeth, whisker-like skin extensions or fins. They also make use of a sensory lateral line organ similar to that of fish. After metamorphosis, these organs become redundant and will be reabsorbed by a controlled apoptotic process. The variety of adaptations to specific environmental circumstances among amphibians is wide, with many discoveries still being made to this day.[13]

Taking into account the explosive increase in the list of pollutants, the scientific community is facing challenges in protecting humans and wildlife from the potentially negative consequences of their effects. The majority, if not all, of the previous efforts conducted to achieve this goal have revealed the global presence of high levels of various pollutants in amphibians. Laboratory research and epidemiological studies imply that xenobiotics exert a broad spectrum of side effects on amphibians by interfering with several parameters, including their endocrine, immune and neural systems, reproduction, development and growth. Nevertheless, inconsistent results may seem to suggest that the potential effects of pollutant exposure on physiological parameters in amphibians are clearly multifactorial, involving simultaneously several and diverse biological processes, species-specific differences, sex and age, all depending upon the types of xenobiotics involved. Another factor deserving attention is the type of interaction between more than one pollutant acting simultaneously with another one. When the conventional concentration additivity is analysed, several studies have suggested that different classes of pollutants may possibly act synergistically or antagonistically based upon their concentration. However, the risk assessment information is still recorded in a scattered and inadequate partial manner, including a lack of enough anuran species under study, a limited number of xenobiotics investigated, and insufficient geographical representation. For these reasons, our knowledge of the effects of xenobiotic-induced deleterious effects on amphibians still remains rudimentary, although on a more positive note, the mechanistic understanding of their mode of action is progressing. On the contrary, research on wild amphibian populations and their ecosystems under pollutant-induced stress is still unavailable, as far as we know. Accordingly, the aim of this book section is to provide some practical evidence predicting and tracing the effects of several xenobiotics at different biological organization levels, especially starting from the molecular, cellular and individual ones to the population, community and ecosystem levels, due to the limited and scattered information available, as mentioned above.

Chapter 2 provides an overview highlighting that amphibians are unique species among vertebrates in their physiological and life history adaptations, which are particularly challenging given their multiphasic existence that enables some species to exist in aquatic and terrestrial habitats. The use of amphibians in toxicity testing protocols to enable environmental risk assessment applications has enabled the making of management decisions regarding hazardous substances released to the environment. Toxicity testing protocols that use larval amphibians are predominantly performed through exposures to test compounds administered to the water. However, terrestrial forms (*e.g.*, frogs, toads and salamanders) are provided where test compounds were mixed into soil and sediments. Oral exposure routes for adult amphibians are complicated by practical considerations in exposing prey and inadvertent soil ingestion. In the chapter, the authors discuss the available science in controlled testing protocols and their potential risk assessment applications in larval and adult species of different amphibians, especially employing salamanders as biotic matrices, while recognizing some of the advantages, disadvantages and uncertainties.

In Chapter 3, authors introduce the concept of ecotoxicology and underpin the emerging field of nanoecotoxicology. The work with three amphibian's very well-known model species in evaluating the potential risks associated with exposure to nanoparticles in the context of ERA is reported and discussed. The African clawed frog *Xenopus laevis* (Anura, Pipidae), the Iberian ribbed newt, also called "gallipato", *Pleurodeles waltl* (Urodela, Salamandridae) and the paedomorphic salamander, also called "*axolotl*", *Ambystoma mexicanum* (Urodela, Ambistomidae), were employed, in which varying levels of toxicity to nanoparticles have been exhibited. Genotoxic effects were observed during the exposure of the amphibian larvae to certain nanoparticles, which makes amphibians a suitable model for assessing their genetic hazards in ERA. The study proposes using larval growth inhibition as a parameter to evaluate the chronic effects of exposure to nanoparticles. To conclude, the chapter proposes the use of a combination of ecotoxicological tools, such as monospecific and standardized batteries of tests and studies in microcosms or mesocosms, to manage and develop a regulatory approach to nanoparticles in the environment. Overall, the authors propose as an example the demonstration of the significance of amphibians in ERA and their potential to contribute to a "Safer by Design" approach to product development.

Following this chapter, Chapter 4 highlights that agrochemical residues affect the sustainability of agroecosystems and compromise the health of many amphibians and other living species, including humans. To monitor the environment, biomarkers of genotoxicity, such as the comet or single-cell gel electrophoresis (SCGE) assay, and mutagenicity, such as the micronucleus (MN) test, can be efficient tools to estimate the early effects of agrochemical pollution. Thus, the present study has evaluated the factors that indicate genotoxicity and mutagenicity in three Neotropical amphibian species that inhabit agroecosystems. Reviewed data from South American studies conducted on agroecosystems used *in situ* approaches with Neotropical anuran species. Two hundred fifty-six manuscripts indexed in the Web of Science, Scopus and Google Scholar databases were analysed, out of which 19 papers were shortlisted. Studies were mainly led by Argentina, Brazil and Colombia. Notably, the three most frequently studied Neotropical species were the lesser frog *Dendropsophus minutus* (Hylidae), the barker frog *Physalaemus cuvieri* (Leptodactilidae) and the South American common frog *Rhinella arenarum* (Bufonidae). Overall, the species found in agricultural areas showed a higher frequency of DNA damage evaluated by CA and MN compared to reference areas that served as controls. Fifty-six formulations of agrochemicals were reported, including insecticides, herbicides and fungicides, as well as fertilizers (*e.g.*, nitrate and phosphorus). Given the sensitivity of these anurans to the genotoxicity and mutagenicity of chemical compounds, new studies should be encouraged in the agroecosystem areas of interest.

Chapter 5 focuses on the detection and quantification of contaminants of emerging concern, such as human and veterinary pharmaceuticals in aquatic environments from mid-eastern Argentina (the Parana River wetlands, one of the largest water basins in the world), using different analytical techniques. In addition, these environmental screenings have been used to test how tadpoles from native amphibian species (*e.g.*, *Rhinella arenarum*, *Leptodactylus luctator*, *Physalaemus albonotatus*, *Trachycephalus typhonius* and *Scinax nasicus*) respond to recorded concentrations of pharmaceutical residues (*e.g.*, diclofenac, dexamethasone, ciprofloxacin, enrofloxacin, dipyrone and streptomycin) and their metabolites. These are widely used medications for the treatment of common ailments in humans. The summary of different studies demonstrated that the presence of these contaminants in aquatic environments poses a deleterious environmental risk for amphibians.

Chapter 6 describes the usefulness of the studies of xenobiotics on the African continent, a region where not much has been published on the subject. These agents are regularly being released into the environment due to increasing human exploitative activities. Individual actions and/or complex interactions among these xenobiotics, *e.g.*, radiation, chemical contaminants, parasites/pathogens, toxic metals and climate change, among others, adversely affect amphibian populations. These stressors either kill or induce sublethal effects in amphibians. The use of adult African toads as bioindicators of xenobiotic-induced geno- and systemic toxicity was considered. Some xenobiotics also generate reactive oxygen species that elicit systemic toxicity and genotoxicity. Xenobiotics can also directly induce sublethal toxicity in the organ systems. Alterations in body biochemistry, haematological indices, body morphology and histopathology were common findings in systemic toxicity-affected African toads. Micronucleus and abnormal nuclear malformations were the common biomarkers of this anomaly. Damage induced by xenobiotics may also lead to pathophysiological, inflammatory and genotoxic disorders, while reproductive abnormalities and morphological deformities are present as well. These alterations may incapacitate the toad and eventually lead to death. Certainly, these factors constitute a threat to amphibian health and are the hallmarks of population decline. There is a clear need for a better understanding of the causal relationship between environmental toxicant impacts and the decline in populations.

The following chapter, Chapter 7, illustrates the fact that the development of massive sequencing technologies has led to an impressive increase in data acquisition capacity and analysis. The study of differential expression of genes (DEG) in species exposed to contaminants using whole transcriptomics enables a response pattern as a massive molecular biomarker fingerprint. This biomarkomic approach has been recently applied to the autochthonous, non-model amphibian *Rhinella arenarum* exposed to organophosphorus pesticides. The whole transcriptome study enabled the annotation of approximately 57 000 gene sequences, an excellent starting point considering that the *R. arenarum* genome remains unknown. In parallel, massive biomarkomic analysis has confirmed that a few top-regulated genes are commonly affected by contaminant exposure, while more genes are specifically affected by organophosphorus pesticides. Interestingly, none of the target or responding genes typically recognized for organophosphorus pesticides at the biochemical or molecular levels were found among the top-regulated genes in the biomarkomic analysis. This fact was also confirmed by a hypothesis-driven DEG analysis focused on these recognized targets for organophosphorus agents' effects and responses, except for a very few genes related to detoxification, metabolism and regulation. Efforts are now directed to long-noncoding RNA biomarkomics and the extension of our transcriptomic studies to other contaminants and aquatic autochthonous species.

Continuing with practical examples of non-conventional amphibian species as experimental models, Chapter 8 highlights that about 41% of the amphibian species are threatened with extinction according to the International Union for Conservation of Nature (IUCN) Red List, undeniably an alarming situation that deserves priority actions. One important environmental factor commonly associated with the amphibian decline phenomenon is solar ultraviolet (UV) radiation due to its genotoxicity. Other stressors present in the environment may potentiate the UV-induced effects in amphibians. It is a case of an emergent topic in the research fields of ecotoxicology and conservation, opening up new lines of study. In this chapter, the authors summarize the state-of-the-art knowledge about the genotoxic risk of solar UV radiation for amphibian species and its potential implications for the population decline phenomenon. The relevance of UV radiation for amphibians in a changing environment is discussed, describing the UV-induced DNA lesions and their lethal and sublethal consequences for amphibians, as well as exploring amphibian strategies to avoid and deal with UV damage, such as behavioural changes and activation of DNA repair mechanisms. Finally, certain considerations about the link between the genotoxicity triggered by solar UV exposure and the amphibian decline and ecological factors are analysed.

Chapter 9 remarks on the fact that amphibian populations have been declining for the last 30 years worldwide, a phenomenon that can be partially explained by the extension of agriculture and, more particularly, the high rate of pesticide application that modern agriculture requires to produce high yields. Moreover, pesticides are often applied in the form of mixtures, which interact with and affect non-target organisms in different ways by altering both their survival and/or fitness. The Pampas region of Argentina (one of the breadbaskets of the world) has not been exempted from this situation and studies involving amphibians and mixtures of pesticides have increased in the last decades, providing valuable information on the matter. Most of these studies have been performed in laboratory conditions employing the South American common frog *Rhinella arenarum* (Anura, Bufonidae), a ubiquitous toad of the region. As far as we know, only one work has been performed in field conditions evaluating mixtures of commonly employed pesticides in four species. Different interactions were detected among the mixture studies, with variations among the formulations employed and/or developmental stages of individuals exposed and time of exposure. These facts highlight the importance of extending the studies to the most employed mixtures of agrochemicals together with native fauna to minimize negative effects on the native biota.

Chapter 10 illustrates that, although considered a globally declining vertebrate taxon, amphibians are often ignored and replaced in ecological risk assessments by surrogate species from other vertebrate groups with striking biological disparities (*e.g.*, fish or birds and mammals for aquatic and terrestrial forms, respectively). This mismatched surrogacy may introduce strong biases in assessing environmental risk for amphibians. A great deal of variation in amphibian life history and morphology further challenges efforts to identify an ideal amphibian model species candidate to represent the whole group in ecological risk assessment. Here we provide an overview of the relative sensitivity of the direct-developing frog *Eleutherodactylus johnstonei* to chemical and physical stressors in the context of amphibian risk assessment. To achieve this goal, this chapter is divided into three sections exploring the role of this species in ecotoxicology—pesticide toxicity, chemical cyto-genotoxicity and UV radiation cyto-genotoxicity—based on existing and novel data. The outcome of this chapter reflects the fact that *E. johnstonei* represents a promising yet underused model for approaching different screening levels in ecotoxicology. Finally, the authors hope that the overview presented in this chapter will encourage other researchers to capitalize on the enormous potential that this species has to offer in ERA.

The following chapter, Chapter 11, stresses that morphological and histological traits are good biomarkers of intermediate responses of organisms to contaminants as they directly correlate with both biochemical and physiological responses. Morphological biomarkers describe changes and abnormalities from subcellular and cellular levels up to individual ones. Morphological changes in organs and tissues can be defensive or compensatory to maintain functions against degenerative changes. New morphological tools used in both field work and laboratory experiments to evaluate the effects of contaminants on frogs are introduced in this chapter. Also provided are rare, detailed tool descriptions that include characteristics of tissue morphometric techniques, cell density and colour-difference analysis. In addition, the authors demonstrate their applicability in different organs and tissues of Neotropical frog species, *e.g.*, the analyses in melanocytes and melanomacrophages. In summary, the authors pinpoint that the application of histomorphometric techniques and colour-difference analyses is a useful tool in the armamentarium for evaluating different tissues, which have proven to be applicable to the liver and skin of different species, demonstrating that the effects of environmental stressors are permanent and tissue responses homogeneous.

In the last chapter of this frame section, *i.e.,* Chapter 12, authors clearly underpin the existence of two key simultaneous features of Neotropical amphibians' biodiversity: one that it is amazing and the other that it is disappearing. This region has the highest diversity of species in the world, containing more species of amphibians than tropical Africa and Southeast Asia combined. Examination of assessment data for all Neotropical anuran species shows that modern agricultural practises top the list of threats, posing risks to up to 30% of all the assessed species. This type of agriculture makes use of a wide range of agrochemicals for genetically modified tolerant seeds. In this context, biomarkers understood as cellular, biochemical, physiological or histological quality in living organisms to assess the exposure or effect of xenobiotics, including agrochemicals, can provide connection clues on the exposure and biological responses in amphibians such as anurans. The information obtained from these sources can be later validated in the field for the implementation of environmental risk assessment strategies. The chapter aims to describe the use of biomarkers at the genetic, biochemical and physiological levels to demonstrate their usefulness as endpoints in Neotropical anuran species to be used as validating tools indicating genotoxicology and ecotoxicology due to exposure to xenobiotics.

The third frame section, *i.e., Amphibian Species as Experimental Models in Environmental Risk Assessment Studies. Epilogue and Final Remarks,* is integrated into one chapter (Chapter 13). In this chapter, the editors of the current book summarize how the growing impact of anthropogenic activities generates huge quantities of toxic residues that can have direct or indirect detrimental effects on the quality of our environment. The increase in their jeopardizing effects can exert short-, medium- and long-term consequences, affecting not only human health but also environmental health in general. Furthermore, these disturbances are usually capable, in most cases, of reducing the biodiversity of native flora and fauna, which will, in turn, encourage resistance and the emergence of new pests and diseases. This book intends to provide an outline and relevant examples, and stimulate practical discussions on the use of this non-conventional group of vertebrates, the amphibians. It includes relevant tools that may be of use in the analyses prior to the implementation of decisions leading to actions, which will hopefully reduce environmental health risks against environmental factors. Our aim was to compile information from various sources into a single volume. The rationale is to give some real-life examples to widen the concept that the use of non-conventional animal models, far from being a scientific conundrum, may instead provide useful avenues to be explored and further developed as a way to provide help and real answers to some of the actual problems the whole planet is dealing with currently. Concomitantly, these concrete examples extend concepts of hazardous factors to living species worldwide, which may stimulate new research ideas and trends in their respective important fields.

In spite of having to deal with many diverse topics, we have tried to compile this "wealth of information" to take into consideration whether different amphibian species of one or more populations can be selected as experimental matrices.

This book is part of a series of four volumes along similar lines to be published in Issues in Toxicology by the Royal Society of Chemistry (see Figure 13.1) utilizing different species rather than amphibians (mammals, birds and reptiles). The present volume is the third one. First in the list is "*Marsupial and Placental Mammal Species in Environmental Risk Assessment Strategies*" (see Figure 13.1A)[14] and is followed by "*Bird and Reptiles Species in Environmental Risk Assessment Strategies*" (see Figure 13.1B).[15]

Without running the risk of being repetitive, we would like to recap some important concepts as mentioned erstwhile in the two volumes previously published in Issues in Toxicology by the Royal Society of Chemistry in 2017 (see Figure 13.2), *i.e.*, "*Ecotoxicology and Genotoxicology – Non-traditional Terrestrial* Models" (see Figure 13.2A)[16] and "*Ecotoxicology and Genotoxicology*

Figure 13.1 Front book covers of the books published in Issues in Toxicology, The Royal Society of Chemistry, *Marsupial and Placental Mammal Species in Environmental Risk Assessment Strategies, 2022* (A), followed by *Fish and Reptiles Species in Environmental Risk Assessment Strategies, 2023* (B) and the current volume *Amphibian Species in Environmental Risk Assessment Strategies, 2023* (C). Images kindly provided by The Royal Society of Chemistry, Cambridge, United Kingdom.

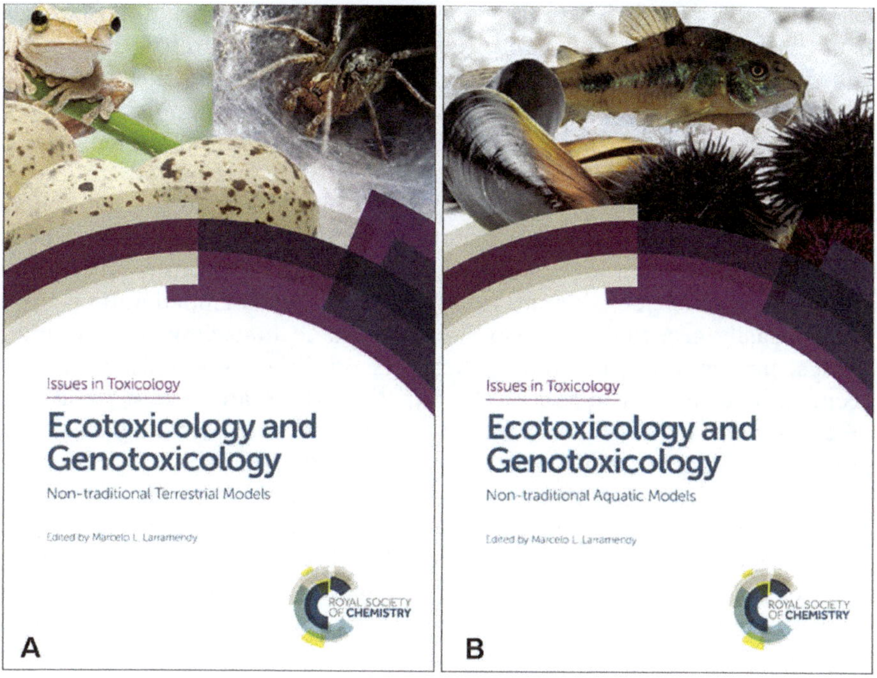

Figure 13.2 Front book covers of the books *Ecotoxicology and Genotoxicology – Non-traditional Terrestrial Models* (A) and *Ecotoxicology and Genotoxicology – Non-traditional Aquatic Models* (B), previously published in Issues in Toxicology, The Royal Society of Chemistry by the same editor group in 2017. Images kindly provided by The Royal Society of Chemistry, Cambridge, United Kingdom.

– *Non-traditional Aquatic Models*" (see Figure 13.2B),[17] by the same editor group. We strongly recommend the perusal of both volumes, which are not overlapping subjects, to gain the full benefit of this series and have a holistic and stimulating approach to the matter.

New emerging studies in several reputable publications of common worldwide human diseases such as arterial hypertension, among others, seem to suggest that not only classical and lifestyle factors should be addressed in the search for adequate treatment.[18] From a medical perspective, it is not only a disease of affluence. One in eight deaths worldwide is due to this condition and its corresponding factors (heart disease, kidney disease and stroke). These studies seem to suggest that a closer look at other, up until now unrelated factors, is required. Early-life nutrition and exposure to air pollution, heavy metals and even noise are factors implicated in pushing blood pressure up later in life. Thinking out of silos and cross-linking medical with applied science will lead to a faster pace of understanding all the factors related to the emergence of a disease and thus finding a cure.

Many scientists feel that time has come for a shift in emphasis away from basic towards applied research. While basic research is needed to shed more light on the fundamentals, applied research is required to find solutions to the many problems the world faces, *e.g.*, overpopulation, excessive use of the Earth's natural resources, and pollution.

The chapters included in this book are just a mere enumeration of some practical examples. There are many more species that can be used as experimental models and the list should be expanded by different research groups and academics from all over the world. Quite recently, among the list of species to be considered, some scientists have included animals that can be easily accessed and also share the same household with human beings, namely dogs, cats and other pets. While on one side, they may seem to be a faithful mirror of the situation, there are certain pitfalls to be considered. One of them is nutrition as they mostly now sadly consume only industrial feeds and not fresh produce. Their metabolic systems are also not closely related to humans, with the exception of pigs. We shall certainly hear more on the matter as this approach is being explored.

We hope that many more scientists realize that it is more important to tackle subjects related to the status of the environment using autochthonous, non-target species that are truly exposed to locally used xenobiotic agents. On the other hand, this research plan would also increase the chances of independent scientists and research institutions in getting access to grants and attracting the attention of local authorities, who would be more interested in financing projects that really are within their sphere of political interest and pride.

Globally, 41% of amphibian species and 19% of reptile species are threatened or dying out. This is the result of their natural habitat destruction due to land use, disease, pollution and climate change. In many species, these factors have added so much pressure that they have been completely wiped out. It is our hope that with the emergence of the new legal figure of Ecocide, governments, politicians, business corporations and producers will have to parse more carefully the consequences of their actions and activities. It is certainly the first important step towards their accountability and public pressure is mandatory to be successful, as actions by agents and politicians will be hard to fathom. Another step forward in this direction is the recently launched "Banana Index" by *The Economist*, a weekly financial publication. It tries to give a hint at the environmental impact of foodstuffs produced, giving consumers, economists and the public in general a rough idea of the matter. As in statistics in general, this indicator reveals a lot but hides the essentials. Nevertheless, it is a trailblazing intent to raise awareness at the time we pick up our supplies from groceries and markets. There are certainly some certification schemes already in place (Rainforest Alliance, FSC, *etc.*) with their labels inserted in the packages. Most of them will have to be dramatically improved to lighten or reduce as much as possible their overload on the environment.

Clearly, the problem here is twofold. On the one hand, the production of foodstuffs and, on the other hand, its packaging and disposal of waste. All this should be factored in at the till. Food safety, as the present war in Ukraine has clearly shown, is a phantom still hanging around. Ultra-processed foods labelled as organic, with a basic "organic" commodity (that can be debated), make use of manufacturing processes that are clearly misleading consumers into making healthy choices and nothing more far away from the truth. All these situations would deserve another volume and clearly escape the purpose of this series.

The UN Biodiversity Conference Intactness Index (BII) shows a world average of 75%; this is a value certainly lower than the 90% required to be considered the safe limit for not falling into "ecological recession".

Stone Age, Copper Age, Iron Age, Medieval Age, Modern Age are just a few of the developmental stages of mankind. The anthropocene will probably be followed by the Extinction Age, if we remain quiet bystanders. While the clock is ticking, we still have some valuable time left to be used diligently.

Many researchers have contributed to the publication of this book. We hope that it serves as a herald to bolster enthusiasm for the use of native or easily available local species to widen our knowledge on the subject. Last but not least, we would like to especially thank the authors who have responded to our call for their time, contribution and feedback, which made the compilation of this book possible. Let us hope that the anthropocene will not be the last age of humanity.

References

1. W. W. Skeat, in *A Concise Etymological Dictionary of the English Language*, Clarendon Press, Oxford, 1895.
2. D. Frost, American Museum of Natural History: Amphibian Species of the World 5.6, Online Reference, 2013, Available from: https://amphibiansoftheworld.amnh.org/. Accessed 19 April 2023.
3. D. C. Blackburn and D. B. Wake, in *Animal biodiversity: An outline of higher-level classification and survey of taxonomic richness,Class Amphibia Gray, 1825*, ed. Z.-Q. Zhang, Zootaxa, 2011, vol. **3148**, pp. 39–55.
4. K. D. Wells, in *The Ecology and Behavior of Amphibians*, Chicago University Press, 2010.
5. D. L. Levy and R. Heald, Biological scaling problems and solutions in Amphibians, *Cold Spring Harb. Perspect. Biol.*, 2016, **8**, a019166.
6. R. R. Schoch, in *Amphibian Evolution. The Life of Early Land Vertebrates*, Wiley Professional, Reference & Trade, Hoboken NJ, 1st edn, 2014.
7. M. Laurin, The evolution of body size, Cope's rule and the origin of amniotes, *Syst. Biol.*, 2004, **53**, 594–622.

8. E. N. Rittmeyer, A. Allison, M. C. Gründler, D. K. Thompson and C. C. Austin, Ecological guild evolution and the discovery of the world's smallest vertebrate, *PLoS One*, **20, 127**, e29797.
9. B. Nguyen and J. Cavagnaro, University of California, Berkeley, CA, Available from: https://amphibiaweb.org/species. Accessed 19 April 2023.
10. L. I. Price, Um anfibio Labirinthodonte da formação Pedra de Fogo, Estado do Maranhão, *Bol. Div. Geol. Mineral., Dep. Nac. Prod. Miner.*, 1948, **24**, 7–32.
11. Y. Uno, Inference of evolution of vertebrate genomes and chromosomes from genomic and cytogenetic analyses using amphibians, *Chromosome Sci.*, 2021, **24**, 3–12.
12. S. Nowoshilow, S. Schloissnig, J.-F. Fei, A. Dahl, A. W. C. Pang, M. Pippel, S. Winkler, A. R. Hastie, G. Young, J. G. Roscito and F. Falcon, The axolotl genome and the evolution of key tissue formation regulators, *Nature*, 2018, **5549**, 50–55.
13. R. A. Newman and A. Robert, Adaptive plasticity in amphibian metamorphosis, *BioSci.*, 1992, **42**, 671–678.
14. M. L. Larramendy and G. E. Liwszyc, in *Marsupial and Placental Mammal Species in Environmental Risk Assessment Strategies*, Royal Society of Chemistry, Cambridge, 2022.
15. G. E. Liwszyc and M. L. Larramendy, in *Bird and Reptile Species in Environmental Risk Assessment Strategies*, Royal Society of Chemistry, Cambridge, 2023.
16. M. L. Larramendy, in *Ecotoxicology and Genotoxicology: Non-traditional Terrestrial Models*, Royal Society of Chemistry, Cambridge, 2017.
17. M. L. Larramendy, in *Ecotoxicology and Genotoxicology: Non-traditional Aquatic Models*, Royal Society of Chemistry, Cambridge, 2017.
18. NCD Risk Factor Collaboration (NCD-RisC), Worldwide trends in blood pressure from 1975 to 2015: A pooled analysis of 1479 population-based measurement studies with 19·1 million participants, *Lancet*, 2017, **389**, 37–55.

Subject Index